*Mineral Deposits and
Global Tectonic Settings*

Academic Press Geology Series

Mineral Deposits and Global Tectonic Settings—A. H. G. Mitchell and M. S. Garson—*1981*

Frontispiece. Photograph taken by Dr Dudley Foster of Woods Hole Oceanographic Institution from the research submersible Alvin at a depth of about 2500 m on the crest of the East Pacific Rise at 21°N, showing a 350°C hot spring, termed a "black smoker", jetting out mineral-laden water at a rate of a few metres per second from a chimney about 15 cm across (MacDonald *et al.*, 1980). This water issues as a clear solution but immediate mixing with ambient seawater results in a rapid precipitation primarily of pyrrhotite plus pyrite, sphalerite and Cu–Fe sulphides. These build up to form chimneys superposed on basal mounds with honeycomb structure situated on fresh basalt pillows or flows. Chimney spire samples show radial changes in composition with outer walls made of anhydrite, gypsum and magnesium sulphate, and innermost walls of alternating bands of sphalerite, pyrite and chalcopyrite. The basal mounds consist of similar material, together with brittle white precipitates of baryte and amorphous silica coating tube walls of honeycomb samples. Part of the mounds and adjacent areas support dense benthic populations as at the active hydro-thermal vents at the Galapagos spreading centre (Corliss *et al.*, 1979).

Manganese, largely absent from the hydrothermal mounds, precipitates hydrogenously, rather than hydrothermally, on basalt surfaces within a few metres of sulphide structures.

This type of hydrothermal activity at the spreading centre, presently under intensive study in terms of chemistry, heat-flow and thermal balance, tectonic setting and related geo-physical and metamorphic conditions (Rise Project Group, 1980), is considered to be closely related in form and origin to that which produced fossil deposits of massive sulphides overlying basal pillow lava successions in the Cyprus and Semail ophiolites and at similar older settings, and also recently-formed metalliferous sediments at brine pools in the Red Sea spreading centre.

Mineral Deposits and Global Tectonic Settings

A. H. G. MITCHELL

Chief Technical Adviser, UNDP/UNDTCD-BMG Project
(Strengthening Geological Survey Division, Bureau Mines and Geosciences)
PO Box 7285 ADC, Mia Road, Pasay City, Metro Manila, Philippines

M. S. GARSON

Chief Technical Adviser, UNDTCD, UN-assisted Project—Mineral
Exploration in Northern Tanzania
PO Box 641, Arusha, Tanzania

1981

ACADEMIC PRESS
A Subsidiary of Harcourt Brace Jovanovich, Publishers
London   New York   Toronto   Sydney   San Francisco

ACADEMIC PRESS INC. (LONDON) LTD
24/28 Oval Road,
London NW1

United States Edition published by
ACADEMIC PRESS INC.
111 Fifth Avenue
New York, New York 10003

British Library Cataloguing in Publication Data

Mitchell, A.H.G.
  Mineral deposits and global tectonic settings.
  1. Mines and mineral resources
  2. Geology, Structural
  I. Garson, M.S.
  533'.1   TN263

ISBN 0-12-499050-9

LCCN 81-68011

553.1
M681m
217613

Typeset by Gilbert Composing Services, Leighton Buzzard, Bedfordshire
Printed in Great Britain by T.J. Press (Padstow) Ltd, Padstow, Cornwall

Foreword

There have been numerous books published which classify ore deposits either genetically or chemically, or they have been assembled on a regional basis or in what has been termed as metallogenic provinces. In this book the authors have related ore deposits to structure as they emphasize in their Introduction they have not attempted to formulate a structural classification. In so doing the authors have compiled a very valuable synthesis of the current knowledge and distribution of all types of ore deposits and, in relating them to structure, have also performed a valuable task of collating the various structural hypotheses presently favoured. Inevitably when writing about structure and oregenesis all opinions are open to debate and this book, in presenting the various aspects, certainly stimulates thought. The contents of this work will be invaluable to students and of great help to the professional exploration geologist in the search for new potential target areas.

R. RICE
Rio Tinto Finance and
Exploration Limited,
London

Preface

The revolution in the earth sciences consequent on the formulation of the plate or global tectonic hypothesis some 14 years ago is familiar to all concerned with the earth sciences. The effects of this revolution on our understanding of the relationship of ore body genesis and distribution to major tectonic settings can be seen from the voluminous literature on the subject which has appeared in the last decade, compared to the few papers, other than in the USSR, which were published previously. Extensive though the literature is, it is virtually confined to individual papers in journals, to proceedings of meetings, seminars and workshops, and to books consisting of chapters by individual contributors. Since the terminology applied to tectonic settings is as yet ill-defined, and since individual authors differ in their interpretations of the relationship of many smaller-scale settings to major plate boundaries, these publications do not always present material in a form which can be readily assimilated either by geologists concerned largely with mineral deposits or by those unfamiliar with the mode of occurrence of economic deposits. The preparation of this book was stimulated to a large extent by the encouraging response to our review on "Mineralization at Plate Boundaries" published in 1976. This review was adopted as a basis for courses on plate tectonics mineralization at many American universities, and many Heads of earth science departments requested at that time that we should amplify the review into a textbook for use by senior undergraduate and post-graduate geology students, economic and exploration geologists and other earth scientists with an interest in recent advances in understanding of global patterns of tectonics and mineralization.

The present book attempts to discuss the relationship of mineral deposits, other than petroleum, to global tectonics in a manner easily digestible by anyone having some familiarity with the basic principles of geology. While the tectonic content is largely restricted to aspects which have become generally accepted in the West in the last few years, parts of both the tectonic framework and relationship to ore genesis reflect the author's views; this is to some extent inevitable in a rapidly developing subject on some aspects of which agreement is by no means unanimous. We emphasize the treatment of ores as rocks, part of the stratigraphic succession or intrusive body in which they occur, a concept followed in the now classical textbook by Stanton (1972). As many types of mineral deposit can occur in more than one type of tectonic setting, and in fact relatively few are diagnostic of a particular setting, a framework for discussion is provided by consideration of the various types of major settings which can be recognized on the earth's surface today (Fig. 1) and which, not without speculation, can be identified from the geological record of the past 2500 Ma. While many of these settings are beside plate boundaries, others lie within plates and are defined by their position relative to continental margins, hot spot tracks, or other major intraplate crustal features. The tectonic settings are thus global rather than related exclusively to plate boundaries.

In the first chapter we discuss briefly why tectonic settings are a major control on the nature of the minerals deposited in economic concentrations, and review the pre-plate concepts of the relationship of mineralization to geosynclinal settings. We then introduce the plate tectonic hypothesis, and indicate the major developments in ideas on the relationship of mineral deposition to plate processes. The next six chapters, comprising the bulk of the book, are concerned with the brief description of each of the major types of tectonic settings recognizable today followed by an account of the main kinds of economic deposit found in modern settings and inferred ancient equivalents. We concentrate on aspects of the deposits' genesis related to the regional tectonic setting, and no attempt is made to review features such as temperature of formation or mineralogy which can be found in textbooks concerned exclusively with mineralization and ore bodies.

We consider that analogues of most modern types of tectonic setting can be recognized in the geological record throughout the Phanerozoic and Proterozoic, and so examples of mineralization in each setting can include deposits of any age from early Proterozoic to late Cenozoic. However, global tectonics in the Archaen were undoubtedly different from those developed subsequently, and Archaen mineralization is hence beyond the scope of this book.

Finally we summarize the evolution of the individual tectonic settings with their associated mineral deposits throughout an orogenic cycle, and discuss briefly the possible value to mineral exploration of relating economic deposits to the tectonic settings in which they formed. The types of deposit discussed in most detail are those which are either hydrothermal or magmatic in origin. This is because the generation of these types of ore bodies tends to be more dependent on the tectonic setting than does the deposition of other metallic and nonmetallic ores unrelated to igneous activity or geothermal gradient. Nevertheless, we include discussion of some metasomatic deposits and of major types of sedimentary ore. Throughout the book many of the postulated relationships between tectonic setting and mineralization are necessarily speculative. Were discussion of this relationship to be restricted to aspects on which there is international acceptance throughout both the western world and Soviet bloc, a publication of a few pages would have sufficed.

We wish to thank all those who have helped directly or indirectly in preparation. In particular we thank Dr H.G. Reading of Oxford University for his influence in emphasizing the importance of time and stratigraphy in all aspects of geology, and for his encouragement and appreciation of geological activities not confined to his own field; Dr T. Deans formerly of the Institute of Geological Sciences, London for discussions and advice over many years on the geology and setting of carbonatites and kimberlites; Dr T.C. Hart of the Regional Remote Sensing Facility in Nairobi, Kenya and Dr G.C. Tolbert, Chief of the Branch of Latin American and African Geology, US Geological Survey for their assistance and provision of Landsat imagery; Professor Dr F.M. Vokes of Trondheim University, Norway for advice on mineralization in the Oslo Graben and the Scandinavian Caledonides; Miss Catherine Collingbourn of the Institute of Geological Sciences, London for painstakingly copying essential and relevant literature and mailing it to us in many far-flung corners of the world; and Dr D.A. Harkin, Technical Adviser, United Nations Department of Technical Co-operation and Development in New York for constant encouragement and for arranging permission to publish the content of this book.

At all stages of the preparation of this book we have been surprised and gratified by the immediate response of scientific colleagues, too numerous to list individually, who contributed photographs, copies of diagrams and recent reprints. We thank all of these for their generosity and also all those authors and publishers acknowledged in the text who gave permission to reproduce their diagrams. We are grateful to Mr T.O. Gama, Mr P. Liwime and Mr J.A. Massawe of the Drawing Office, Arusha, Tanzania in the UN-assisted Project-Mineral Exploration in Northern Tanzania for preparation of the bulk of the diagrams in their spare time, and to Mrs Betty Travas of the same Project who typed drafts of various sections of the book. The staff of Academic Press are thanked for their patience and help.

Finally special thanks are due to our respective wives, Orawin Mitchell and Jean Garson for their patience, encouragement and assistance.

September 1981   A.H.G. Mitchell
                                                          Kathmandu, Nepal

                                                              M.S. Garson
                                                          Arusha, Tanzania

Contents

CHAPTER 1

Introduction

I TECTONIC SETTINGS AS A CONTROL ON MINERALIZATION

The tectonic setting, in which a particular suite of rocks formed, can conveniently be defined as the location, relative to major features of the crust, within which
the rocks including associated economic minerals of similar age were deposited,
intruded, or less commonly formed as a result of deformation amd metamorphism. The crustal features usually have a morphological expression
affecting sedimentation or erosion and reflect variations in the nature, thickness
or temperature of the crust or upper mantle which may result in magmatism.
The life of any particular tectonic setting may be no longer than the period of
formation of the rocks which define it, in some cases only a few million years or
less, and therefore the present tectonic setting of all but the youngest rocks
bears no direct relationship to the tectonic setting in which they were formed.

The main aims of relating the environment of formation of mineral deposits
to tectonic settings are to obtain an indication of the mineral potential of rocks
formed in various settings and of the preferred locations within the setting for
mineralization. Tectonic settings exert a major control on the type of mineralization, its deformation and preservation potential in several ways. First, types of
magmatism, each associated with particular types of hydrothermal alteration
and mineralization, whether pluton-related or submarine exhalative, are each
characteristic of one or more modern tectonic settings. For example in general
calc-alkaline volcanic and plutonic rocks are erupted and intruded in subduction-
related magmatic arcs, alkaline rocks in intracontinental rift zones and transform
faults, and tholeiitic lavas at oceanic spreading centres.

Secondly, the nature of the earth's major sedimentary successions, their
geometry, thickness, composition and to some extent their detailed facies,
and associated formation of syngenetic and diagenetic or early epigenetic
mineral deposits, is controlled at least partly by the tectonic setting. For
example, extensive prisms of shallow marine carbonates with associated lead–
zinc deposits are characteristic of anorogenic or passive continental shelves and
epeiric seas, and thick elongate belts of folded feldspathic greywackes lacking
stratabound sulphides are typical of orogenic fore-arc basins.

Thirdly, the rocks in many tectonic settings undergo subsequent deformation

1

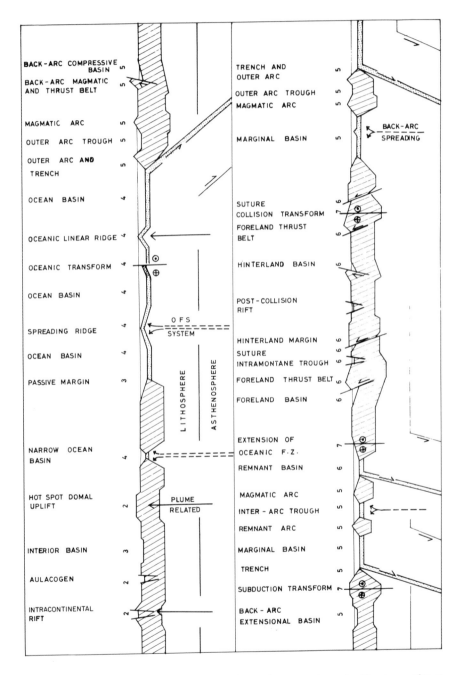

Fig. 1. Diagrammatic illustration of tectonic settings, not to scale. Numbers refer to chapters.

and in some cases, metamorphism, uplift and erosion, the nature and degree of which is often dependent on the tectonic setting in which they formed. The preservation potential of the rocks, and hence of the associated mineral deposits, is therefore determined partly by their tectonic setting at the time of formation.

Fourthly, the geothermal gradient is dependent on the tectonic setting. This is not only of particular significance to the generation and maturation of hydrocarbons, and in determining the rank of coals, but also exerts a major control on the circulation of mineralizing geothermal brines. The movement of hot brines and in some cases hydrocarbons is essential to the formation of many types of epigenetic sediment-hosted metal sulphide deposit.

Finally, because major faults can control circulation of mineralizing fluids and hence ore deposition, they are considered as a type of tectonic setting, although their control on formation of rocks other than hydrothermal deposits is indirect, and confined to erosion and sedimentation in adjacent areas.

Although relatively few types of mineral deposit can be observed forming in modern tectonic settings at the earth's surface, the association of many ore bodies with older volcanic and sedimentary rocks identical to those erupted or deposited at the surface today is well established, and hence the ancient equivalent of the modern tectonic setting can be inferred. With regard to pluton-related deposits, e.g. the predominantly magmatic hydrothermal porphyry copper, tin-tungsten and uranium deposits, erosion in young tectonic settings reveals the relationship of plutons to volcanic rocks, facilitating recognition of the tectonic setting in which these and similar older plutons elsewhere were emplaced.

Tectonic setting is thus a more fundamental control on mineralization than the local sedimentary or volcanic environment in the usual rather restricted sense of the term. For example, heavy mineral placer deposits in channel lag facies may accumulate within fluvial deposits formed in very many of the recognizable types of tectonic settings; whether the fluvial deposits formed in a foreland basin, outer arc trough or rift zone can be determined only from interpretation of the mineralized successions' regional depositional environment and of the surrounding source rocks.

II TECTONIC SETTINGS AND CLASSIFICATION OF MINERAL DEPOSITS

The relationship of mineral deposits to tectonic setting does not result in a strict classification of ore bodies as some types of deposit can form in more than one type of setting. For example, considering classifications according to the predominant ore metal, economic deposits of lead occur in both carbonate-hosted lead–zinc deposits, considered here to be characteristic of continental shelves and epeiric seas, and in Kuroko-type deposits widely accepted as restricted to volcanic arc environments. Similarly copper ore bodies are formed in rift, magmatic arc and foreland basin settings, and major deposits of tin are known from granites emplaced in outer arcs, foreland thrust belts and back-arc magmatic belts as well as above mantle hot spots. There is also no direct correspondence between either a genetic classification of mineral deposits (e.g. Lindgren, 1933) or even a classification in terms of associated host rocks (e.g. Stanton, 1972), and the tectonic setting in which they formed. Thus, submarine exhalative sedimentary ores may form in magmatic arcs and at oceanic spreading centres, and diagenetic stratabound ore bodies in sedimentary rocks can form in intracontinental rifts, in foreland basins, and in back-arc basins. However, consideration of the tectonic setting in which mineral deposits formed does lead to a broad "geotectonic" grouping of types of deposit as attempted by Russian geologists (Bilibin, 1955) and the need for which was stressed by Petersen (1970); this geotectonic grouping is analogous to the tectofacies concept (van Houten, 1974) applied so successfully to sedimentary rocks.

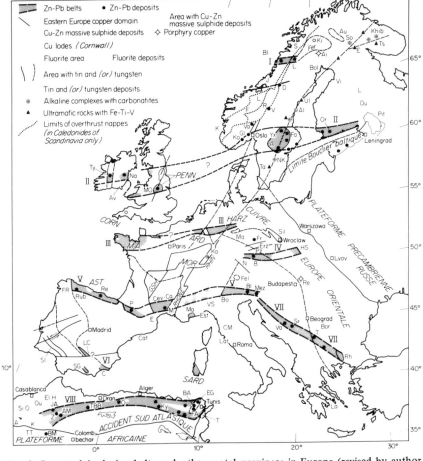

Fig. 2. Proposed lead–zinc belts and other metal provinces in Europe (revised by author, from Routhier, 1980). Note this is a metal province rather than metallogenic map. By permission of the author and publisher, Mem. Bur. Recherche Geol. Min. No. 105.

III METAL AND METALLOGENIC PROVINCES

In discussing tectonic settings for mineralization, most authors have considered, where convenient, not only individual ore bodies but also metal and metallogenic provinces, the concept of which was gaining support by the 1950s.

Metal provinces, a term which includes the more local mining districts of Park and MacDiarmid (1964), are commonly defined as areas characterized by an abnormal concentration of large deposits of a particular metal or metals, by numerous occurrences of a metal, or by both. A metal province map includes one or more provinces; an example is that of the western United States produced by Noble (1970). A province can include deposits irrespective of age, with different modes of formation, which in some cases had formed in a variety of tectonic settings. For example, a single copper province might include porphyry ores formed in a late Mesozoic magmatic arc, stratiform volcanic exhalative ores formed in early Mesozoic ocean floor and tectonically emplaced in an outer arc, and late Tertiary stratabound deposits formed in a foreland basin during continent-arc collision. Provinces are thus heterochronous and multitypic in the sense of Routhier, whose Ardennes-upper Harz lead-zinc belt or province (Fig. 2) is defined to include a stratiform volcanogenic deposit (Rammelsberg), a stratiform sedimentary deposit (Monback) and fossil karst deposits, with a time range of Devonian to Triassic (Routhier, 1976). The lead-zinc deposits of the Sierra de Carthagena, Spain (Fig. 3) are similarly heterochronous and multitypic (Pavillon, 1973). While of some use in showing regions broadly favourable for copper or lead-zinc mineralization, metal province maps of this type are of limited value either in relating the formation of the deposits to a specific setting or in determining their mode of genesis.

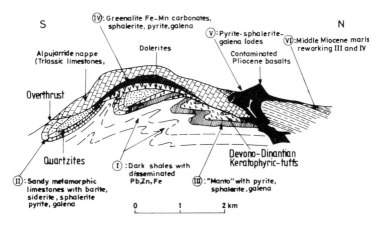

Fig. 3. Schematic cross-section through Sierra de Carthagena, Spain, showing six types of mineralization (from Pavillon, 1973); on a metal province, as opposed to metallogenic map, the Pb-Zn occurrences of various ages and types could be depicted by one symbol. By permission of the author and publisher, *Mineral. Deposita* 8, 237-258, Springer-Verlag, Heidelberg.

Metallogenic maps are more specific than metal province maps. A metallogenic belt shows the types of deposit of a metal or metals formed within a
narrow time range, ideally not more than 10 or 20 million years, and in some
cases much less. It corresponds to the homochronous and monotypic provinces
of Routhier (1976). Metallogenic belts therefore comprise deposits formed
within a particular tectonic setting, to which the origin or genesis of the metals
is related. The term "metallogenic" is therefore not normally applied to metal
province maps, lacking information on the age of the mineralization, for
example those drawn by Petersen (1970) for South America and Guild (1974a)
for North America. A fairly recent discussion of metallogenic maps and
provinces in the USSR, where the concept of metallogeny is particularly popular, is provided by Shcheglov (1976), and metallogenic map preparation in
general was recently reviewed by Guild (1978a).

IV MINERALIZATION AND TECTONIC SETTINGS: EARLY IDEAS

Until the middle of this century the relationship of mineral deposits to tectonic
settings had attracted very little attention. This was largely because the more
influential authors, and particularly those of textbooks on economic deposits,
considered that most ore bodies were either magmatic or magmatic hydrothermal in origin (Lindgren, 1933; Bateman, 1950) and hence related mostly to
intrusive rocks. As the intrusions were considered to be largely unrelated to the
adjacent pre-intrusive host rocks, there was little interest in or awareness of the
relationship of mineral deposit genesis to the structural setting.

Beginning in the early 1950s, the concept of ore formation on the depositional surface of the underlying volcanic rocks or sediments, yielding stratiform
ore bodies of approximately the same age as the enclosing host rocks, began to
gain ground, and has been comprehensively reviewed by Stanton (1972).
Following the work of, for example, King and Thompson (1953) on the Broken
Hill orebody in Australia, of Ehrenberg *et al.* (1954) on the Meggen ores in
Germany, and important papers by Stanton (1955) and Knight (1957), many
deposits long considered to be magmatic hydrothermal and epigenetic in origin
were reinterpreted as syngenetic deposits. The Rio Tinto deposit in Spain, for
example, was reinterpreted as syngenetic in a significant reappraisal by
Williams (1962), and in the last 20 years numerous ore bodies broadly concordant with their host rocks have been accepted as syngenetic.

Although the genetic relationship between many types of ore body and their
stratified host rocks was widely accepted by 1965, only relatively few authors
had shown interest in going a further stage and interpreting the various types of
mineralized stratigraphic successions as characteristic of specific tectonic settings
within a global framework. One of the earliest discussions of the relationship of
mineral deposits and metal provinces to major structural belts was that by
Blondel (1936). The concept was later elaborated on by Turneaure (1955), and
formed a broad working hypothesis utilized by many economic geologists.

Turneaure recognized, in particular: shield areas mostly of Precambrian age; orogenic belts, mostly mountain ranges built largely of Phanerozoic rocks; stable regions of gently folded thick sedimentary successions. Although of value in comparing ore bodies from analogous tectonic settings in different regions, this was an essentially non-dynamic concept and could not explain for example why rocks and associated mineral deposits characteristic of a stable region were often tectonically juxtaposed with and even included within those of an orogenic belt.

V THE GEOSYNCLINAL HYPOTHESES AND MINERALIZATION

To most geologists the geosynclinal concepts of mountain building and orogeny provided at least a partial explanation of geological processes not only in orogens but in the adjacent platforms and shields. Although the relationship of mineralization to geosynclines was considered in detail only by relatively few authors, the geosynclinal hypothesis still has a number of adherents who consider it as a preferred alternative to plate tectonics in explaining the association of some types of ore body with particular types of rock sequence. We shall show later that the geosynclinal hypothesis can be easily reconciled with that of plate tectonics, and that difficulties in the former can be explained by the latter, but first, it is necessary to consider in some detail the various geosynclinal tectonic frameworks to which mineral deposits were considered to be related.

There were major differences in the development of the geosynclinal hypothesis in North America, Europe, and Russia, and later in Japan, and the significance of magmatic rocks and associated mineralization to geosynclines in each region received different emphasis.

A Geosynclinal Concepts in North America, Europe, and the USSR

In North America, following Hall (1859) and Dana (1873), the ideas of geologists were strongly influenced by the position of the major mountain ranges of the Appalachians and Western Cordillera built largely of folded Palaeozoic rocks: consequently geosynclines were considered to develop on a continental margin. Subsequently the American views were influenced by the European concept of a miogeosyncline and eugeosyncline forming parallel paired belts, the miogeosyncline consisting of relatively undeformed shallow-water clastics and carbonates, and the eugeosyncline of very thick tectonized successions of greywackes and volcanic rocks (Fig. 4). This led to elaborations in the terminology and classification, notably by Kay (1947, 1951). In the early 1960s the laterally equivalent late Mesozoic and Cenozoic continental shelf and continental rise successions of the Atlantic coast of North America described by Drake *et. al.* (1959) were compared respectively to the parallel Palaeozoic miogeosynclinal and eugeosynclinal belts of both the Appalachians and Western Cordilleras. The Atlantic coast sequences were considered to be actualistic examples of the Palaeozoic successions, despite differences in the volcanic rocks of the two regions.

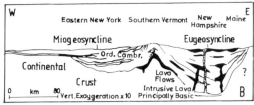

Fig. 4. Palaeozoic miogeosynclines and eugeosynclines on margins of North America. (A) Schematic map (after Kay, 1947); (B) schematic cross-section through Palaeozoic cordilleran geosyncline of eastern North America (after Kay, 1951). By permission of the publisher, *Bull. Am. Ass. Petrol. Geol.* **31**, 1289–1293; courtesy *Mem. geol. Soc. Am.* **48**.

In Europe the geosynclinal concept of Dana and Hall was modified to explain the intercontinental position of the Alpine chains between Africa and Europe (Fig. 5). Subsequently, to most European geologists a geosyncline was located between and on the margins of two adjacent continents, the foreland and hinterland of Suess. The orthogeosyncline of Stille (1936, 1940) included both continental margin and intercontinental geosynclines, provided they had undergone Alpine-type deformation. Each orthogeosyncline comprised a miogeosyncline without igneous rocks, bordered on the ocean side by a eugeosyncline characterized in the case of the Hercynian chains of Europe by pre-orogenic ophiolites or greenstones, with syn- and late-orogenic granitic rocks and late to post-orogenic andesites.

Dutch geologists were strongly influenced by their experiences in Indonesia: Umbgrove (1949), van Bemmelen (1949) and others used Indonesia, and particularly the Sunda Arc, as an actualistic model for mountain building hypotheses. Within the Sunda Arc they recognized the continental margin "inner" or volcanic arc of Sumatra, extending southeastwards as an island arc through Java to Flores, and the "outer arc" consisting of deformed flysch-type sediments with

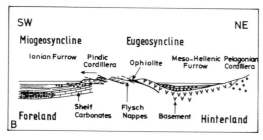

Fig. 5. Alpine geosynclines of Europe. (A) Tertiary Alpine chains between African and Eurasian continents; (B) schematic cross-section through European geosyncline based on Alpine chain in Greece at late orogenic stage of development (after Aubouin, 1965). By permission of the author and publisher, "Geosynclines", Elsevier, Amsterdam.

basic and ultrabasic rocks, bordered on the Indian Ocean side by a submarine trench (Fig. 6).

In the 1950s and early 1960s, geosynclinal models based on the Alps and *circum*-Mediterranean chains became increasingly sophisticated. Aubouin (1965) recognized a development stage with an episode of basic and ultrabasic or "ophiolitic" igneous activity in the eugeosyncline, an orogenic stage and a late orogenic stage (Fig. 5); minor andesitic or trachyandesitic volcanism and granodiorite emplacement were considered typical of the late geosynclinal stage, following flysch deposition and orogeny, and accompanying deposition of clastic molasse sediments. Granitic magmatism and high-grade metamorphism were insignificant in many of the Alpine chains, and thus played a minor role in Aubouin's model of geosyncline evolution.

Aubouin tried to explain the modern volcanically active Sunda Arc by a model based on the Alps and Hellenides, rather than interpreting the latter in terms of the former. He equated the active volcanic arc of Sumutra with the late geosynclinal trachyandesite volcanism in the Hellenides, and hence was forced to conclude that the Sunda Arc geosyncline was at a late stage of development. Consequently he was unable to explain satisfactorily the presence of oceanic

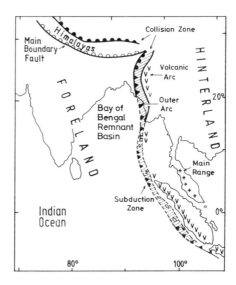

Fig. 6. Cenozoic Sunda Arc subduction zone and adjacent orogen showing Indian Ocean foreland and Bay of Bengal remnant basin (from Mitchell, 1978a). Note oceanic Bay of Bengal foreland, with Himalaya-derived Bengal Fan turbidites, in comparable tectonic position to continental miogeosynclinal foreland (Ionian furrow) in Alpine geosyncline (Fig. 5B); with subduction of Bay of Bengal, India will underthrust flysch and ophiolites of Outer Sunda Arc, resulting in similar situation to Fig. 5B. By permission of the 11th Min. Metall. Cong., Hong Kong.

foreland (Indian Ocean) in the Sunda Arc in a similar tectonic setting to continental foreland (Eurasia) in the Alpine chains (Fig. 6).

European geologists including Aubouin attempted to interpret the Caledonian and Hercynian orogens in geosynclinal terms, but in general were unable to explain the presence of extensive synorogenic deformation and syn- to late- or post-orogenic metamorphism and granite emplacement in terms of the Alpine models.

In Russia ideas on the tectonic setting of geosynclines were influenced by the intracontinental position of the major mountain ranges, especially the Urals and Tien Shan. Peyve and Sinitzyn (1950) recognized primary geosynclines with ultrabasic rocks corresponding to the eugeosynclines of Stille but situated within fractures in continental platforms, and characterized by vertical movements rather than thrusting. The significance of vertical as opposed to horizontal movements in mountain building received additional impetus from the work of Beloussov in the Greater Caucasus.

The consequences of this intracontinental or ensialic development of geosynclines were developed by Beloussov (1968) and others, e.g. Smirnov (1968), who argued that oceanic crust developed from basification of ancient continental basement, with the northwest Pacific back-arc or marginal basis as an example. In contrast, a few authors, e.g. Gnibidenko and Shashkin (1970),

Avdeiko (1971) and Gnibidenko (1973) interpreted the marginal basins as geosynclines developing on a continental margin, leading to oceanward growth of continental crust.

B Metallogenesis and Geosynclines

As discussed above, in both North America and Europe until the late 1950s metal and metallogenic provinces were considered to be mostly related to igneous and especially intrusive rocks, a view reflecting the predominant magmatic hydrothermal theory of ore genesis in the then current textbooks. There were only a few attempts to relate ore genesis to geosynclinal tectonic settings (Cady *et al.*, 1950; Stanton, 1955), partly because most geosynclinal models were conceived by stratigraphers, to whom magmatism and particularly emplacement of plutons were of less significance than sedimentation, and partly because an understanding of tectonic settings, as opposed to the more local environment indicated by the ore host rock, was considered to be of little value as a guide to the discovery of new deposits. Moreover, many economic geologists were unaccustomed to thinking in terms of a global framework. The significance of geosynclinal models in metal deposit genesis, particularly in Europe, is indicated by the fact that the origin of many ore bodies of Palaeozoic age continued to be interpreted until fairly recently (e.g. Baumann, 1970) in terms of the geosynclinal magmatic episodes of Stille.

In the 1960s and early 1970s the distribution of metal and metallogenic provinces (Routhier, 1969) relative to geosynclines attracted increasing interest. A number of authors emphasized the significance of the boundary between copper provinces characteristic of orogens and lead–zinc provinces found in adjacent less deformed carbonates, and Laznicka and Wilson (1972) demonstrated how the two provinces were characteristic respectively of the eugeosyncline and miogeosyncline of Aubouin.

In the USSR the relationship of metallogenesis to geosynclinal tectonic settings and their development received far more attention than in Europe, largely under the leadership of Bilibin (1948, 1955, 1968), and was applied to exploration concepts in the Canadian cordillera by McCartney and Potter (1962). It continues to provide a framework for discussion of many types of mineral deposit in USSR (e.g. Smirnov, 1977).

In the initial stage of the Russian geosynclinal model (Table I) ultrabasic and basic rocks with associated platinum group metals, chromite and titaniferous magnetite are intruded into a subsiding intracontinental trough. Following folding, the early stage is characterized by intrusion of gabbros, and accumulation of thick sedimentary and volcanic successions intruded by porphyries with associated cupriferous or polymetallic pyritic massive sulphides. Smirnov (1968) stressed the relationship of these early geosynclinal stage ores to the magmatic rocks and recognized a basaltophile group of metals (Fe, Mn, Va, Ti, Cr, Pt, Cu, Zn), considered to have a sub-crustal source.

In the middle or orogenic stage uplift is followed by sedimentation and

TABLE I

Generalized scheme of development of a folded belt based on Russian sources (from McCartney and Potter, 1972). By permission of the authors and publisher, *Can. Min, J.* **83,** 83–87.

Phase	Appalachian age	Tectonic movements	Sedimentary and volcanic rocks	Type of intrusive rock	No.	Composition of intrusive rocks	Endogenous deposits	Location	Examples Canadian Appalachians	Exogenous deposits	Examples Canadian Appalachians
Final	Pennsylvanian-Triassic	Weak oscillations, some faulting	Terrestrial sedimentary rocks, pyroclastics, flows	Commonly absent					Not recognized		
			Basaltic, andesitic, and rhyolitic volcanic rocks							Iron oxides and hydroxides	
					F-2	Diabase and ladradorite-porphyry, quartz-porphyry granosyenite-porphyry	Pb, Zn, Cu, F, Ba, Fe (Hg, Sb)	In depressions far from border of mobile zone			Northern NS and Shippigan, NB
			Molasse		F-1	Granodiorite-porphyry, monzonite, diorite, gabbro, pyroxenite, peridotite, alkaline gabbros, nepheline syenites, syenites, granites, trachy-rhyolites, trachy-andesites	Ba, Fe, Ag, Zn, Pb, Ni, Bi, Cu, Co, Au, As F	In activated margins of platform	Walton, NS; much barite occurs elsewhere in areas of Carboniferous folding	Cu and U sandstones	
	Mississippian		Multicoloured continental and lagoonal formations	Small intrusions, commonly sub-volcanic (stocks, sills, dykes, laccoliths)	L-6	Dacites, andesites, rhyolites, (commonly absent)	Hg, Sb, As, (W)	Within blocks and late depressions	Not recognized	Coal	Minto, NB Sydney, NS
		Epirogenic movements (crustal adjustment), formation of great faults	Salt, gypsum		L-5	Alaskite, granosyenite, granite-porphyry, and syenite-porphyry. Diorite, monzonite, granodiorite, granite	Pb, Zn, Ag, (As)			Salt	Malagash etc.
					L-4	Granodiorite-porphyry, monzonite-porphyry	Al, Mo, Cu, Au	Late internal and border depressions	Not recognized	Oil and gas	Moncton, NB
Late			Petroliferous rocks		L-3	Plagiogranite, granite, granosyenite, granite, diorite, gabbro, diabase. In some places alkali-pyroxenites, shonknites and monzonites	Fe, Cu, Co			Coal	
					L-2	Granite, grandiorite, diorite and porphyritic equivalents. In some cases comprise gabbros and diorites. (not accompanied by mineralization)	Pb, Zn, Sn, **Ag,** As (W, Mo)	Late internal and border depressions	Mt Pleasant, NB	Iron oxides Carbonates Hydroxides	Londonderry
			Terrestrial sedimentary rocks, carbonates		L-1	Granites, granodiorites, quartz-syenites, quartz-diorites, and porphyritic equivalents, Quartz-porphyry and albitophyre	Mo, Au, Cu, (Pb, Zn) As, W, Sb Hg, Co	Late internal depressions	Not known		
	Devonian	Uplift and intense folding	Erosion	Batholiths	M-3	Ultra acid potassic granites, alaskites, aplites, pegmatites	W, Mo, Sn, Bi, F Li, Be, Ta, Nb	Within or on the borders of structures of diverse mobility, near borders	Burnt Hill, NB Square Lake, NB		
			Flysch and flychlike rocks in isolated basins, calcareous shales and limestones, acid to intermediate volcanic rocks		M-2	Granodiorite, biotite-granite, plagiogranite, quartz-diorite	W, Mo, Au, (As, Co) Sn, Fe	of central geanticlines	St Lawrence, Nfld		
Middle		Slight subsidence		Small intrusions	M-1	Quartz-diorite, diorite, diorite-porphyry	Au, (W, As, Sb, Mo)		Nova Scotia gold-quartz-scheelite veins	Iron oxides	Torbrook, NS

TABLE I (Cont'd)

Generalized scheme of development of a folded belt based on Russian sources (from McCartney and Potter, 1972). By permission of the authors and publisher, Can.Min.J. 83, 83–87.

Appal-achian age	Phase	Tectonic movements	Sedimentary and volcanic rocks	Type of intrusive rock	No.	Composition of intrusive rocks	Endogenous deposits	Location	Examples Canadian Appalachians	Exogenous deposits	Examples Canadian Appalachians
Ordovician-Silurian	Early	Uplift	Erosion	Small intrusions	E-5	Andesite, trachyte, phonolite, rhyolite	Au, Ag		Buchans, and numerous Cu-pyrite deposits in Notre Dame Bay area of Newfoundland		
		Localized folding	Lithic greywacke, shales. Cherts, (basic volcanic rocks not as abundant as in initial phases)	Small stocks or batholiths	E-4	Granodiorite, diorite, granite, and porphyritic counterparts	Cu, Pyrite, Ag, Pb Zn, Au, Ba				
					E-3	Quartz-albitophyre, albitophyre, quartz-porphyry and syenite porphyry					
		Subsidence			E-2	Gabbro, diorite, tonalite, plagiogranite	Magnetite-Cu skarn (Co, As)		Not known		
					E-1	Gabbro, monzonite, granosyenite, nepheline syenite, syenite	Au, As, W, (scheelite), Cu, Mo				
	Initial	Deformation localized near great fractures	Predominantly thick basic volcanic flows. Sedimentary rocks include siliceous shales, jaspers, tuffs, greywackes, and minor calcareous rocks.	Chains or belts of bodies of various size and shape, commonly sills or dykes	I-2	Gabbro-norite, gabbro-diorite, pyroxenite, peridotite and dunite	Ti, Fe, Cu, Ni, Cu	Junction of marginal and initial depressions with platform. Also on margins of internal geanticlinal uplifts	Numerous small occurrences in the Eastern Townships and Newfoundland	Iron oxides in basins far from mobile zone	Wabana
					I-1	Peridotite, pyroxenite, dunite, serpentinite, gabbro-peridotite, hornblendite, subordinate gabbro, diorite, seldom quartz-diorite, and plagiogranite	Pt, Cr, Fe, Ti, Ni, asbestos			Fe, Mn cherts	Notre Dame Bay Nfld

intrusion of dioritic stocks with associated gold-quartz veins, and granodioritic and adamellite batholiths with gold or scheelite-bearing veins. Subsequently "ultra acid" granites and pegmatites are emplaced, with vein-type deposits containing cassiterite, wolfram, molybdenum and bismuth sulphides and fluorite, and pegmatites with Be, Ta, Nb, Cs, Rb and rare earth minerals as well as certain gemstones.

The late- or post-orogenic stage is characterized by faulting and sedimentation with emplacement of granodiorite and quartz-monzonite with Cu–Mo–Au deposits associated with intense hydrothermal alteration. Smirnov (1968) considered the granitophile group of metals (Sn, W, Be, Li, Nb, Ta), characteristic of the middle and late stages, to be derived from a crustal source.

The final stage of geosynclinal evolution according to Bilibin included tele-thermal Pb–Zn–Ba deposits associated with carbonates. However Smirnov related these deposits to the stable continental platform in which trap-type basalts with associated copper-nickel sulphides, kimberlites, and carbonatites could also occur. The significance of continental non-geosynclinal settings was also emphasized by other authors in USSR, and intracontinental rifts which in northeastern Asia include tin granites were considered to be more important metallogenic belts than geosynclines by Itsikson and Krasnyy (1970).

Bilibin's (1955, 1968) emphasis of the relationship of mineral deposits to intrusive rocks, and of these in turn to the geosynclinal stages, are valuable as an aid to understanding ore genesis, but inevitably when applied in detail to any region lead to major difficulties because of imperfections in the geosynclinal hypothesis itself. An attempt to relate the stages of evolution of the Russian geosynclinal concept to that of Aubouin in Europe is shown in Table II; the relationships of both to the plate tectonic or Wilson orogenic cycle described later are also shown.

VI OTHER GLOBAL TECTONIC HYPOTHESES

In Japan, Matsuda and Uyeda (1971) defined a Pacific-type orogenic belt characterized by pairs of parallel outer and inner belts, the inner belt consisting of granitic to dioritic magmatic rocks and the low pressure–high temperature metamorphics of Miyashiro (1971), and the outer geosynclinal belt of flysch and high pressure–low temperature metamorphic rocks. They related the belts to inclined subduction of ocean floor, and suggested that the North American and European geosynclines resulted from vertical subduction. This concept combined aspects of the geosynclinal and subduction hypotheses, but no attempt was made to relate metallogenic provinces to the tectonic belts.

Among the global tectonic hypotheses other than those related to geosynclines, the undation concept of van Bemmelen (1949, 1978), which included a process of basification of oceanic crust analogous to that favoured by Russian geologists, has attracted most support, especially among Dutch geologists, and

TABLE II

Equivalent stages of development of plate tectonic orogenic cycle, European geosyncline and Russian geosyncline. Position of granodiorite and Cu, Mo, Au porphyry mineralization in Russian geosyncline is anomalous and perhaps indicates repeated collision-ocean floor spreading cycle in Urals (from Mitchell, 1978a). By permission of the publisher, 11th Min. Metall. Cong., Hong Kong.

Plate-tectonic stage of orogeny	European geosynclinal stage (Aubouin, 1965)	Russian geosynclinal stage (Bilibin, 1955, and others)
1. Intracontinental rifting	Post-geosynclinal and early generative	Post-orogenic and final *Carbonate sedimentation Pb, Zn, Ba*
2. Ocean-floor spreading	Late generative and development *Shelf sediments* *Ophiolite formation*	Initial *Ophiolite Formation Pt, Cr*
3. Subduction and development of remnant basin	Early orogenic *Flysch deposition*	Early *Volcanism Cu, Pb, Zn* *Diorite, Au Granodiorite*
4. Continent–continent or continent–arc collision	Late orogenic *Molasse deposition*	Orogenic *Granite and pegmatite Sn, W, Mo, Bi, F*
1. Intracontinental rifting	Post-geosynclinal and early generative	Post-orogenic and final *Granodiorite Cu, Mo, Au*

has been to some extent successfully related to plate tectonic theory. However, as the hypothesis did not include metallogenesis, it will not be considered here.

Walker (1976) recognized world-wide synchronous orogenic cycles—eight in the Precambrian (e.g. Fig. 7) and two in the Phanerozoic—and emphasized their role in controlling metallogenesis. Each cycle was considered to be related to sea-floor spreading, with stages of development and associated mineralization analogous to those of the Russian geosynclinal model of Bilibin (1968); inter-cycle intervals were characterized by sedimentary ores, nickel and chromite.

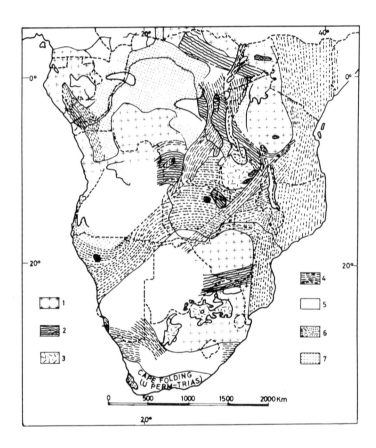

Fig. 7. Structural sketch of the Precambrian of Equatorial and Southern Africa showing provinces of distinct ages of folding, ranging from 1 (end Archaen) to 6 (latest Proterozoic); metallogenic episodes coincide with tectonic episodes. 1, Area folded and metamorphosed 2500-2600 Ma ago or earlier; 2, belts, folded and metamorphosed about 2100–1950 Ma ago; 3, beds, more or less tabular, corresponding in age to 2, Transvaal and Griquatown "System", Bushveld Igneous Complex; 4, belts, folded and metamorphosed about 1300–110 Ma. ago; 5, beds, more or less tabular or folded, which might correspond in age to 4; 6, belts, folded and metamorphosed about 730–600 Ma ago; 7, beds, more or less tabular, corresponding to 6 (after Cahen and Snelling, 1966; Walker, 1976). By permission of the authors and publisher, *Spec. Pap. geol. Ass. Can.* **14**, 517-557.

Difficulties in applying this concept to individual orogens is indicated by the long duration of the cycles, for example the postulated 200 Ma Alpine cycle. Another rather recent scheme of crustal evolution, the tectonospheric model of Tatsch (1976), has apparently been devised specifically to explain the distribution of many types of mineral deposit, but has attracted little interest among economic geologists and less from stratigraphers and structural workers.

Aspects of Chinese concepts of tectonic evolution have recently been summarized by Bally *et al.* (1980). Until recently Chinese geologists were influenced by the hypothesis of meridional and latitudinal horizontal movements of the earth's crust, the former resulting in equatorward movement while the latter led to N-trending rifts and fold belts. Both movements were considered to be related to the earth's rotation, and were applied to mineral exploration concepts. Development of this hypothesis in the 1960s emphasized the importance both of block faulting in the tectonics of China, probably in part reflecting Soviet ideas, and of the influence of older tectonic systems on younger either adjacent or superimposed systems. This latter concept is no doubt related to the overlap of orogenic belts resulting from the progressive accretion of small continental fragments and arc systems which formed mainland China, as recognized increasingly by Chinese geologists (e.g. Chang and Cheng, 1973) in the last decade.

VII PLATE BOUNDARIES AND GEOLOGICAL PROCESSES

At the beginning of the 1960s Hess and later Dietz and Wilson had suggested that ocean floor was created at mid-ocean ridges and lost in seismically active zones around the ocean margins. Subsequent evidence from magnetic anomalies of sea floor spreading (Vine and Matthews, 1963) supported this concept, and preceded the plate tectonic hypothesis by about five years. Wilson (1968) had clearly recognized that orogens were former mountain ranges where ancient oceans had closed, and at the same time several authors suggested mainly on seismic evidence that the earth's surface consisted of a number of rigid lithospheric plates. It was recognized that the ocean floor created at spreading ridges descended along inclined seismic Benioff zones which approached the surface beneath submarine trenches bordering volcanically active island arcs and continental margin mountain chains, thus conserving surface area. The lithosphere generated at ridges and subducted beneath trenches was considered to form a rigid plate comprising ocean crust about 5 km thick overlying upper mantle then thought to be around 80 km thick which extended beneath the continents. The boundary between lithosphere and underlying asthenosphere, now considered to be at least 120 km deep, is defined by a zone of low seismic velocity interpreted as a low viscosity layer. Other plates consisted either of continental lithosphere, or, where a continent lay within the plate, of both continental and oceanic lithosphere. It followed from this that volcanism in island arcs as well as at ocean ridges was directly related to relative movements of plates. Further consequences of the rigid plate hypothesis were that plates slide past one another along transform faults, that continent–continent or continent–arc collision must follow ocean floor closure, as recognized by Wilson (1968), and that most orogens are old collision belts.

In 1969 and 1970 attempts were made (e.g. Mitchell and Reading, 1969; Dewey and Bird, 1970) to relate thick successions of sedimentary and volcanic rock, previously interpreted in terms of geosynclines, to geological processes at or adjacent to plate boundaries. As the understanding of geological processes at or related to present plate boundaries has increased, interpretations of ancient orogens have become correspondingly more sophisticated and included most types and associations of rocks. Consequently the plate tectonic concept has now largely replaced the geosynclinal hypothesis as a tectonic framework for interpretation of ancient rock sequences. Nevertheless, the various stages of development in the geosynclinal models can readily be reinterpreted in terms of continental rifting, ocean floor subduction and continental collision (Table II), and some of the geosynclinal terminology is retained in describing plate tectonic settings.

In attempting to relate mineral deposits to global tectonic settings, each of the four major types of plate boundary (rift, ocean floor subduction, collision and transform) is here divided into a convenient number of sub-settings, some of which are dependent on whether the contiguous plates are continental or oceanic in nature, while others develop only as a result of prolonged plate inter-action. For example in subduction-related settings where ocean floor descends into the mantle, settings extending far into the overriding plate as well as at the actual plate boundary are recognizable, and in collision-related settings, a number of tectonic settings on the underthrusting and overriding plates may be defined. A few settings also occur within a plate and are related either to their position with respect to a continent–ocean boundary or to crustal hot spots, although the evolution of hot spot settings is determined largely by movement of plates relative to the sub-lithospheric upper mantle or asthenosphere.

In each of these tectonic settings, volcanic, plutonic, sedimentary, meta-morphic and deformation processes may occur, which yield sedimentary and volcanic successions, plutons or metamorphic rocks characteristic of the particular setting. In most cases modern settings are defined with respect to present plate boundaries (Fig. 8), rocks forming today in that setting are observed, and similar and older rock sequences are then interpreted as having formed in an ancient setting analogous to one of the modern ones.

Today the formation, deformation and metamorphism of most major belts of plutonic rock and most thick stratigraphic successions can be related to processes occurring in tectonic settings associated with types of plate boundary, whether rift, subduction, collision or strike-slip; others can be related to either crustal hot spots or position with respect to the continental-oceanic crust boundary within intraplate settings.

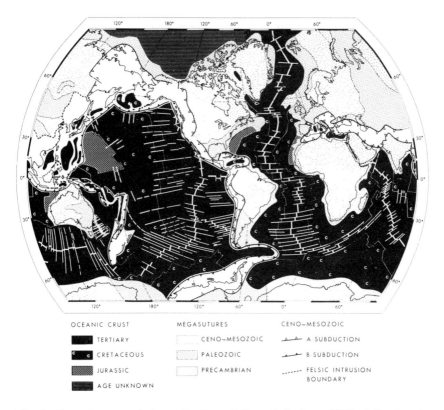

OCEANIC CRUST | MEGASUTURES | CENO–MESOZOIC

TERTIARY — CENO–MESOZOIC — A SUBDUCTION

CRETACEOUS — PALEOZOIC — B SUBDUCTION

JURASSIC — PRECAMBRIAN — FELSIC INTRUSION BOUNDARY

AGE UNKNOWN

Fig. 8. Tectonic map of the earth (from Bally and Snelson, 1980), indicating the distribution of some major plate boundaries of Cenozoic age and older. By permission of the authors and publisher, Can. Soc. Petroleum Geologists Mem. No. 8.

VIII   PLATE TECTONICS AND MINERAL DEPOSITS, 1967-1980

As in the case of the geosynclinal hypothesis, although on a much shorter time scale, at first no attempts were made to relate the formation of mineral deposits to plate tectonic settings despite the obvious success of the hypothesis in explaining the formation of rock successions and associations. The most probable explanation of this is the gulf in reading habits between stratigraphers and volcanologists, accustomed to searching the literature for modern analogies of older rocks, and most exploration geologists, for whom an understanding of present day volcanism and sedimentation was and to a lesser extent still is of limited value.

At the beginning of the 1970s the first attempts were made to relate mineralization to plate boundaries, with papers by Sillitoe (1970), Guild (1971), Pereira and Dixon (1971) and Snelgrove (1971). The number of papers on this theme was still limited in 1972 (e.g. Mitchell and Garson, 1972; Sawkins, 1972; Sillitoe,

1972a, b) but subsequently increased rapidly. It is perhaps significant that almost all the earliest papers, and a high proportion of the later ones, were concerned largely with porphyry copper deposits and hence with the subduction-related magmatic arcs in which they occur (e.g. Fig. 9). Here, as in the case of the various geosynclinal hypotheses for explaining rock successions, ideas were based largely on the present position of a major mountain range, in this case the volcanically active Andes above subducting ocean floor; formation of the numerous porphyry deposits of the Andes, although older than the latest Cenozoic and active volcanoes, were then assumed to be related to a tectonic setting broadly similar to the present one.

The main class of ore bodies next related to plate settings was that of stratiform massive sulphides, both those associated with silicic volcanic rocks, for which the Kuroko ores in the Japanese magmatic arc formed a late Cenozoic example, (e.g. Sawkins, 1972; Guild, 1972), and also those associated with basaltic submarine lavas, such as Cyprus (e.g. Sillitoe, 1972a) for which Quaternary submarine metal-rich muds were considered a modern analogy. The extensive recognition in the previous 15 years of many types of massive sulphide deposits as stratiform and syngenetic, with consequent emphasis on their stratigraphic position and similarity in age to the host rocks, greatly

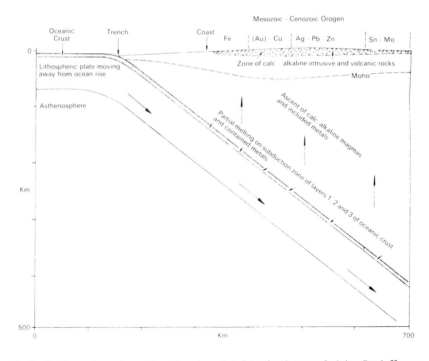

Fig. 9. Metal provinces in western America related to depth to underlying Benioff zone (from Sillitoe, 1972b). By permission of the author and publisher, *Bull. geol. Soc. Am.* **83**, 813–818.

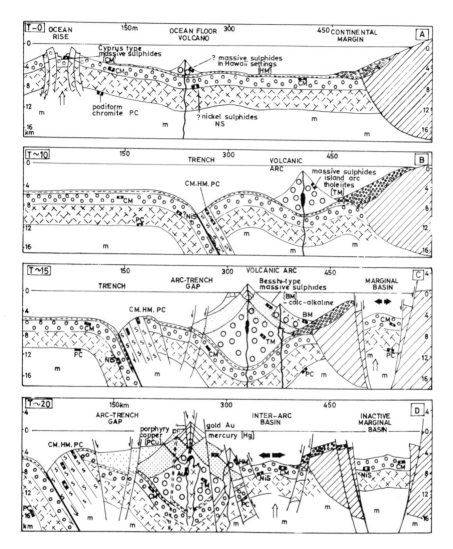

Fig. 10. Mineralization in an evolving island arc subduction system (after Mitchell and Bell, 1973). By permission of the authors and publisher, *J. Geol.* **81**, 381–405.

facilitated interpretation of the environment of formation of the ores together with the rocks enclosing them in terms of tectonic setting. Other deposits interpreted in terms of plate settings included those of the tin–tungsten association, Cenozoic examples of which were known from continental margins in Asia and South America, and which could thus readily be related to a Benioff zone (Mitchell and Garson, 1972; Mitchell, 1973).

Since 1972 the types of deposit related to plate settings, and the number of tectonic settings considered, have increased rapidly, some authors explaining the

TABLE III

Proposed relationship of some ore-deposit types to lithospheric plates (from Guild, 1974a). By permission of the author and publisher, "Metallogenetische und geochemische Provinzen" 10–24, Springer-Verlag, Vienna.

Deposits formed	Types, possible examples
1. *At or near plate margins*	Orientation of deposits, districts, and provinces tends to parallel margin
(a) Accreting (diverging)	Red Sea muds. Ancient analogues? Certain cupriferous-pyrite (massive sulphide) ores, Cyprus? Newfoundland? Podiform Cr (may be carried across ocean and incorporated in island arc or continental margin)
(b) Transform	Podiform Cr, Guatemala? Cu and Mn, Boleo, Baja California
(c) Consuming (converging)	Chiefly of continent/ocean or island arc/ocean type; deposits formed at varying distances on side opposite oceanic, descending plate Podiform Cr, Alaska FeS_2-Cu-Zn-Pb stratabound massive sulphide, New Brunswick, Japan (Kuroko ores), California, British Columbia Mn of volcanogen type associated with marine sediments, Cuba, California, Japan Magnetite-chalcopyrite skarn ores, Puerto Rico, Hispaniola, Cuba, Mexico, California, British Columbia, Alaska Cu (Mo) porphyries, Puerto Rico, Panama, SW USA British Columbia, Philippine Is., Bougainville Ag-Pb-Zn, Mexico, western USA, Canada Au, Mother Lode, California; Juneau Belt, Alaska Bonanza Au-Ag, W USA, W, Sn, Hg, Sb western and S America
2. *Within plates*	Deposits tend to be equidimensional, distribution of districts and provinces less oriented (may be along transverse lineaments)
(a) In oceanic parts	Mn-Fe (Cu, Ni, Co) nodules Mn-Fe sediments in small ocean basins with abundant volcanic contributions? Evaporites in newly opened or small ocean basins
(b) At continental margins of Atlantic (trailing) type	Black sands, Ti, Zr, magnetite, etc. Phosphorite on shelf
(c) In continental parts	Au (U) conglomerates, Witwatersrand Mesabi and Clinton types of iron formation Evaporites, Michigan Basin, Permian Basin; salt, potash, gypsum, sulphur Red-bed Cu; Kupferschiefer and Katangan Cu-Co U, U-V deposits, Colorado Plateau Fe-Ti-(V) in massif anorthosite, Canada, USA Stratiform Cr, Fe-Ti-V, Cu-Ni-Pt, Bushveld Complex Carbonatite-associated deposits of Nb, V, P, RE, Cu, F Kimberlite, diamonds Kiruna-type Fe(P), SE Missouri Mississippi Valley type deposits, Pb-Zn-Ba-F-(Cu, Ni, Co)

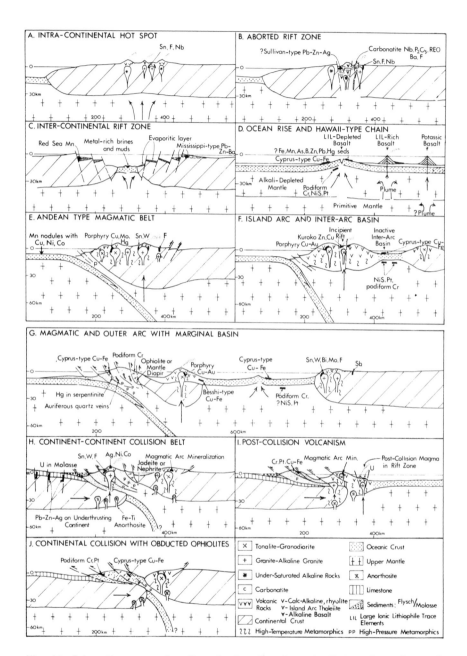

Fig. 11. Schematic cross-section through plate boundary-related tectonic settings and associated mineralization (from Mitchell and Garson, 1976). By permission of the authors and publisher, *Minerals Sci. Engng* 8, 129–169.

types of possible mineralization characteristic of a particular setting (Fig. 10) or settings (Table III, Fig. 11), and others concerned with the types of setting in which a particular type of ore body could form. While in general the emphasis has been on magmatic arc and to a lesser extent on ocean rise settings, intra-continental rift zones have attracted wide interest as possible sites for formation of stratiform copper ores in particular (e.g. Sawkins, 1976a). In the last five years the possibility of pluton-related mineralization in continental collision belts has received increasing attention, and a few authors have emphasized the importance of continental transform faults and fracture zones in controlling the location of a variety of types of ore body (Mitchell and Garson, 1976; Kutina, 1980a).

IX TECTONIC SETTINGS IN THE PRECAMBRIAN

Many authors now accept that tectonic processes analogous to those well-documented for the late Mesozoic and Cenozoic took place throughout the Phanerozoic. We consider that the existence of Lower Palaeozoic ophiolites, magmatic arcs and belts of imbricate flysch wedges interpreted as outer arcs requires that plate tectonics operated since the early Cambrian, and that further discussion here is superfluous. Similarly, the presence of magmatic arcs and ophiolitic belts in the Pan-African Belt and Saudi Arabia (e.g. Rogers *et al.*, 1978a) indicates that Wilson-type plate tectonics operated in the late Protero-zoic, from about 1000Ma ago onwards.

Concerning the early and middle Proterozoic, the presence by about 2700 Ma ago of the extensive and thick largely non-marine deposits of the Witwatersrand Basin in South Africa and similar successions in Australia and South America demonstrates that major continents have existed since the end of the Archaen. Most metallic mineral deposits of early and middle Proterozoic age occur in intracontinental settings, many of which can be interpreted as failed rift zones, and since we are concerned here with mineral deposits rather than crustal evolution we are fortunately not obliged to discuss whether extensive areas of oceanic crust existed and were or were not subducted prior to the later Proterozoic. Evidence from the mineralized successions indicates that failed rifts of early and middle Proterozoic age were broadly similar to Phan-erozoic rift systems which failed to develop into ocean basins, and whether these rifts occurred within a single major super-continent, as suggested by palaeo-magnetic evidence (e.g. Piper, 1974) and the distribution of Proterozoic linea-ments (Davies and Windley, 1976: Glikson, 1976), or within a number of con-tinents which underwent at least limited relative movement, does not materially affect the discussion.

Nevertheless, evidence for mid-Proterozoic collision in the Coronation Geo-syncline and formation of the Athapuscow aulacogen (Hoffman, 1973), and for similar collision in the Labrador Geosyncline (Dewey and Burke, 1973), suggests that ocean floor spreading, subduction and collision were operating by 1800 Ma ago. We consider that many of the early and mid-Proterozoic "ensialic" orogenies can be interpreted as back-arc magmatic and thrust belts analogous to those of Cretaceous age in North America or the Tertiary Eastern Cordillera in Bolivia, developed in continental interiors landward of magmatic arcs during subduction along a shallow-dipping Benioff zone.

In view of the similarity in tectonic features between Proterozoic and Phanerozoic rift-related successions, we include examples of Proterozoic mineral deposits within the tectonic framework established for the late Proterozoic and Phanerozoic. However, recognizing the difficulty in identifying Phanerozoic analogies for the host-rock sedimentary succession for some types of mineral deposit of early or early and middle Proterozoic age (i.e. quartz–pebble conglomerate uranium–gold and unconformity vein-type uranium deposits), we group these in a separate section on intracontinental deposits of uncertain tectonic setting.

There is widespread evidence that, before the Proterozoic, tectonic settings and major global tectonic processes were different from those existing and operating today: in particular, in contrast to the succeeding Proterozoic era, major intracontinental rifts did not exist, and while there is evidence for collision of small continental areas, it is unlikely that subduction of ocean floor along Benioff zones took place. Discussion of Archaen mineralization in terms of tectonic settings comparable to those today is therefore impractical, and we make no further reference to it in this book.

CHAPTER 2

Deposits Formed in Continental Hot Spots, Rifts and Aulacogens

The tectonic settings and associated mineral deposits described in this chapter respectively developed and accumulated within continental environments unrelated to orogeny. The hot spot settings can be explained by thermal processes in the asthenosphere, while the rift settings are clearly directly related to incipient plate boundaries. Descriptions of these anorogenic intracontinental and intraplate settings precede those of passive continental margins, oceanic settings and the orogenic subduction and collision settings because it has become customary to consider that a plate tectonic cycle of rifting, ocean floor spreading, subduction, collision and again rifting starts with the rift stage, although the plate tectonic hypothesis itself developed largely from the concept of spreading ocean floor.

In geosynclinal terms the rift stage corresponds to both the early generative stage of the European geosynclinal model of Aubouin, and to the post-geosynclinal stage at the close of Aubouin's cycle; it also is equivalent to the post-orogenic and final stages of the Russian geosynclinal model (Table II).

I INTRACONTINENTAL HOT SPOTS AND HOT SPOT TRACKS

A Tectonic Setting

The concept of hot spots in the upper mantle, fixed relative to the earth's spin axis and hence to overlying lithosphere, was developed by Wilson (1963, 1965) who suggested that oceanic linear island chains were the tracks formed by movement of oceanic lithosphere over hot spots in the upper mantle. Subsequently Morgan (1972) argued that the hot spots were the sites of rising mantle plumes.

The surface effects of hot spots beneath continental plates has been discussed largely with reference to Africa. Burke and Dewey (1973) considered that rifts develop on thermal domes above hot spots, as discussed below, and Burke and Wilson (1976) argued that magma formed in mantle plumes can

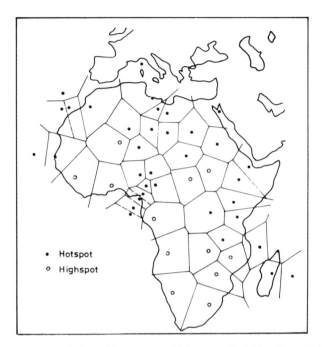

Fig. 12. Suggested population of hot spots and high spots in Africa (from Thiessen *et al.*, 1979). By permission of the authors and publisher, *Geology* 7, 263–266.

penetrate continental crust only where the lithosphere is more or less stationary with respect to the overlying mantle. Thiessen *et al.* (1979) have postulated the presence of about 36 mantle plumes beneath Africa, each overlain either by a hot spot with active volcanism or by a high spot, a crustal dome lacking volcanism (Fig. 12). They followed earlier suggestions that hot spot magmatism develops where the mantle plume underlies an older zone of crustal weakness, but suggested that where mantle plumes underlie stable cratonic areas only upward doming occurs, forming a high spot. Thiessen *et al.* also argued that major basins of Africa are situated between hot spots or high spots, and are underlain by zones of mantle downwelling. Although their existence has been disputed by some geologists, mantle plumes can explain many features of ocean basins and spreading ridges as well as of continents, and it is becoming evident that their control on crustal processes is second only to that of lithospheric plate accretion and consumption.

Descriptions of volcanism and plutonic activity in hot spot settings are based largely on rocks formed in inferred ancient hot spots of Mesozoic age or older, in particular from the Jos Plateau in Nigeria described below. In general hot spots are characterized by alkaline magmatic activity with well-developed ring structures. Eruptive rocks are predominantly rhyolitic with minor trachyte and in some cases basalt; intrusive rocks include carbonatite, per-alkaline granites and undersaturated alkaline rocks together with per-aluminuous granites. Initial

Sr^{87}/Sr^{86} ratios are highly variable, with basalts and some plutonic rocks showing low ratios indicative of a mantle source, while granites mostly have high ratios attributed to crustal anatexis and possibly to zone refining of rising mantle melts.

The concept that doming is related to mantle hot spots underlying continental lithosphere stationary relative to the underlying asthenosphere (Burke, 1977) implies that, where a continent is moving relative to the asthenosphere, mantle plumes are generally unable to penetrate the continental crust. However, a few authors have suggested that in some continental areas volcanism without associated rift or graben development was related to the track of a mantle plume formed by the continent drifting across it, for example the Neogene volcanism in the Massif Central of France (Burke *et al.*, 1973) and basaltic lavas of Cenozoic age in eastern Australia (Wellman and McDougall, 1974). It follows that any mineralization associated with the volcanic rocks was similarly related to the hot spot, and hence in some cases largely mantle-derived. One possible example of a late Cenozoic hot spot track and eruptive centre, apparently unmineralized despite intense hydrothermal activity, is provided by the basalts of the Snake River Plain in Idaho and the Yellowstone resurgent caldera and ring complex at its northeastern end (see Fig. 119). Southwestward drift of North America relative to an underlying hot spot now beneath Yellowstone has resulted in a rhyolitic track in and beneath the Snake River Plain overlain by subsequently erupted basaltic lavas (Smith and Christiansen, 1980).

B Mineral Deposits

Ore bodies associated with continental hot spots consist almost entirely of deposits formed in or adjacent to alkaline intrusives and lava flows and peraluminous granites (Table IV). They comprise some deposits of tin and associated minerals, rare economic deposits or uranium and some deposits of apatite, magnetite, vermiculite and pyrochlore associated with carbonatites.

The genesis of very few mineral deposits has been related to hot spot tracks or plume traces, despite the large world population of hot spots indicated by the number postulated beneath Africa. This suggests that in general mantle plumes have little effect on the crust of moving continental lithosphere, and it may be that hot spot tracks with significant expression in the continental crust are very rare and possibly non-existent. Nevertheless we include a brief description of two mineral deposits for which a hot spot track origin is at least a possibility.

1 Tin and uranium mineralization associated with anorogenic hot spot granites

Burke and Dewey (1973) suggested that many anorogenic granites were emplaced above mantle plumes prior to development of intracontinental rifts, and Sillitoe (1974) pointed out that economic deposits of tin were associated with some of these granites. The most widely quoted example of tin mineralization in plume-

TABLE IV

Mineral deposits characteristic of intracontinental hot spots.

Tectonic setting	Association	Genesis	Type of deposit/metals	Examples
Hot spots and hot spot tracks	Per-aluminous and per-alkaline granites	Magmatic and meteoric hydrothermal	Sn, Nb	Jos Plateau, Nigeria (Jurassic); St Francois Mts, Missouri (U Proterozoic); Rondônia, Brazil (U Proterozoic)
	Per-alkaline granites	Magmatic and meteoric hydrothermal	U	Bokan Mt., Alaska (Mesozoic); Appalachians (M Palaeozoic-Mesozoic)
	Carbonatites	Magmatic–metasomatic	Apatite, magnetite, vermiculite, pyrochlore	Kola Peninsular (U Proterozoic–Palaeozoic); Gebel Uweinat, Egypt (U Mesozoic–L Cenozoic)
	Basanitoid basalt		Sapphire, ruby	Kampuchea, Thailand (Quaternary)

related granites is that of the Jos Plateau in Nigeria; other examples include the early Upper Proterozoic tin mineralization in the Saint Francois Mountains of Missouri, and the later Proterozoic deposits of Rondônia in Brazil.

(a) Tin deposits of the Jos Plateau, Nigeria

The Jurassic mineralized granites of the Jos Plateau are the southernmost group of a 1300 km long discontinuous chain of ring structures including the Carboniferous complexes of southern Niger and the mid-Palaeozoic complexes of northern Niger (Figs 13 and 14A), both of which also contain cassiterite (Bowden *et al.*, 1976). The Jos Plateau granites lie immediately west of the NE-trending Benue Trough, interpreted as a failed rift in which subsidence began in the Albian, but there is no obvious genetic relationship between emplacement of the granites and subsequent rifting.

According to Bowden and Kinnaird (1978) more than 40 ring complexes, each up to 25 km diameter, are known in the Jos Plateau, within a 200 km wide N-trending belt. Plutons include per-alkaline albite–riebeckite granites, sub-alkaline hastingsite, and biotite granites, the last occurring as both concentric intrusions and stocks or cupolas. The granites intrude Precambrian metamorphic and igneous rocks, and in some places coeval silicic volcanics and minor basalts. The per-alkaline granites contain high concentrations of zinc and several other trace metals, while tin is confined to the biotite granites, erosion of which has given rise to the important alluvial workings. In the biotite granites two episodes of mineralization have been noted: an early post-magmatic phase that resulted in dispersed mineralization with columbite, xenotime, thorite, and cassiterite, and a later vein-controlled episode with formation of greisen and associated cassiterite, sphalerite, chalcopyrite, galena and pyrite.

Initial Sr^{87}/Sr^{86} ratios of the mineralized granites exceed 0.720, indicating a crustal source for the magmas and hydrothermal fluids, and providing some support to those favouring an origin of metals from remobilization of older mineralized rocks at depth. This hypothesis is also supported by the location of the richest area of tin mineralization on the Jos Plateau, where the N-trending belt of Younger Granites intersects the belt of tin mineralization associated with the Older Granites of Precambrian age (Hunter, 1973). Turner and Bowden (1979) suggested that in the Ningi-Burra caldera complex of Nigeria, an initial mantle-derived melt underwent crystal fractionation, and that zone refining with crustal anatexis took place during trapping of the magma in the absence of crustal extension or rifting.

Bowden *et al.*, (1976) suggested that the ring complexes in Niger and Nigeria could have formed while Africa was more or less stationary with respect to an underlying heat source, the progressive southward decrease in age of each group of complexes recording the northward movement of Africa relative to the sub-lithospheric mantle from the mid-Palaeozoic to Jurassic. A possible relationship of mineralization to lithospheric plate movement is provided by the suggestion of Turner and Bowden (1979) that plate movement leads to removal of the

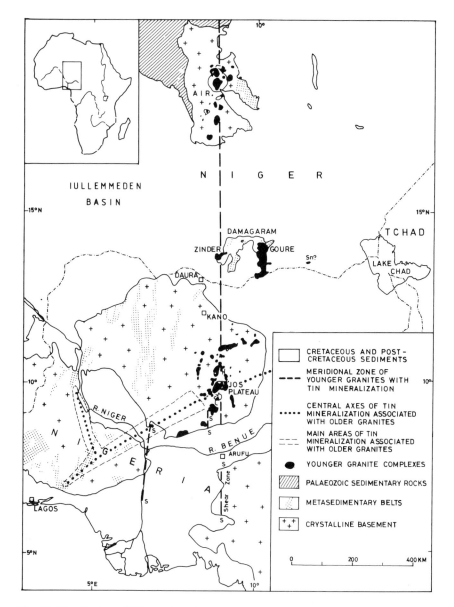

Fig. 13. Distribution of Palaeozoic and Mesozoic ring complexes with tin mineralization in Niger and Nigeria (from Turner and Webb, 1974). The zone of mineralization, associated with the Older Granites (after Hunter, 1973), intersects the 9° meridional zone of Younger Granite ring complexes in the area of high tin concentration at the Jos Plateau. By permission of the authors and publishers, *J. geol. Soc. Lond.* **130**, 71–77 and *Minerals Sci. Engng* **5**, 53–57.

Fig. 14. (A) Landsat image of the Aire area in Niger, published by permission of NASA. The dark-coloured elliptical or circular mountains are tin-bearing alkali granites of upper Palaeozoic age. These from north to south are Agalak, Ashkout, Baguezans and Taraouadji. They lie on a meridional zone of postulated hot spot activity extending south-wards through to the Mesozoic ring-complexes with tin mineralization at the Jos Plateau, Nigeria (see Fig. 13). (B). Landsat image of the northern part (about 120 km long) of the Great Dyke of Zimbabwe, published by permission of NASA. The Great Dyke which averages 6 km in width is postulated to lie within the root-zone of a Precambrian rift (see Fig. 26). Associated mineralization includes chromite, asbestos, copper, nickel, gold and platinum. The synclinal features in the southern part of the image are lopolithic subcom-plexes with layering dipping inwards at shallower angles than the main dyke.

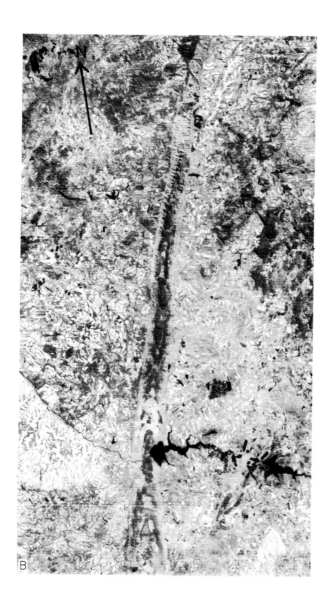

source of mantle-generated magma from beneath a ring centre: this results in an isolated magmatic system which does not undergo post-caldera resurgence, and facilitates development of highly evolved granites at depth. Much of the mineralization is probably associated with final stages of emplacement of these evolved magmas.

(b) Tin mineralization in the St Francois terrain, Missouri

The late Proterozoic granitic ring complexes and tin mineralization of southeastern Missouri, considered to be related to a mantle plume, have been described by Lowell (1976) and Kisvarzanyi (1980). Silicic and minor basaltic volcanic rocks are intruded by biotite granite forming sub-volcanic massifs, by amphibole granite emplaced along ring fractures related to cauldron subsidence, and by later two-mica granite with which the tin is associated. The tin granites, which are cylindrical and up to 20 km in diameter, form central plutons and their emplacement is considered to have been related to resurgent doming. The style of tin mineralization, which occurs entirely as cassiterite, reportedly resembles that of sub-volcanic tin deposits of Japan, although the sulphides present in the latter are absent. Concentrations of W, Nb and U as well as topaz are associated with the tin.

Kisvarzanyi (1980) has argued that the absence of significant volumes of basalt associated with the silicic volcanics indicates that only very limited rifting took place, although Lowell (1976) implied that the late Precambrian rifting in central North America was related to hot spot magmatism including that in the St Francois Mountains.

(c) Tin deposits in Rondônia, Brazil

Generation of tin-bearing granite complexes in Rondônia, Brazil (Kloosterman, 1967) has been related (Sawkins, 1976b) to a mantle hot spot. About 30 ring-like granitic intrusions, ranging in size from 5 km to 20 km across, form elongate, possibly fault-controlled belts within high-grade metamorphic rocks (Priem et al., 1971). In the northern part of the area, several granitic complexes intrude metasedimentary rocks of pre-Grenville age which infill a rift valley trending roughly easterly.

The intrusions are strongly differentiated with chemical compositions ranging from those of biotite granite to quartz monzonite and syenite. Their syenitic affinities and field relations were believed by Priem et al. (1971) to be indicative of anorogenic emplacement in a stable cratonic environment. Rb-Sr data provide a computed isochron age of 977 ± 20 Ma, with an initial Sr^{87}/Sr^{86} ratio of 0.710 ± 0.008, indicating considerable assimilation of crustal rocks. The tin mineralization, in greisen-like zones and topaz-quartz veins cutting both complexes and country rocks, is characterized by a topaz/cassiterite association and locally wolframite and columbite–tantalite are present.

Sawkins (1976b) has suggested that the age of emplacement of the Rondônia granites coincided with a major widespread episode of hot spot activity associated with the fragmentation of a postulated proto-Pangea.

(d) Uranium mineralization of Bokan Mountain type

Rogers *et al.* (1978b) recognized a type of granite-related uranium mineralization, distinct from that of Rossing-type (Ch. 6, IV. B.1. (f)), and defined by its association with post-tectonic commonly sodic and per-alkaline granite rich in Th, Nb and F, rocks which are characteristic of the hot spot settings referred to here.

The type-example of Rogers *et al.*, is the Bokan Mountain deposit of southeastern Alaska, within a Mesozoic peralkaline riebeckite granite intruding Upper Proterozoic and Lower Palaeozoic sediments. Uranium (MacKevett, 1963) occurs as primary disseminations and segregations in the granite and in associated aplites and pegmatites, and as epigenetic or secondary mineralization in veins and fractures within the granite and in pore spaces within the sedimentary host rocks. The two episodes of mineralization are interpreted respectively as the results of magmatic fluids and later circulation of post-magmatic hydrothermal solutions. Rogers *et al.* (1978b) consider the magma and associated uranium to be mantle-derived, because of probable absence of sialic source rocks at depth, a Th/U ratio greater than 1, and low initial Sr^{87}/Sr^{86} ratios in post-tectonic granites elsewhere which are geochemically similar to the Bokan pluton.

Areas identified by Rogers *et al.* in the eastern USA with plutons comparable to those of Bokan Mountain and hence with a possisble uranium potential are the Upper Palaeozoic pluton belt of Georgia, South and North Carolina, and Virginia, part of the Mesozoic White Mountain Magma Series of New England, and the Palaeozoic molybdenum-copper province of Maine. In New England the post-tectonic ring dykes and plutons have been interpreted as a Mesozoic plume track, and comprise one of a number of mid-Palaeozoic and Mesozoic anorogenic per-alkaline granite complexes in the northern Appalachians related by Taylor (1979) to episodic cessation of lithospheric movement relative to underlying mantle plumes.

2 Apatite, magnetite, vermiculite and pyrochlore mineralization associated with hot spot carbonatites and related alkaline and ultrabasic rocks

Carbonatites are rare rocks of limited areal extent and hence relatively unfamiliar to most geologists; a general definition is therefore useful here.

Carbonatites may be defined as intrusive and extrusive igneous rocks, of deep-seated origin, composed largely of the carbonate minerals calcite (sövite, coarse-grained, alvikite, fine-grained), dolomite or ankerite (beforsite, dolomite-carbonatite or ankeritic carbonatite), strontianite and calcite (strontianitic sövite or carbonatite), and rarely of Na–K–Ca carbonates (natrocarbonatite), together with other minerals such as Na and K feldspars, sodic pyroxenes and amphiboles, apatite, magnetite, baryte and zircon.

Carbonatite complexes are generally characterized by fenitic aureoles of alkali metasomatism, with, at deeper levels, country rock desilication concomitant with introduction of soda and potash feldspars, aegirine and soda amphiboles, and at higher subvolcanic and volcanic levels a potassic metasomatism of the fenites and country rocks to feldspathic rocks and breccias, and ultrapotassic trachytic feldspar rocks. Mobilization of syenitic and nepheline-bearing fenites may result in formation of intrusive equivalents although some geologists consider these to be differentiates of alkali basalts. Experimental evidence, however, suggests that kimberlite–carbonatite magmas and related olivine-melilites and nephelinites can be generated, depending on the H_2O/CO_2 ratio and P/T conditions, by partial melting of the peridotite mantle (Brey and Green, 1976). Nepheline syenites as well as the ijolitic series (urtite-jacupirangite) and alnoites may therefore be derived from the nephelinitic fraction.

Compared with carbonate rocks of sedimentary origin, carbonatites are generally enriched in Nb, Zr, Ti, U, Th and Ce/La rare earths; Sr^{87}/Sr^{86} ratios are typically between 0.702 and 0.7065, the higher values possibly having been influenced by wall-rock contamination (Deans and Powell, 1968).

Carbonatites have been recognized in the last 25 years as important economic sources of apatite, magnetite, pyrochlore, rare earth, titanium and vanadium minerals, zircon, baddeleyite, fluorite, uranium minerals and exceptionally copper and lead. Of these, apatite, magnetite, vermiculite and pyrochlore may be of particular importance in hot spot carbonatites and alkaline rocks.

Two possible hot spot areas characterized by carbonatite and alkaline rocks are described briefly below. Many other areas, including the Lake Natron–Ngorongoro–Kilimanjaro area of Tanzania and much of the Chilwa Alkaline Province of southern Malawi, are probably also related to underlying hot spots which resulted in formation of the East and South Africa rift system.

(a) The Kola Peninsula–northern Karelia alkaline province

This area exhibits all the anticipated features of hot spot activity with an extended period of alkaline activity from 610 Ma at the Kovdor carbonatite-ultrabasic complex to 270 Ma at the Turiy cluster of alkali and carbonatite dykes (Fig. 15), supporting Sawkins' (1976a) suggestion that hot spots initiate protracted periods of igneous activity. The presumed triple junction is centred approximately on the alkaline massifs of Khibiny (40 km x 35 km) and Lovozero (30 km x 20 km) with two arms within a fault zone trending NW, outlined by widespread basic and ultrabasic dykes. Within a SW-trending area, interpreted as an aborted Caledonian rift, there are large clusters and zones of basic and ultrabasic dykes and carbonatite, including the Ozernaya Varaka, Afrikanda, Kovdozero, Vuorijarvi and Salantava centres (Ginzburg, 1962). Belyayev and Uyad'yev (1978) show igneous centres and massifs of alkaline, basic and ultrabasic rocks and carbonatites with a remarkable system of arcuate and associated radial dykes and faults related to an updomed area some 450 km across centred on the Khibiny massif (Fig. 15), and similar in area to the Hawaiian hot spot.

The position of the Kola alkaline rock province was probably controlled to

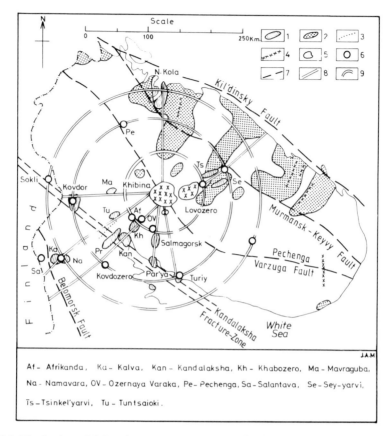

Fig. 15. Distribution of dykes, igneous centres and massifs of alkaline, basic and ultrabasic rocks and carbonatites at the Kola Peninsula hot spot (after Belyayev and Uyad'yev, 1978). 1, Large clusters and zones of dykes of alkaline rocks and carbonatites; 2, areas of wide-spread basic and ultrabasic dykes; 3, large alkaline dykes; 4, large basic dykes; 5, alkaline massifs; 6, central intrusions of carbonatite and associated rocks; 7, major fractures; 8, radial faults; 9, idealized position of arcuate faults. By permission of the authors and publisher, *Int. Geol. Rev.* **20**, 273–280.

some extent by the Kandalaksha deep fracture zone (Paarma and Talvitie, 1976), which possibly dates back to the late Archaen or early Proterozoic (Petrov, 1970). This fracture is seismically active and is clearly visible on Landsat imageries as a lineament over 150 km long and 10 to 20 km wide (Vartiainen and Paarma, 1979).

Important mineralization at the Kola hot spot is associated with the nepheline-syenite/ijolite massif of Khibiny, the ultrabasic/carbonatite centres and the carbonatite centres.

At Khibiny (Fig. 16) reserves of fluor-apatite rock have been estimated at 2700 million tonnes averaging 18% P_2O_5 (Notholt, 1979), and annual production of apatite is about 15 million tonnes per annum from four large

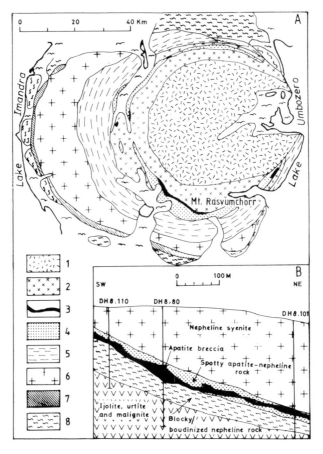

Fig. 16. Khibiny Complex, Kola Peninsula (after Ivanova, 1963). (A) Simplified geological map: 1, foyaite; 2, nepheline–syenite (rischorrite); 3, apatite–nepheline deposit, 4, ijolite, urtite and malignite; 5, trachytoid nepheline syenite (khibinite); 6, non-trachytoid khibinite; 7, hornfels; 8, country rocks. (B) Cross-section through apatite–nepheline deposit at Mt Rasvumchorr. By permission of the author and publisher, "Apatite Deposits of the Khibiny Tundra", Gosgeoltechizdat, Moscow.

mines. The Khibiny ring complex comprises steep inward-dipping layered intrusions of various type of nepheline syenite and ijolite (Gerasimovski *et al.*, 1974). The apatite-nepheline ore bodies occur in a 11 km long zone of layered ijolitic rocks up to 100 m thick, dipping towards the centre of the complex. Deposits consist of apatite breccia, and apatite-rich ijolites and urtites comprising 15-75% apatite, 10-80% nepheline, 1-25% sodic pyroxene, 5-12% sphene and smaller amounts of other minerals (Ivanova, 1963). Contents of up to 11% SrO and nearly 5% Re_2O_3 have been found in the fluor–apatite. Nepheline concentrates are also worked for the production of alumina.

The Kovdor alkalic/ultrabasic complex with minor dykes and veins of carbonatite is typical of several of the ring structures in the Kola Peninsula, located in most cases at or near the intersection of radial and arcuate faults probably produced by updoming above a mantle plume. Iron ore and apatite reserves of about 700 million tonnes and 110 million tonnes respectively are being open-cast mined to produce nearly a million tonnes of apatite concentrate per annum. Baddeleyite (ZrO_2) is a by-product mineral, and vermiculite is

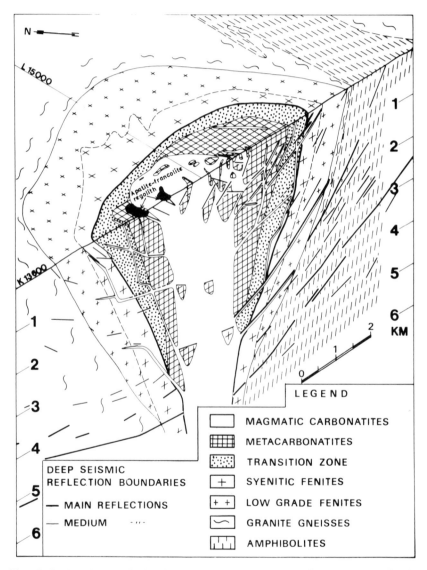

Fig. 17. Section through the Sokli carbonatite complex showing fracture zones (infilled by dykes) outlined by seismic soundings, and the position of apatite-francolite regoliths (from Vartiainen and Paarma, 1979). By permission of the authors and publisher, *Econ. Geol.* **74**, 1296–1306.

worked separately in olivinites. The Kovdor complex is a ring structure consisting of an outer ring of ijolitic rocks, an inner zone of melilite rocks and an ultrabasic core of pyroxenites, nepheline-pyroxenites and olivinites cut by veins and dykes of carbonatite (Rimskaya-Korsakova, 1964). The core rocks resemble the pyroxenites and phoscorites of the Palabora carbonatite in South Africa in containing abundant vermiculite and pegmatoid diopside-forsterite-phlogopite rocks with apatite. The apatite is obtained from veins and lenses consisting mainly of apatite, forsterite, magnetite, calcite and phlogopite.

The 18 km^2 Sokli carbonatite complex in Finland (Paarma, 1970; Vartiainen and Paarma, 1979) is one of the world's largest carbonatites, perhaps due to its position at an intensely fractured intersection of the Kandalaksha fracture zone with an arcuate fault of the Kola hot spot stress system. Mineral occurrences of economic importance include: reserves of nearly 100 million tonnes of phosphate rock in the form of an apatite-francolite regolith; hydromica-apatite residues with up to 35% apatite; early stage pyrochlore with U-Ta enrichment (UO_2 up to 25%) and late stage hydrothermal pyrochlore with thorium (ThO_2 1-7%), both in magmatic carbonatites; and apatite-magnetite mineralization in metaphoscorites with 15-30% magnetite.

The Sokli complex lacks alkalic intrusives and consists of four carbonatitic units surrounded by a large fenite aureole comprising syenitic fenites and low-grade fenites of shock zone type (von Eckermann, 1948). These units (Fig. 17) comprise a magmatic core of sovite, a metacarbonatite area of metasilico-sovites, a metaphoscorite zone of primary rocks such as magnetite-bearing olivinites and pyroxenites and a disrupted transition zone adjacent to the fenites consisting of fenite and amphibolitic rock breccia in banded silicosovites together with thin dykes and veins of carbonatite. Calcitic, dolomitic and numerous kimberlitic dykes and veins form an irregular stockwork. Seismic soundings indicate that the magmatic core tapers to 1 km diameter at a depth of 5 km. A general slight tapering in depth seems to be a feature of carbonatites, e.g. Palabora in South Africa, and it is probable that in extreme depth carbonatites, like kimberlites, are dyke-like bodies within deep fracture zones tapping the mantle.

(b) Gebel Uweinat, SW Egypt-Libya border

The Gebel Uweinat oasis area provides an isolated example of magmatic rocks, in a stable cratonic area in Africa remote from any recognizable rifting, which closely matches the expected surface expression of pre-rift hot spot activity. A cluster of ring-like features consists of Upper Mesozoic or Lower Cenozoic rocks including syenite, nepheline syenite, periodotite and carbonatite intrusive into basement gneisses (A.A. Hussein, pers. comm.). However no economic mineralization has so far been reported.

3 Mineral deposits possibly associated with continental hot spot tracks

The postulated relationship of the two types of deposit described below to migration of continental lithosphere over hot spots should be regarded as highly speculative; in both cases the hot spot track hypothesis was put forward in the absence of any other satisfactory explanation.

(a) Porphyry copper deposits of southwestern North America

While most deposits of porphyry-type copper are considered to form in magmatic arcs above subducting ocean floor, Livingstone (1973) proposed that an important belt of porphyry coppers in the southern Basin and Range Province of Arizona and Mexico was related to movement of the North American plate relative to an underlying hot spot. Livingstone showed that the age of the calc-alkaline pluton host rocks to the mineralization increased systematically throughout the NW-trending belt of deposits, from 54 Ma in the southeast to 70 Ma in the northwest. He suggested that this age distribution could be explained by northwestward migration and clockwise rotation of the North American plate at a rate relative to the hot spot of around 3.5 cm/yr, and that the mantle hot spot had a maximum dimension of 450 km, comparable to that of the oceanic Hawaiian hot spot. The postulated hot spot source for these porphyry deposits has not been widely accepted, and other hypotheses of origin are discussed briefly later (Ch. 5, II.B. 1.(d)).

(b) Gemstone deposits of Kampuchea and Thailand

Placer deposits of gem-quality sapphire, zircon and spinel, and reportedly ruby (Berange and Jobbins, 1975), are derived from deeply weathered Quaternary alkali basalts near Pailin in western Kampuchea and adjacent areas of southeastern Thailand.

The gemstone-bearing basalts lie near the southeastern end of a chain of scattered late Cenozoic mostly barren basalts flows and plugs extending to northern Thailand (Fig. 18). Barr and Macdonald (1978) showed that the basalts in southeastern Thailand are of two types, olivine basalts and basanitoid basalts. The basanitoid basalts are the hosts to the gemstones, which are interpreted as high-pressure megacrysts formed with the basalt in the upper mantle and transported rapidly to the surface.

The basalts in Thailand and more extensive basalts in Indochina were related by Barr and Macdonald (1978) to tensional conditions associated with late Mesozoic to Cenozoic opening of the China Basin. However, as the southeastern Thailand basalts are largely or entirely Quaternary in age, and the sapphire-bearing Denchai basalt in northern Thailand is Upper Miocene (Barr and Macdonald, 1979), there is no obvious necessity for them to be related to a much earlier event. Mainland southeast Asia is generally considered to be more or less stationary with respect to underlying lithosphere, although a former position to the southeast may have been required to allow the Cretaceous-

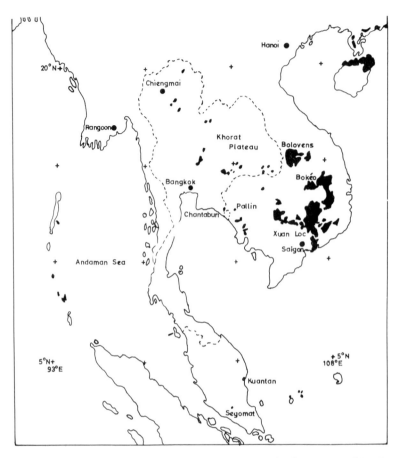

Fig. 18. Distribution of late Cenozoic basalts in Southeast Asia (from Barr and MacDonald, 1979). By permission of the authors and publisher, *Earth planet. Sci. Lett.* **46**, 113–124.

Palaeocene northward migration of Greater India. This suggests that, despite the absence of rhyolites so characteristic of, for example, the Yellowstone hot spot, the alkali basalts and associated gemstones could be related to north-westward migration over an underlying hot spot or mantle plume; nevertheless, why most of the basalts are no older than Quaternary remains a problem.

4 Preservation potential

Mineral deposits associated with hot spots should have a rather high preservation potential, as they tend to form within continents remote from orogenic belts on and near the continental margins. However, whether the rather limited number of economic deposits associated with ancient hot spots reflects the scarcity of hot spot magmatism in the geological record, or the common erosion of the upper mineralized portions of intrusives, is uncertain.

II INTRACONTINENTAL RIFTS AND AULACOGENS

A Tectonic Setting

Graben, widely interpreted as extensional zones of rifting, have long been recognized as a major type of structural feature with associated sedimentary and volcanic fills within continents, and were commonly referred to in geosynclinal terminology as taphrogeosynclines. Shatski (1947) noted that many graben in the USSR pass at one end and at a high angle into orogenic belts and Burke and Dewey (1973) described a large number of graben of Proterozoic and Phanerozoic age which they considered had a similar origin to the East African Rift System, and represented rift zones which failed to develop into ocean basins.

The more recent literature indicates that many graben or rift zones can be grouped according to their tectonic setting at the time of formation: those which developed as a "trilete" system of three or more rifts radiating from a crustal dome centred on a hot spot, during incipient ocean floor spreading as described by Burke and Dewey (1973); and collision-related rifts which developed following, and were related to, continent–continent or continent–arc collision (Sengor *et al.*, 1978).

1 Hot spot-related rifts

We describe here active rifts, failed rift arms, and half-graben or rifts now separated by ocean floor, as well as aulocogens or failed rifts terminating at one end in a younger orogenic belt, development of which is considered to be related to prolonged hot spot activity.

(a) Trilete rift systems, failed arms and half-graben

Recognition of the numerous rift systems described by Burke and Dewey (1973) was based largely on analogy with late Cenozoic rifts in Africa (Burke and Whiteman, 1973) and with Mesozoic rifts formed during the break-up of Gondwanaland. For example in the eastern rift system of East Africa, the intracontinental North and South Gregory Rifts meet the Kavirondo Rift in a triple junction, while in the Afar area the intracontinental Ethiopian Rift joins the Red Sea and Gulf of Aden which are underlain by oceanic crust.

Burke and Dewey (1973) following Burke and Whiteman (1973) suggested that the Ethiopian Rift and similar structural features elsewhere intersecting a continental margin at a high angle each represented the failed arm of an original three-armed or "trilete" intracontinental rift system, two arms of which developed into an ocean basin while the third became inactive before it reached the stage of ocean floor spreading. They argued that the rifts develop over a hot spot in the continental crust, probably related to an underlying mantle plume (Morgan, 1972), and undergo a distinctive sequence of

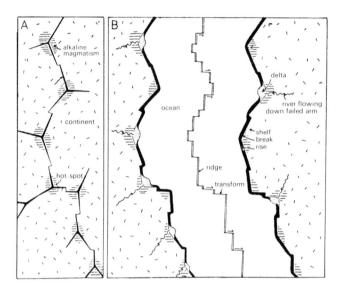

Fig. 19. Diagrammatic representation of continental break-up by (A) linking of hot spot-related rift arms to form (B) a zone of subsequent ocean floor spreading and transform faults (from Dewey and Burke, 1973). By permission of the authors and publisher, *J. Geol.* **81**, 683–692.

development (Fig. 19). Updoming of the crust above the plume is followed by formation of three or rarely more rifts radiating from near the centre of the dome; emplacement of basaltic dykes is succeeded in two of the rifts by ocean floor spreading, while the third rift, intersecting the ocean at a high angle, either becomes inactive, or may develop into a "leaky" transform fault, e.g. the Dead Sea Rift. The spreading rifts join those radiating from adjacent plumes, forming a continuous network of spreading rift systems.

The presence of the exceptionally well-developed Neogene rift system in East Africa is considered to be related to the lack of relative movement between Africa and the underlying mantle for the last 25 Ma, allowing heat from mantle plumes to penetrate the overlying continental lithosphere (Burke, 1977). With continued development Africa could split along the rift system, in which case the failed arms would become intracontinental rifts extending to the margin of the opening ocean, as in the case of the Ethiopian Rift.

Many of the recognized rifts of Phanerozoic and Proterozoic age are now considered to have been initiated at trilete junctions, and numerous examples were described by Burke and Dewey (1973). Examples of failed arms are known from the east African and eastern India coasts, and the western and north-western coasts of Australia, as well as the Atlantic, while Mesozoic graben beneath the North Sea provide an example of a failed rift system. Most authors emphasize the role of tension in the lithosphere in generation of intracontinental rifts, but Falvey (1974) has attempted to explain rift

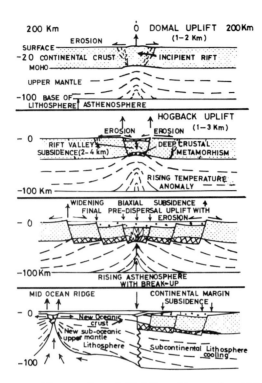

Fig. 20. Development of intracontinental rift followed by continental break-up and ocean floor spreading in response to rising thermal anomaly or hot spot. Dashed lines indicate approximate isotherms (from Falvey, 1974). By permission of the author and publisher, *Aust. Petroleum Explor. Ass. J.* **10**, 95–106.

development entirely by thermal processes which result in uplift, erosion, and finally in subsidence (Fig. 20).

Rift origins unrelated to hot spot and trilete junctions have been put forward by some authors. For example, Turcotte and Oxburgh (1973) suggested that membrane stresses generated within a plate moving longitudinally on the earth's elliptical surface could result in rifting, but this hypothesis has not been widely accepted.

Trilete rift successions typically lie in marked unconformity on older rocks and may include evaporites, fanglomerates and arkoses, although shallow marine sediments are also often present; failed rift successions commonly terminate at their oceanic end in a major delta. Volcanic rocks are usually interbedded with the sediments in the lower part of the succession, and include alkaline basalts, as in the Benue Trough and bordering the Red Sea, and undersaturated alkaline rocks as in the East African Rifts (see Fig. 24). Rift successions are usually only gently folded: highly deformed rocks are anomalous and perhaps occur only in rifts which began to develop into an ocean basin and subsequently closed as a result of subduction. Volcanoes associated with rifting may form major intracontinental features, e.g. Tibesti, associated with only a poorly-defined rift structure.

Intrusions of mantle-derived undersaturated alkaline and per-alkaline rocks emplaced beneath volcanic centres are typical of many rifts, and seismic refraction and gravity data suggest that major basaltic dykes of probable tholeiitic composition at least several kilometres thick underlie parts of the East African Rift system. Predominantly alkaline granite provinces related to hot spots are associated both with failed rifts, for example the Nigerian granites adjacent to the Benue Trough, and with successful rifts on half-graben e.g. the Tertiary granites of northwest Scotland.

It is now realized that failed arms and failed rift systems are often subject to rejuvenation or reactivation. This is best illustrated in the southerly rifts of East Africa, involved in Mesozoic rifting during the break-up of Gondwana long before the onset of Neogene rifting which continues today (Burke and Dewey, 1973). Bailey (1977) has strongly questioned the validity of intra-continental magmatism resulting from deep mantle plumes on the grounds that well-documented multiple episodes at rift junctions (centres of alleged hot spots) extend back to the Precambrian, e.g. Rungwe–North Nyasa area in Africa. During the whole Precambrian to Recent period of alkaline-carbonatite activity the African plate shows a polar wander shift of over 110° of latitude rendering it highly improbable that the Rungwe area, and many other similar alkaline areas, happened to alight on several different occasions on several entirely different sub-lithosphere plumes. Alternatively it has been suggested that hot spot activity sets in motion a sequence of geological events that can continue for an extended period (Sawkins, 1976a), and that mantle "hot lines" postulated by Bonatti and Harrison (1976) rather than hot spots would allow more ready access of plume activity into an area of ancient crustal fractures.

There is some evidence that at a late stage in development of a failed rift the zone of subsidence changes from that of the initial relatively narrow rift to a broad basin which includes the rifts. Examples are the development of the North Sea over a pre-existing graben system at the end of the Palaeocene, the increase in area of sedimentation with time in the Proterozoic Athapuscow aulacogen and subsidence of the Neogene Chad Basin overlying Cretaceous rifts (Burke, 1976, 1977).

The arms of trilete rift systems into which ocean floor was emplaced are split longitudinally by the ocean crust, and form half-graben which subsequently lie on the opposite sides of an ocean basin. The half-graben successions, resembling those of failed rifts, are eventually largely buried by the miogeosynclinal successions of the continental shelf and slope deposited across the subsiding continental margin (Ch. 3, I.A.).

(b) Aulacogens

Elongate intracratonic troughs or graben terminating at one end in an orogenic belt were first recognized as distinctive geological features by Shatski (1947,

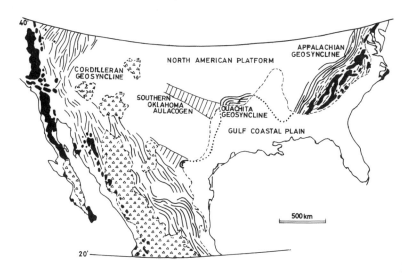

Fig. 21. Location of Southern Oklahoma aulacogen in structural framework of North America (after Hoffman *et al.*, 1974). Mississippi Embayment shown by dashed line northeast of Ouachita Geosyncline. By permission of the authors and publisher, *Spec. Publs. Soc. econ. Palaeont. Miner.* **19.**

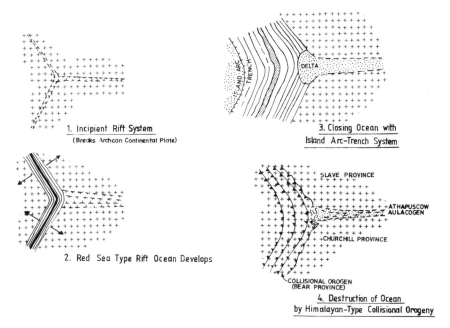

Fig. 22. Evolution of the mid-Proterozoic Athapuscow aulacogen (after Hoffman *et al.*, 1974, modified by Morton, 1974). By permission of the authors and publisher, *Spec. Publs. Soc. econ. Palaeont. Miner.* **19.**

1955) who applied the term "aulacogen" to troughs of this type in the Russian platform west of the Urals, and to the Southern Oklahoma trough in the USA (Fig. 21). Much later, aulacogens were identified elsewhere in North America, where Hoffman (1973) recognized that the structure and Proterozoic stratigraphy of the East Arm of Great Slave Lake resembled that of aulacogens, and termed it the Athapuscow aulacogen. Hoffman (1973) argued that the aulacogen formed one arm of a trilete rift system, the two other arms of which opened and subsequently closed in the mid-Proterozoic collisional orogeny of the Coronation Geosyncline (Fig. 22). Although Shatski's definition of aulacogen was independent of the cause of the initial rifting which formed the trough, the term is now commonly restricted to rifts which strike at a high angle into orogens and which originated at the hot-spot related trilete rift junction (Burke, 1977).

Failed rifts extending into the continent should be equally abundant on both sides of a spreading ocean, and hence when the ocean has closed aulacogens might be expected on the former subducting and also overriding continental plates. However, for some reason which is not entirely clear, aulacogens appear to be restricted to the subducting foreland plate with respect to the collision belt, and none has been recognized on overriding continental plates.

In general aulacogen successions are similar to those of failed rifts which pass into oceans described above, as their early genesis is identical, and the collision required to form an aulacogen post-dates the volcanism and most of the sedimentation in the rift. Despite this broad similarity, it is significant that in the Palaeozoic Southern Oklahoma aulacogen a major rhyolite unit near the base of the succession, and associated granites, are sub-alkaline in character, resembling silicic magmatic rocks of volcanic arcs rather than intracontinental rifts (Hanson and Al Shaieb, 1980).

2 Collision-induced rifts

Sengor (1976) proposed that some rifts terminating at one end in a collision orogen post-dated the collision and were genetically related to it. These rifts resemble aulacogens in plan and cross-section, but Sengor *et al.* (1978) suggested some significant differences in their stratigraphy. Unlike aulacogens, formation of a collision-related rift is not preceded by updoming, and the rift fill is younger than the deformed successions, other than those of the foreland basin, within the collision belt in which the rift terminates.

The best-known example of a collision-related rift is the Upper Rhine Graben (Fig. 23) extending from the Jura Mts to the Vogelsberg, and considered to have formed during major Eocene collision in the Alps (Sengor *et al.*, 1978); Illies and Greiner (1978) also suggested that the Upper Rhine Graben was initiated in the Eocene, just before the main episode of nappe formation in the Alps, and was modified by late Cenozoic sinistral faulting. Volcanism took place in both the Eocene and mid-Tertiary episodes of rift development.

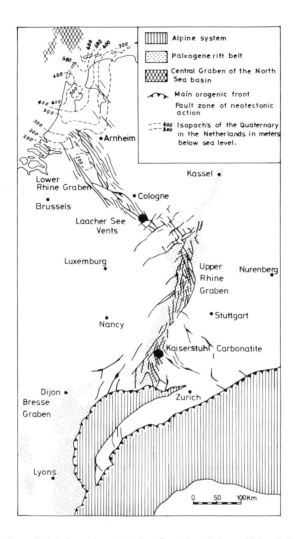

Fig. 23. An active subplate boundary stretches from the Alpine collision belt to the extinct Central Graben of the North Sea basin (from Illies and Greiner, 1978). The Kaiserstuhl carbonatite was intruded into the Upper Rhine Graben during its formation as an extensional rift in mid-Eocene to Miocene times, this Graben has since been affected by sinistral shearing produced by the per-Alpine stress regime with its staggered normal faults behaving as Ridel Shears. The Quaternary Laacher See Vents with carbonatite ejectamenta are associated with extensional rifting in the Lower Rhine Graben. By permission of the authors and publisher, *Bull. geol. Soc. Am.* **89**, 770-782.

While the Upper Rhine Graben lies on the former subducting foreland, another example of a collision-related rift, that of Lake Baikal, and poorly-known N-trending rifts on the Tibetan Plateau, referred to later (Ch. 6, VII. A), are situated on the overriding plate of the Himalayan collision belt. Strike-slip

TABLE V

Mineral deposits characteristic of hot spot and collision-related intracontinental rifts and aulacogens.

Tectonic setting	Association	Genesis	Type of deposit/metals	Examples
Intracontinental rifts and aulacogens	Carbonatites	Magmatic–metasomatic	Apatite + vermiculite, Cu–U–baddeleyite, pyrochlore, rare earth ± strontianite	Palabora, S Africa (Proterozoic) Oka, Canada L Cretaceous, Chilwa, Kangankunde, Malawi L Cretaceous
	Undersaturated alkaline complexes	Magmatic	Apatite	Baikal rift (U Palaeozoic); Oslo Graben (Permian)
	Kimberlites associated with carbonatites	Magmatic	Diamonds	Tanzania (Mesozoic?) and S Africa (Proterozoic and Cretaceous)
	Basic–ultrabasic intrusions	Magmatic	Cr–Ni–Pt–Cu	Great Dyke, Zimbabwe (L Proterozoic); Bushveld, S Africa (L Proterozoic); Duluth Complex, Minnesota (U Proterozoic)
	Biotite granite	Magmatic–meteoric hydrothermal	Porphyry Mo	Glitrevann, Oslo rift (Permian)

TABLE V (Cont'd)

Tectonic setting	Association	Genesis	Type of deposit/metals	Examples
	Shales, commonly calcareous and bituminous, above unconformity and beneath evaporites	Diagenetic or early epigenetic meteoric hydrothermal	Stratabound Cu	Atlantic margin Africa (Aptian); Kupferschiefer Europe (Permian); Zambia, Zaire (U Proterozoic)
	Shales, commonly bituminous, within terrigeneous sequence	Diagenetic or early epigenetic meteoric hydrothermal	Strarabound Ag-Pb–Zn 'Sullivan type'	Sullivan Mine, Br. Columbia (U Proterozoic); Mt Isa, Queensland (U Proterozoic); McArthur River (mid-Proterozoic); Gamsberg S Africa (U Proterozoic)
Intracontinental rifts and aulacogens	Terrigeneous clastics	Diagenetic or epigenetic	Stratabound sandstone type U	Athapuscow aulacogen Canada (M Proterozoic)
	Magnesian carbonates	Chemical sedimentary	Evaporites	Zechstein (Permian)
	Lacustrine brines and evaporites	Sedimentary	Na and K salts, magnesite and phosphate	E. Africa Rift lakes (Quaternary)
	Veins in black shale (Benue Trough)	Magmatic–connate hydrothermal	Pn–Zn veins	Benue Trough and Amazon fracture zone (Cretaceous)
	Veins in faults and lineaments	Meteoric, magmatic and connate hydrothermal	F veins	W North America (Cenozoic); Illinois (U Cretaceous)
	Veins in granite and basement	Epigenetic hydrothermal	Mo-quartz veins Ag Co–Ni arsenides	Oslo Graben (Permian); Keweenawan Rift (U Proterozoic)

faults related to collision (Molnar and Tapponier, 1977) are important in both the Baikal rift and the Upper Rhine Graben.

Whether collision-induced rifts can develop into ocean basins is uncertain. If they can, it is possible that some aulacogens were initiated as rifts in a collision orogenic belt, remained as failed arms during ocean opening, and now terminate in a younger collision belt as a result of ocean closure.

B Mineral Deposits

Intracontinental rifts, and aulacogens in particular, have long been of economic interest because of their petroleum potential, and to a lesser extent because of their evaporites. Carbonatites, kimberlites and alkaline granites, within or adjacent to rifts, provide a major source of metallic and other minerals (Table V), although except in East Africa (Figs 24 and 25) it was only in the last decade that the association of these with rift zones was fully appreciated. The common occurrence of stratabound copper deposits within intracontinental rifts was suggested by Burke and Dewey (1973), and re-emphasized by Sawkins (1976a) who suggested that rift successions deposited in the late Proterozoic about 1000 Ma ago were a preferred host for copper. Other types of stratabound deposit are also known from some rift successions.

It is not yet clear whether there are significant differences in mineralization between hot spot related rifts and aulacogens, and collision-induced rifts; mineralization is therefore here discussed independently of the inferred mode of origin of the rift, although in some cases the rift's origin is reasonably well understood. Mineral deposits associated with rifts can be conveniently divided into those within or related to magmatic rocks, stratabound deposits in sedimentary successions, and vein-type deposits.

1 Deposits associated with magmatic rocks

These include important magmatic mineral deposits of apatite, vermiculite, pyrochlore, rare earths, strontianite and copper associated with carbonatite intrusions, deposits of apatite associated with alkaline and alkaline/ultrabasic intrusions, diamond emplaced with kimberlites, and magmatic chromite, copper and nickel in major ultrabasic dykes and associated intrusive complexes, e.g. the Bushveld Complex of Zimbabwe. A few rifts contain porphyry molybdenum deposits associated with granitic intrusions.

(a) Carbonatites

In eastern Africa (Figs 24 and 25) alkaline centres and carbonatite complexes lie within or adjacent to the rift system and associated areas of swells (Baker *et al.*, 1972), although considerable sections of the rift valleys lack these rock types and in places several carbonatite centres cluster preferentially either at

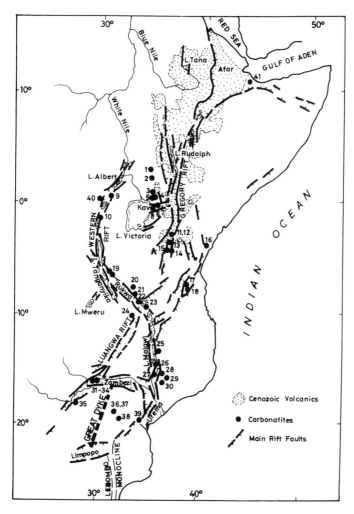

Fig. 24. Distribution of carbonatites in relation to rift faults in eastern Africa (from Mitchell and Garson, 1976, after McConnell, 1972). 1, Toror; 2, Napak; 3, Bukusu; 4, Tororo; 5, Sukulu; 6, 7, Ruri Homa Mt; 8, Rangwa; 9, Fort Portal; 10, Lueshe; 11, 12, Oldoinyo Lengai, Kerimasi; 13, Oldoinyo Dili; 14 Galappo; 15, Hanang; 16, Mrima; 17, Wigu; 18, Maji ya Weta; 19, Karema Depression; 20, Ngualla; 21, Songwe Scarp; 22, Mbeya; 23, Nachendezewaya; 24, Nkombwa; 25, Lipuche; 26, Lake Malombe Vents; 27, Kangankunde; 28, Chilwa Island; 29, Tundulu; 30, Nkalonje; 31–34, Kaluwe, Nachomba Mwambuto, Chasweta; 35, Katete; 36, 37, Shawa, Dorowa; 38, Chishanya; 39, Chiluvu; 40, Bingo; 41, Darkainle. By permission of the authors and publishers, *Bull. geol. Soc. Am.* **83**, 2549–2572 and *Minerals Sci. Engng.* **8**, 129–169.

intersections of rift troughs or along transverse fault structures (King and Sutherland, 1960).

Most of the carbonatites occur as ring complexes of central and arcuate carbonatite intrusions associated with alkaline rocks, and at deeper levels with

ultrabasic/alkaline rocks. These complexes are comparable to the alkaline centres
in the Kola Peninsula (see I. 2.A) and have been termed core carbonatite by
Heinrich (1966). Major examples are Dorowa in Zimbabwe (Johnson, 1962),
Chilwa Island and Tundulu Hill in Malawi (Garson, 1965a), Rangwa and Homa

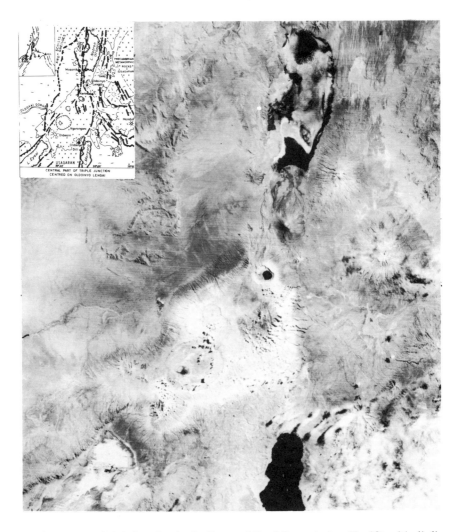

Fig. 25. Hot spot/triple junction in the Eastern Rift of Tanzania (see Fig. 29), with alkalic
and carbonatitic volcanoes and flows of Pliocene to present-day age (Landsat image published
by permission of NASA). The triple junction is approximately centred on the active
carbonatite/nephelinite volcano Oldoinyo Lengai, with two arms forming the Eastern Rift
and a third arm to the southeast forming a young rift within which are the major volcanoes
of Mount Meru and Kilimanjaro (see inset). Mineralization includes trona in Lake Natron,
magnesite on the eastern shores of Lake Natron, niobium and phosphate at the Kermasi
and Oldoinyo Dili carbonatites, and sedimentary phosphate and limestone on the eastern
terraces of Lake Manyara.

Mountain in Kenya (McCall, 1963) and Sukulu and Bukusu in Uganda (King and Sutherland, 1966). Vent types of carbonatite with little or no alkaline rock include Kangankunde in Malawi (Garson, 1965b) and the Rufunsa vents and sills of Zambia (Bailey, 1960). Flows and tuffs of carbonatite, nephelinite etc. occur in Tanzania at Kerimasi (James, 1958) and Oldoinyo Lengai (Dawson, 1962; see Fig. 31A) and in Uganda at Fort Portal (von Knorring and DuBois, 1961). In some cases, elongate dykes or sheets of intrusive carbonatite are localized in major rift faults, e.g. Sangu Complex, Tanzania (Coetzee, 1963) where three tabular carbonatite bodies extend for 25 km along the shore of lake Tanganyika.

Ages of intrusion tend to decrease northwards, from Zimbabwe with Triassic carbonatites, through the Rufunsa area of Zambia and southern Malawi both with Lower Cretaceous complexes, to northern Tanzania along the eastern edge of the East African swell where carbonatites range in age from Cretaceous in the south to intermittently active in the north (Oldoinyo Lengai). Possibly the Precambrian carbonatites in South Africa such as Palabora form the southern extension of the rift system.

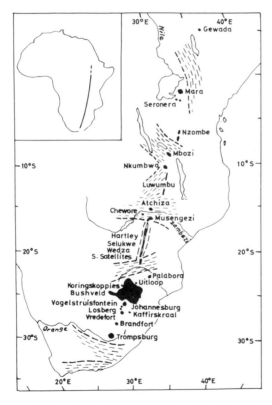

Fig. 26. Alignment of Precambrian igneous centres and major dykes in southern and eastern Africa (from Vail, 1978). Inter-cratonic mobile belts are shown schematically by dashed lines. By permission of the author and publisher, _Trans. geol. Soc. S. Afr._ **62**, 87–92.

Garson and Mitchell (1981) suggested that the formation of the East Africa rift and the persistence of alkaline and carbonatitic activity have been largely controlled by a remarkable 3800 km long Proterozoic lineament described by Vail (1978). Along this lineament, approximating a great circle, are situated the Trompsburg centre, the Vredefort Dome, The Bushveld Complex, and the Great Dyke of Zimbabwe (e.g. McConnell, 1980). The lineament possibly continues northwards through the Luangwa Rift of Zambia and numerous other basic/ultrabasic centres (Fig. 26), although its truncation by the Zambezi Belt, the probable eastward continuation of the latest Proterozoic Damara collision belt described later, implies that the Precambrian lineament was cut by an opening ocean which subsequently closed without lateral displacement. Deep-seated alkaline, carbonatitic and ultrabasic magma evidently penetrated the lineament as the African continental lithosphere intermittently came to rest over mantle hot spots from late Precambrian time onwards.

The simultaneous origin of several intrusions at the southern end of Vail's lineament has been ascribed to meteorite impact (Dietz, 1961). The Vredefort Dome and the Bushveld Complex exhibit some impact structures compatible with this theory (Hargraves, 1970; Hamilton, 1970), suggesting triggering of the Precambrian igneous activity in the south by impact which produced a deep-seated fracture accessible to later cycles of magmatism. Dietz (1964) also postulated an impact origin for the Sudbury Complex (1720 Ma) in Canada, which is aligned with younger carbonatite centres to the southeast in

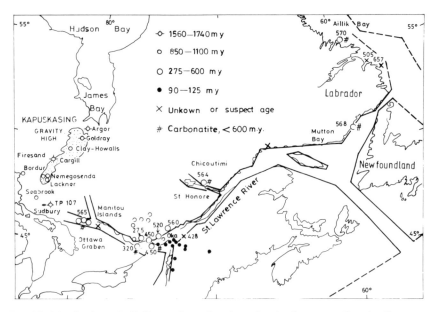

Fig. 27. Distribution of alkaline rocks and carbonatites in the eastern Canada rift system (after Doig, 1970, and Erdosh, 1979). By permission of the authors and publishers, *Can. J. Earth Sci.* **7**, 22–28 and *Econ. Geol.* **74**, 331–338.

the Ottawa Graben (Kumarapeli, 1976) within the St Lawrence rift system. However, as this rift shows several features considered typical of aborted spreading centres, acceptance of the meteorite impact hypothesis would imply a genetic relationship between impact and subsequent ocean floor spreading, for which there is little unambiguous evidence.

The now inactive St Lawrence rift system and related major structures in Ontario rival the African rift system in complexity (Fig. 27). The associated carbonatites are of four main ages, 1740-1560 Ma, 1100-850 Ma, 600-275 Ma and 125-90 Ma, correlated with major periods of orogeny (Gittins *et al.*, 1967; MacIntyre, 1977). Doig (1970) linked Lower Palaeozoic carbonatites of the St Lawrence rift system with carbonatites of similar age (565 Ma) in Greenland and Scandinavia, and suggested that the Kola Peninsula centres (515-290 Ma) may also lie within an interconnected North Atlantic rift system.

Many of the Precambrian carbonatites including the apatite-rich Cargill Complex, are associated with the Kapuskasing gravity and magnetic high which coincides with a rift zone stretching southwest from James Bay to the Seabrook carbonatite. This and the Township 107 carbonatite are aligned with the Ottawa Graben to the east, along which are at least nine carbonatites mainly of Lower Palaeozoic age. The earliest tensional structures here developed around 600-700 Ma (Kumarapeli, 1976), perhaps during rifting and post-rift sedimentation in the Appalachian geosyncline. The mid-Mesozoic activity which included the niobium-rich Oka carbonatite (Gold *et al.*, 1967) can be correlated with the opening of the North Atlantic Ocean.

In the collision-related Upper Rhine Graben, the carbonatite-nephelinite centre at Kaiserstuhl possibly formed above a mantle diapir in the Miocene. Later, faults in the lower Rhine were reactivated, and explosion vents with associated carbonatite and phonolite formed at the southeastern end of an extensional rift system linked to the central graben of the North Sea. Within the related Central European Rift Zone cutting the Bohemian Massif, fenites, nepheline syenites and carbonatites occur as xenoliths in young alkaline volcanics, and baryte-fluorite mineralization is associated with the major rift faults (Kopecky, 1971).

In India, six Precambrian carbonatites occur within an *en echelon* NE-trending fault system interpreted by Borodin *et al.* (1971) as an Upper Precambrian trough or rift valley. The Koratti (Sevathur) pyrochlore-rich carbonatite, the largest occurrence, is an elongate body some 2000 m by 200 m in area, flanked to the east by syenite and to the west by pyroxenite. Quartz-baryte veins with galena occur within the fault-zone to the southwest.

(i) Carbonatites with apatite ± vermiculite mineralization. Examples of these are Dorowa and Shawa in Zimbabwe, and Palabora in South Africa with medium-grade primary apatite ores; Glenover in South Africa, Bukusu in Uganda, Tundulu in Malawi and Cargill in Canada with secondary supergene phosphate; and Sukulu in Uganda with apatite-rich residual soils which infill broad valleys.

The Precambrian Palabora complex (Lombard *et al.*, 1964; Hanekom *et al.*, 1965) is an elongate plug-like intrusion of pyroxenite measuring 6.5 km x 2.5 km pierced by a central carbonatite plug surrounded by a coarse-grained apatite–magnetite–olivine rock known as phoscorite (Fig. 28). The latter is rimmed by a serpentinized pyroxene–vermiculite–olivine pegmatoid which also forms two other centres along the axis of the pyroxenite intrusion. Comparison with the Sokli Complex (see I.B.2. (a)) indicates that the similar Palabora pyroxenites, phoscorites and pyroxene–apatite–magnetite–olivine pegmatoid rocks are probably also country rocks metasomatically replaced by CO_2-rich fluids derived from the carbonatite magma.

There are three separate open-pit mines at Palabora: the carbonatite and phoscorite are worked for copper (see iv) and the olivine–vermiculite pegmatoid for vermiculite, a weathering product of phlogopite; an apatite-rich pyroxenite

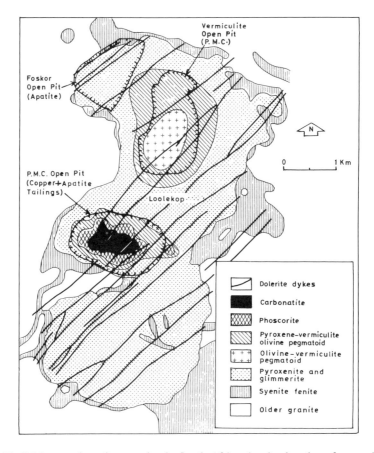

Fig. 28. Palabora carbonatite complex in South Africa showing location of open pits for apatite, vermiculite and copper plus apatite (from De Jager and Fourie, 1978). By permission of the authors and publisher, Am. Inst. Min. Metall. Eng.

forms the world's largest igneous phosphate deposit. Reserves in the quarry area alone are calculated at 3000 million tonnes of apatite concentrates (36.5% P_2O_5).

The Dorowa deposit in Zimbabwe (Deans, 1966) consists of weathered (decalcified) pipe-like bodies in fenite and ijolite. The Cargill carbonatite-pyroxenite complex in Ontario (Erdosh, 1979) is similar to the Sokli carbonatite, in containing a high-grade residual phosphate deposit up to 170 m thick associated with karst topography beneath glacial clays; widespread vermiculite occurs near the pyroxenites. Large residual phosphate deposits also occur at the Glenover carbonatite in South Africa (Verwoerd, 1966) and at Tundulu in Malawi (Garson, 1965a).

(ii) Carbonatites with pyrochlore mineralization. Typical examples of these are Oka (Gold *et al.*, 1967) and St Honoré (Ferguson, 1971) in Canada, Chilwa Island in Malawi (Garson, 1965a) and Sevathur in southern India (Borodin *et al.*, 1971).

At Oka, the ore-bodies are steeply-inclined tabular sövitic bodies up to 60 m wide consisting largely of calcite with pyroxene, biotite, magnetite, apatite, pyrite and pyrochlore. Annual production until recently has been above 3000 tonnes of pyrochlore with 0.45% Nb_2O_5. At St Honoré, pyrochlore is worked in a dolomitic section of a large carbonatite plug rich in apatite (Vallée and Dubuc, 1970). Forty million tonnes of ore averaging 0.76% Nb_2O_5 have been proved.

Dolomitic apatite-rich tremolite carbonatite on Chilwa Island contains up to several per cent pyrochlore. The ore bodies are arcuate intrusions in the sövitic zone of steeply-dipping carbonatite ring-belts. The Chilwa complex, which together with several smaller carbonatite complexes is situated in a down-warped zone roughly parallel to and about 60 km east of the main rift valley, is believed to have been emplaced progressively upwards in a series of very high pressure pulses (Garson, 1965a), possibly comparable to the emplacement mechanism of kimberlites.

The Precambrian ankeritic carbonatite at Sevathur in Tamil Nadu is an elongate arcuate body some 2 km long, flanked to the east by syenite and to the west by pyroxenite with vermiculite. The carbonatite is unusually rich in zircon and uranium pyrochlore with a high tantalum content, and has been investigated as a source of uranium.

(iii) Carbonatites with rare earth ± strontianite mineralization. Rare earth carbonatites occur at several complexes in Malawi (Garson, 1965a), and high contents of rare earth minerals are also recorded in several Zambian carbonatites, in the Wigu Hill carbonatite in Tanzania and the Mrima carbonatite in Kenya. Elsewhere in the world rift systems rare earth carbonatites are not particularly abundant.

At the Kangankunde vent within the main rift valley in southern Malawi three ore-bodies consist of ankerite-strontianite carbonatite which has invaded and partly replaced high-potash feldspathic breccia (Garson, 1965b). These ore-bodies carry from 4 to 7% of virtually thorium-free monazite, 14% strontianite and minor baryte, sphalerite and other rare earth carbonate minerals, and constitute one of the world's largest rare earth/strontianite deposits. Elsewhere in Malawi and Zambia and also at the Wigu carbonatite in Tanzania (Deans, 1966) the rare earth carbonatites mainly contain rare earth carbonates and phosphates.

(iv) Carbonatites with copper, uranium and baddeleyite mineralization. The Palabora carbonatite in South Africa is unique in its content of economic concentrations of copper sulphides comprising chalcopyrite, bornite, chalcocite and valleriite (Palabora Mining Company Ltd, 1976). Dyke-like intrusions of carbonatite in the centre of the complex contain about 1% Cu, the grade decreasing radially from the centre, with reserves of more than 300 million tonnes averaging 0.69% Cu. By-products are magnetite, apatite, sulphuric acid, gold, silver and platinum-group metals, together with baddeleyite and uranothorianite in gravity concentrates.

Small amounts of chalcopyrite are present at a few other carbonatites, notably Bukusu in Uganda (Baldock, 1969) but no other economic deposits have been found.

(b) Alkaline complexes with apatite in the Baikal Rift and Oslo Graben

The Synnr ring complex is one of the seven alkaline complexes of Upper Devonian to Upper Carboniferous age which form a 400 km long belt in the Baikal Rift System, USSR. The complex, 25 km in diameter (Nechayeva, 1965), consists of concentric zones of pulaskites, potash-rich nepheline syenites and pseudo-leucitites, but in contrast to the comparable Khibiny Complex in the Kola Peninusla, no carbonatites are reported. Apatite-bearing rocks form narrow arcuate bodies up to several metres thick composed of melanocratic nepheline syenites, with up to 10% apatite together with nepheline, analcite, aegirine and feldspar (Notholt, 1979).

Several apatite-bearing jacupirangite and related alkaline intrusions of Permian age are associated with the Oslo Graben. Of these the most important is the steeply-inclined Kodal jacupirangite dyke, which can be traced for 1.9 km (Bergstøl, 1972). Proved reserves are about 30 million tonnes of ore with over 15% fine-grained apatite (with 1.1% Re_2O_3) and about 40% titaniferous magnetite.

(c) Kimberlites with diamonds

Intrusive rocks termed kimberlites or kimberlitic dykes have been reported from numerous carbonatite centres in various tectonic settings including rifts, e.g. at Fen in Norway, Alno in Sweden, Rangwa in Kenya and Ngualla in Tanzania. Dawson (1967) described these rocks as central complex kimberlites, distinct from true kimberlites with a potential for diamonds. True kimberlites have been defined by Mitchell (1979) as inequigranular alkalic peridotites containing rounded and corroded megacrysts of olivine, phlogopite, magnesian olivine, magnesian ilmenite and pyrope, in a fine-grained groundmass of second generation euhedral olivine and phlogopite together with primary and secondary serpentine after olivine, perovskite, carbonate and characteristic Mg–Ti–Fe spinels. Chrome diopside is also present in some kimberlites.

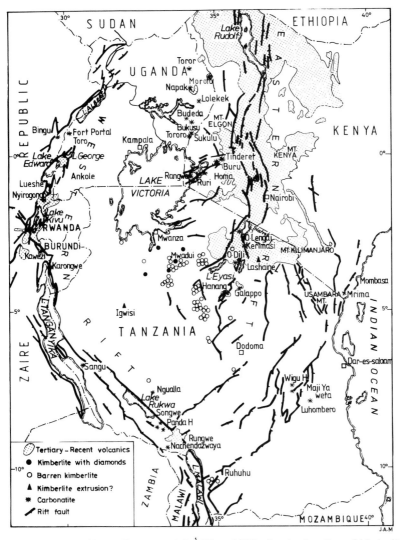

Fig. 29. The East African rift system (after King, 1970) showing location of kimberlites (from Dawson, 1970) and carbonatites. By permission of the authors and publishers, "African Magmatism and Tectonics", 263–283, 321–335, Oliver and Boyd, Edinburgh.

Despite Mitchell's assertion to the contrary we consider that kimberlites are closely related spatially and also in age to clusters of carbonatites and alkaline centres in certain rift systems of East and South Africa and in associated continental extensions of transform faults in western and southwestern Africa. While large stretches of rift systems, e.g. in southern Malawi, lack kimberlites, and the central complex-type kimberlitic dykes associated with a few carbonatite centres are typically devoid of diamonds and true kimberlitic minerals, elsewhere for example, in Tanzania, diamondiferous kimberlites occur within the main swell of the rift zone.

In Tanzania, south of Lake Victoria (Fig. 29), the diamondiferous kimberlites apparently occupy the central and presumably thickest part of the ancient Tanzania Craton, tectonically stable since the mid-Proterozoic. "Barren" kimberlites occur adjacent to the diamondiferous areas, in places where the main rift zones either disappear into the shield or pass into zones of multiple smaller fractures. However about 50% of these so-called barren kimberlites, with typical true kimberlitic minerals, reportedly contain a few diamonds.

In South Africa the distribution of kimberlites and carbonatites is more complex (Fig. 30). Broadly there are two intersecting zones in both of which the diamondiferous pipes lie within the long-stable Transvaal Craton. One zone with both diamondiferous and "barren" kimberlites, carbonatites and alkaline rocks stretches from the Sutherland area through the famous Kimberley area to

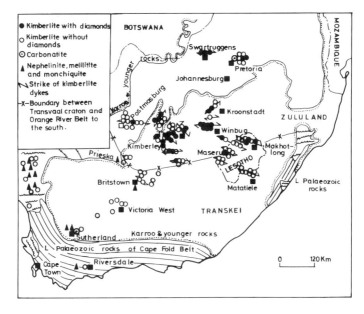

Fig. 30. Tectonic setting of the South African kimberlites (from Dawson, 1970). Trends in the basement of Namibia and Bushmanland shown by wavy lines. By permission of the author and publishers, "African Magmatism and Tectonics" 321–335, Oliver and Boyd, Edinburgh.

Fig. 31. (A). Active carbonatite Oldoinyo Lengai in the Eastern Rift of Tanzania (see Figs. 24, 25 and 29). The crater, in which there is fumarolic activity at present, is infilled with sodium carbonatite lava with up to 4% F, the upper parts of the volcano are coated with soda ash from a recent eruption, and on the slopes are layers of ash and mudflows. The carbonatite in the neck of Oldoinyo Lengai is probably sovite with Nb and P_2O_5 mineralization as in similar but more eroded volcanic centres to the south. Photo by Mr A. Nyblade. (B) Letsing kimberlite pipe and satellite pipe (left) in the Transvaal/Zimbabwe craton within a WNW–ESE fracture zone extending through to Kimberley (see Fig. 30). Photo by Professor J. B. Dawson.

Pretoria (Dawson, 1970) and the Bushveld Complex. The second broad zone extends WNW from Lesotho through the first zone at Kimberley to the Postmasburg area (Fig. 31B).

South African kimberlites are of two ages. Precambrian diamondiferous kimberlites at the Premier Mine can be linked with the nepheline syenites, trachytes and carbonatites of the NNW-trending Franspoort Line (Verwoerd, 1966). Most of the other kimberlites were intruded in the Cretaceous during updoming with related peripheral downwarping or faulting (Dawson, 1970), which accompanied the break-up of Gondwanaland, formation of the Lebombo monocline and intensified rifting in Malawi and Mozambique. In the peripheral cratonic areas in South Africa, as in Tanzania and Yakutia in the USSR (Arsenyev, 1962), the most highly diamondiferous diatremes lie at the intersections of fundamental fracture trends.

The petrogenetic relationship of kimberlite to carbonatite and alkaline rocks is controversial, despite their close association in space and time (Cox, 1980). A direct petrogenetic relationship is favoured by the presence in many kimberlites of an apparently immiscible primary carbonate residuum, in the form of carbonate dykes, ocelli of carbonate in a silica–oxide matrix and patches of silica–oxide groundmass enveloped by amoeboid patches of carbonate.

In seeming contradiction, Boctor and Boyd (1979) emphasized that Fe-Ti oxide minerals in Benfontein carbonate–oxide layers are typically kimberlitic, differing from those found in sovite, and that this carbonate–kimberlite association does not necesarily imply a genetic relationship with carbonatites. Also, Mitchell (1979) stressed that the carbonate residuum of kimberlites consists largely of calcite with lesser serpentine and magnetite and minor apatite, and lacks typical carbonatite minerals. However several carbonatites consist largely of such assemblages which are not necessarily cumulate phases, and Heinrich (1966) showed that kimberlites have similar trace elements to those of carbonatites. Significantly also, high-pressure experimental studies have shown that progressive melting of carbonated model mantle compositions will form silica-undersaturated and CO_2-rich liquids, termed haplo-carbonatites (Wyllie and Huang, 1975, 1976), and that the H_2O/CO_2 ratio is important in yielding kimberlite–carbonatite magmas and related rocks at deep crustal levels.

According to Perchuk and Vaganov (1980) the presence of diamonds in kimberlites is related primarily to the depth of magma formation, but also to the magma ascent velocity, the oxidation degree of liquids and their post-magmatic transformation. They consider than an increase in temperature with depth prevents ultrabasic magma from being carbonatized and liquated into carbonate and silicate melts, the most favourable depth range being from 200 km to 220 km. A very suitable environment for kimberlite emplacement with preservation of diamonds in rapid upward movement seems therefore to be a relatively stable craton which has been updomed gradually under the influence of a plume until pressure causes kimberlitic magmas, produced by plume-related metasomatism, to perforate the crust by "gas drilling" at preferential points. More extensive

rifting in less thick marginal cratonic crust would favour the formation of carbonatitic end-products together with nephelinites and phonolites.

(d) Rift-related basic/ultrabasic intrusions with Cr–Ni–Pt–Cu mineralization

Most Phanerozoic rifts are characterized by tholeiitic dykes and/or alkaline and
carbonatite rocks, and large mineralized basic/ultrabasic complexes are uncommon in such settings at present erosion levels. Seismic and magnetic studies
indicate however that basic/ultrabasic bodies have been intruded into deeper
parts of several rift zones during limited extension. Some mega-fracture zones
in the Proterozoic can be considered as early rifts infilled in places by major
basic/ultrabasic layered complexes, e.g. the Great Dyke of Zimbabwe and the
Bushveld Complex in South Africa, and the Duluth Gabbro in Minnesota.

(i) Great Dyke of Zimbabwe and Bushveld Complex, South Africa. These two
types of intrusion occur within the Proterozoic mega-fracture system in southern
Africa (Vail, 1978), which is possibly the root-zone of an aborted intracontinental rift (Burke, 1977; Garson and Mitchell, 1981).

The Great Dyke of Zimbabwe (Figs. 14B and 26), about 480 km long and
nearly 6 km wide, includes large lopolithic sub-complexes with layering dipping
inwards at shallower angles than the main dyke contacts (Worst, 1960). The
complexes consist of cyclic sequences of ultrabasic rocks overlain by a thick
gabbroic cap (Bichan, 1970); ultrabasic cycles have basal chromite seams
overlain by peridotite, pyroxenite, anorthositic gabbro and norite. In addition
to chromite, there is disseminated and cumulate Ni–Pt–Cu mineralization.
Bichan considered that the dyke formed in successive magmatic surges over the
thermal updraft of a mantle convective cell, and that a waning of the heat flow
pattern resulted in slumping of the dyke into its graben.

The Bushveld Complex is centred on the continuation to the south of the
great circle mega-fracture through the Great Dyke recognized by Cousins (1959).
The Complex, about 66 000 km² in area and up to 8 km thick, is one of the
world's largest igneous bodies and repositories of magmatic ore deposits. The
central part is underlain by acidic rocks and there are two lobate marginal bodies
consisting of rocks ranging from dunite to norite, anorthosite and ferrodiorite.
Magmatic ore deposits in the ultramafic parts include vanadiferous and
titaniferous magnetite layers and pipe-like bodies in the upper layered sequence
(anorthosite); sulphides with Pt, Au, Ni and Cu in the Merensky Reef and in
bronzitite pipes; and chromite in chromitite layers averaging about 0.90 m thick
in the lower layered sequence of pyroxenites, norites and anorthosites. Reserves
are estimated at over 1000 million tonnes of vanadiferous magnetite with an
average grade of 1.6% V_2O_5; 1350 million tonnes of chrome ore with at least
45% Cr_2O_3; and more than 12 000 tonnes of ore with recoverable platinum
group metals, Au, Ni and Cu (von Grueneweldt, 1977).

(ii) Duluth Complex, Minnesota. The late Precambrian Duluth complex consists

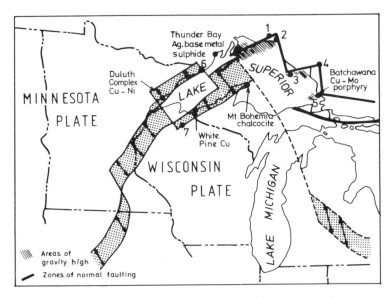

Fig. 32. Mineralization associated with postulated zones of Keweenawan rifting in the Lake Superior area, Canada (after Norman, 1978). Alkalic complexes and regions of extensive felsic magmatism apparently related to intersections of transform faultig and rifting are: 1, Prairie Lake; 2, Coldwell; 3, Michipicoten Island; 4, Fireside River; 5, Batchawana area; 6, Hoveland; 7, Mellon. By permission of the author and publishers, "Petrology and Geochemistry of Continental Rifts", 245–253, Riedel, Holland.

of troctolitic-gabbroic intrusives believed by Weiblen *et al.* (1978) to be related to rift tectonics. The best Cu–Ni sulphide mineralization seems to be concentrated along offsets in the basal contact which follow transverse (transform?) fault directions in the Midcontinent Rift. The spreading rate in this rift has been calculated at about 0.3 mm/yr, i.e. comparable to that of the East Africa, Rhine Graben and Rio Grande rift systems, so that the sequence of intrusion, crystallization and equilibration of the 10 km wide Duluth intrusions occurred within about 40 Ma (Fig. 32).

(e) Porphyry molybdenum and copper–molybdenum deposits

(i) Porphyry-type molybdenum in the Oslo Graben. Possible porphyry-type molybdenum mineralization has been found associated with porphyritic phases of biotite granite at several places within the Oslo Graben (Vokes and Gale, 1976), part of the post-Hercynian rift system which extends beneath the North Sea. The evolution of the Oslo Rift which culminated in the Permian consisted of initial basaltic volcanism, followed by fissure eruptions of latitic rhombporphyry lavas and the formation of central volcanoes and cauldron subsidences. Finally there was emplacement of composite batholiths of monzonite and syenitic composition. Biotite granites were intruded throughout the entire period of the magmatic activity.

The largest area of molybdenite mineralization in the Oslo Rift (Fig. 33) occurs in the central part of the large Glitrevann cauldron in a late, stock-shaped intrusion of aplitic biotite granite and quartz-feldspar porphyry (Geyti and Schønwandt, 1979). Sericitic alteration, resembling that at the Climax porphyry molybdenite deposit, is the dominant feature, below which occurs a stock-work of molybdenite mineralization with K-feldspar alteration, considered to be the uppermost part of a possible orebody.

The location of the Oslo Graben mineralization indicates that porphyry molybdenum deposits occur not only in magmatic arcs, and in back-arc rift settings (Ch. 5, VI. B. 2) as in the case of the mid-Cenozoic deposits of the Colorado Mineral Belt (Guild, 1978b), but also in rifts related to either hot spots or continental collision. Sillitoe (1980a) has emphasized that the world's major porphyry molybdenum orebodies with grades of 0.15–0.6% Mo are characteristic of rift settings. Nevertheless the absence of porphyry molybdenum deposits in the extensive East African Rift system suggests that their generation may require tectonic conditions not realized in most intracontinental rifts, especially those related to hot spots.

(*ii*) *Porphyry-type copper–molybdenum of the Keweenawan Rift.* Four disseminated Cu–Mo deposits in the Batchawan area of Ontario within the Keweenawan Rift are similar to Cu–Mo porphyry deposits in western American Cordilleran belts in style of mineralization, high tonnage-low grade disseminations, hydrothermal alteration and association with highly altered felsic porphyries and breccia pipes (Norman, 1978). Fluid inclusion studies indicate similar fluid inclusion morphology and depth and temperature of mineralization to those of Cordilleran porphyry deposits, while stable isotope measurements indicate that the initial mineralization resulted from typical porphyry-type magmatic hydrothermal solutions (see Fig. 32).

2 Stratabound deposits in sedimentary successions

Syngenetic and diagenetic or epigenetic mineral deposits in sedimentary successions deposited in rifts and aulacogens include evaporites, major stratabound and in some cases stratiform sandstone or red-bed copper ore bodies, stratabound Ag–Pb–Zn and Pb–Zn deposits, and rare stratabound uranium deposits.

(*a*) *Stratabound copper deposits*

Copper mineralization confined to clastic sedimentary successions which commonly include red beds is widespread in the mid-Cretaceous sediments of the Atlantic margin of Africa, and also occurs in Tertiary sandstones bordering the Red Sea, providing indisputable evidence that deposits of this type can form in rift environments. Older deposits considered to be related to ancient failed rifts include the Kupferschiefer of Europe and the African Copperbelt, described below, and the sub-economic deposits of the Adelaide Geosyncline in Australia (Lambert *et al.*, 1980) discussed in terms of a rift or graben environment by

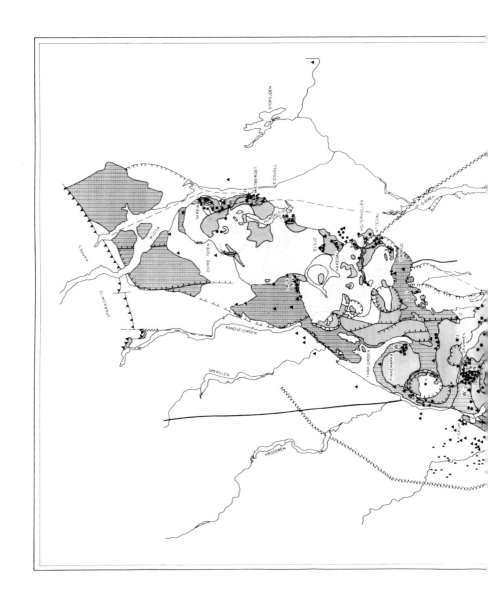

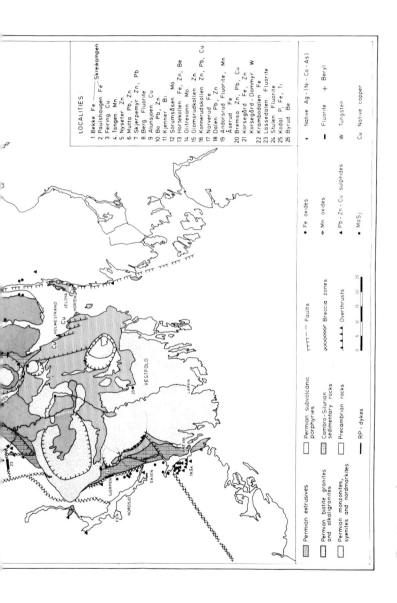

LOCALITIES

1 Bekke   Fe ⎯ Skrekampen
2 Paulshaugen Fe
3 Feiring   Cu
4 Tangen   Mn
5 Nyseter   Zn
6 Mutta   Pb, Zn
7 Skjerpemyr Zn, Pb
8 Berg   Fluorite
9 Alunsjoen   Cu
10 Bø Pb, Zn
11 Kjenner   Bi
12 Sorumåsen   Mo
13 Hortekollen   Fe, Zn, Be
14 Glitrevann   Mo
15 Glomsrudkollen   Zn
16 Konnerudskollen   Zn, Pb, Cu
17 Narverud   Fe
18 Dalen   Pb, Zn
19 Andorsrud   Fluorite, Mn
    Åserud   Fe
20 Bremsa   Zn, Pb, Cu
21 Korsegård   Fe, Zn
    Korsegård-Dammyr   W
22 Krambodalen   Fe
23 Lassedalen   Fluorite
24 Stulen   Fluorite
25 Kodal   P, Fe, Ti
26 Byrud   Be

• Fe oxides   • Native Ag - (Ni - Co - As)

✦ Mn oxides   ǀ Fluorite + Beryl

▲ Pb - Zn - Cu sulphides   ǀ Tungsten

■ MoS₂   Cu Native copper

━━━ Faults

✕✕✕✕ Breccia zones

▲▲▲ Overthrusts

0 5 10 15 20 25

Permian extrusives

Permian subvolcanic porphyries

Permian biotite granites and alkaligranites

Cambro-Silurian sedimentary rocks

Permian monzonites, syenites and nordmarkites

Precambrian rocks

━━ RP - dykes

Fig. 33. Map of the mineral deposits of the Oslo Graben prepared by Professor F.M. Vokes and Dr P.M. Ihlen.

Rowlands *et al.* (1978). Deposits possibly formed in similar settings elsewhere include those of Carboniferous age in the Dzhezkaggan region and the very large Proterozoic deposits of Udokan, USSR, described by Samonov and Pozharisky (1977), and the stratabound copper ores of the Upper Proterozoic Keweenawan succession in Michigan (Raybould, 1978).

The ore in many deposits of this type is confined to carbonaceous shale horizons, but in others it occurs in arenaceous sediments, and Rayner and Rowlands (1980) have suggested that in the Udokan and some of the Adelaide Geosyncline deposits the host rocks are shallow marine deltaic sandstones. In general the influence of sedimentary facies on the location of deposits of this type is becoming increasingly apparent, as a control either on sedimentary deposition of metal, or more probably on Eh/pH boundaries in migrating hydrothermal solutions during diagenesis. However, as some deposits are hosted by deltaic sediments, it is clear that the apparent favourability of rifts for stratabound copper mineralization is related to the presence of adjacent metal source rocks and possibly geothermal gradient in and adjacent to the host rocks, as well as to the availability of suitable sedimentary facies in the depositional basin. Moreover, while some deposits undoubtedly formed in rift environments, it is not always evident that rift structures adjacent to the ore body are not appreciably younger than the mineralization.

(i) Copper mineralization in the Atlantic margin basins of Africa. A valuable summary of the Cretaceous copper deposits of Africa has been provided by Caia (1976) who emphasized the similarity of deposits in Morocco, Gabon and Angola. The mineralization is everywhere concentrated in non-marine sediments of Aptian or approximately Aptian age, deposited during initial subsidence of the intracontinental rift which in the Albian was subjected to marine sedimentation associated with the initial opening of the South Atlantic. Mineralization, mostly of copper with varying amounts of lead and zinc, is largely confined to sandstones and conglomerates, usually grey or white in colour but associated with red beds. The metal mostly occurs in ancient fluviatile channel deposits, and is commonly stratiform but also forms "clouds" or irregular disseminations, and is associated with the carbonate cement of the host rocks. Most deposits are overlain by evaporites formed during the start of marine incursions into the basin as subsidence and marine transgression proceeded. Caia (1976) favoured a diagenetic origin for the copper, which he suggests was derived from the continent, and deposited as sulphides during compaction of the host rocks. Whether the transport and precipitation of metals required by this process was related to the sabkha process of Renfro (1974) is uncertain.

The copper mineralization in beds of Aptian age in Angola has been described in some detail by van Eden (1978). The deposits are scattered over a distance of more than 400 km along the eastern margin of the basin, and occur in transitional continental–lagoonal deposits and underlying Lower Aptian red continental sandstones and conglomerates which rest unconformably on Precambrian basement. The host rock ranges from mudstones to reddish conglomeratic sandstones. Sulphides predominate at the largest known deposit, with 7 million tonnes of 2% Cu, but at some deposits carbonates and oxides are abundant. The mineralized beds are overlain by evaporites up to 2000 m thick (Ch. 3, I.B.1), which pass seawards into a carbonate reef facies (Fig. 34).

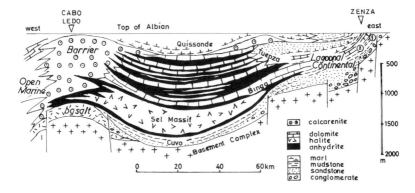

Fig. 34. Schematic cross-section through Cuanza basin, coastal Angola, at end of evaporite stage showing facies relationship and position of copper (1) and zinc (2) mineralization (from Van Eden, 1978). By permission of the author and publisher, *Econ. Geol.* **73**, 1154–1161.

Van Eden (1978) has emphasized the relationship of mineralization to the thick sedimentary sequence which accumulated as a result of subsidence associated with step faulting and consequent marine transgression. Although clearly stratiform, as in the Kupferschiefer described below, mineralization post-dated sedimentation and was probably diagenetic in origin. It is probable that copper was carried in solution in meteoric waters from the eroding continent, through the oxidizing environment of the basal sandstones, and deposited in reducing environments at greater depth nearer the basin centre. Subsequently loading by younger sediments resulted in expulsion of connate water which flowed back through the Cretaceous conglomerates and sandstones and deposited copper at the boundary between red beds and overlying sediments of euxinic lagoonal facies in a reducing environment. The presence of native copper in Miocene sediments at a shallow depth in the thick offshore succession of the Angola Basin (Siesser, 1976) suggests the possibility that transport and deposition of copper by connate or meteoric brines continued into the late Tertiary.

Although the host rocks to the copper are considered to pre-date ocean floor emplacement, the overlying evaporites form part of the South Atlantic salt province referred to below and were probably deposited in the early stages of ocean floor spreading. Deposition of the copper itself thus took place in the initial stages of development of the passive continental margin, rather than in an intracontinental rift.

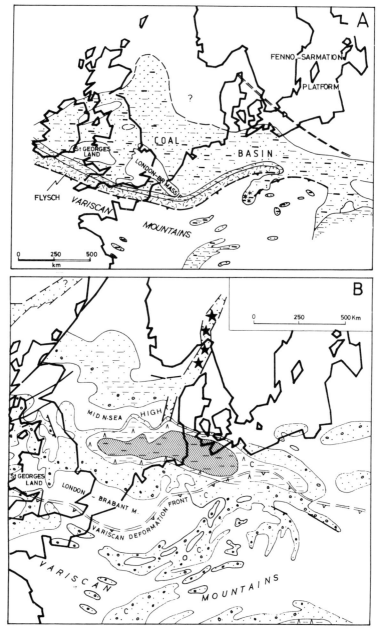

Fig. 35. Basins in northwestern Europe related to the Hercynian orogeny (modified from Ziegler, P. A., 1975 and 1978). (A) Upper Carboniferous coals north of the flysch belt and possibly occupying a pre-collision remnant basin (see Fig. 143); (B) Middle Permian Rotliegendes basin immediately prior to deposition of Kupferschiefer: Λ gypsum, anhydrite; cross-hatch, halite; * volcanics; hollow circles and dots, sandstones and conglomerates. By permission of the author and publishers, "Petroleum of North West Europe" Vol. 1, 131–148, Applied Science, Essex and *Geol. Mijn.* **57**, 589–626.

(ii) The Kupferschiefer of northern Europe. The Kupferschiefer (copper slate) or Mansfield ore lies within the very extensive succession of the southern Permian Basin (Fig. 35B) extending from northern England, where it is termed the Marl Slate, eastwards to Russia. The gently dipping mineralized horizon, which has yielded more than 2 million tonnes of copper, is up to 60 cm thick and consists of a black bituminous dolomitic shale with sulphides of copper, and of lead and zinc which tend to increase upwards through the formation, and with anomalous uranium content; the ore-grade deposits, averaging about 1.5% Cu with Pb, Zn, Ag, Co and V are restricted to parts of the basin in East Germany and Poland mostly lying adjacent to low topographic swells.

The Kupferschiefer forms the basal unit of the Upper Permian Zechstein succession and lies with a local basal conglomerate on the Lower Permian non-marine Rotliegende comprising red sandstones including dune sands, and fan-glomerates grading basinwards into sabkha deposits (P.A. Ziegler, 1975). It is overlain by a dolomitic limestone which forms the base of the lowest of four major Zechstein carbonate-evaporitic cycles.

The Kupferschiefer ore was for many years considered one of the best examples of syn-sedimentary copper deposits. However, more recent work especially that by Rentzsch (1974), indicates that the copper is concentrated outside the margins of a local unmineralized red "Röte Faule" facies developed over offshore sand bars. The Röte Faule transgresses the top of the Rotliegende, the Kupferschiefer and the overlying Zechstein Limestone, with lead and zinc overlying the copper mineralization (Fig. 36). This distribution, together with evidence for replacement of syn-diagenetic pyrite by copper sulphides, suggests that the mineralization post-dates sedimentation and is syn-diagenetic.

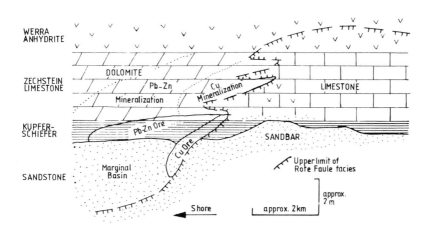

Fig. 36. Typical stratigraphic sequence of basal Zechstein with Rote Faule alteration zone transgressing bedding above sandbars. Copper is concentrated in adjacent unoxidized zone and bordered by lead–zinc (after Rentzsch, 1974, and Brown, 1978). By permission of the authors and publishers, Soc. Geol. Belgique, Liege and *Minerals Sci. Engng.* **10,** 172-181.

Deposition of the Rotliegende is considered to have been contemporaneous with initial development of intracontinental rifts in the North Sea area. While the rifting may have been related to a system of mantle hot spots which subsequently resulted in opening of the Central and North Atlantic (Whitehead *et al.*, 1975) there is an absence of pre-rift doming and scarcity of related igneous rocks other than in the Oslo Graben, and it is perhaps more probable that the incipient rifting was consequent on the collision which terminated the Hercynian orogeny. The position of the rifts on the formerly underthrusting continental plate relative to the end-Carboniferous collisional orogen to the south would then have been analogous to that of the Rhine Graben to the Alpine collision in the Tertiary.

It is uncertain why, if Kupferschiefer sedimentation is related to an intracontinental rift setting, deposits similar to the Kupferscheifer are not more common in association with other rift environments. One possibility is that deposition of the Kupferschiefer facies was dependent on the sudden marine incursion and transgression of the Zechstein Sea as postulated by Smith (1979), during which the sub-sea-level inland drainage basin with Rotliegende facies was flooded to a depth of at least 100 m, forming an almost enclosed sea in which black carbonaceous shales accumulated. Similar flooding has been suggested for the Mediterranean in the Miocene, but comparable events are probably sufficiently rare to explain the uniqueness of the Kupferschiefer in phanerozoic rocks.

(iii) Deposits of the Central African Copper Belt. The stratiform copper-cobalt deposits of Zambia (Fig. 37A) and Zaire account for nearly a fifth of the western world's copper and at least two thirds of its cobalt production; the latter proportion is projected to rise to more than 80% with future production from Zambia. The mineralization occurs near the base of the Katanga Supergroup, a thick succession of Upper Proterozoic sedimentary and metasedimentary rocks lying in angular unconformity on the basement complex, and is confined to shallow marine sediments of the transgressive sequence (Fig. 37B). In Zambia the sedimentary facies are predominantly detrital, but carbonates predominate in Zaire. The sequence is tightly folded and locally metamorphosed to amphibolite facies.

The Zambian deposits, recently described by Fleischer *et al.* (1976), are mostly within sandstones or carbonaceous greywackes and in argillites, which contain carbonate and sulphate. They are underlain by conglomerates and sandstones and overlain by dolomites with anhydrite beds.

Mineralization, originally considered to be epigenetic and related to magmatic hydrothermal solutions, was subsequently interpreted as syn-sedimentary in origin (e.g. Garlick, 1961). More recently a syn- to late-diagenetic origin has become increasingly accepted, and it has been suggested that in some of the Zambian deposits the sulphides formed in areas of increased bacterial activity as a result of upward movement of metal chloride-bearing brine into local

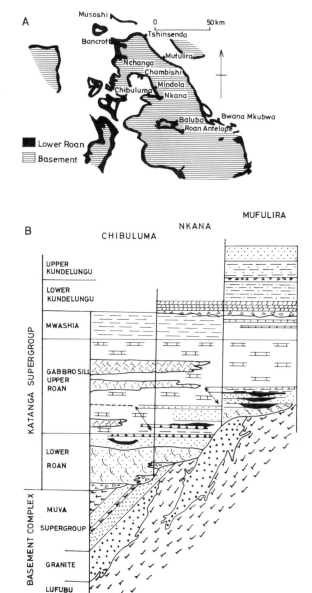

Fig. 37. Zambian copperbelt. (A) Geological map showing distribution of copper deposits and lower Roan calc-arenite sequences (from Jacobsen, 1975); (B) stratigraphic section and facies diagram, Zambian copperbelt (from Fleischer *et al.,* 1976). By permission of the authors and publishers, *Minerals Sci. Engng* **7**, 337–371, and "Handbook of Stratabound and Stratiform Ore Deposits". Vol. 6, 223–352, Elsevier, New York.

structural traps beneath impermeable dolomite beds where hydrocarbons had accumulated (Annels, 1979).

The Copperbelt succession has been interpreted in general terms as the result of deposition in an initial intracontinental rift zone (Burke and Dewey, 1973), an hypothesis supported by similarities between the ore host rock sequence and that in the Kupferschiefer and Cretaceous basins of the west coast of Africa. Similarities include the concentration of sulphides in dolomitic carbonaceous mudstones and sandstones above an angular unconformity and the presence of dolomites and evaporites above the ore horizon. Some authors (e.g. Rowlands, 1974, 1980) have suggested that the Copperbelt mineralization formed during

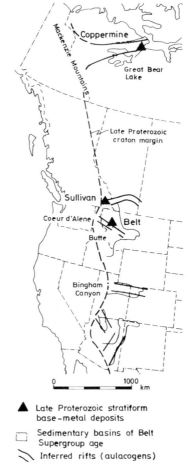

Fig. 38. Upper Proterozoic stratabound copper and lead–zinc deposits related to inferred rifts, western North America (after Burke and Dewey, 1973, and Raybould, 1978). By permission of the authors and publishers, *J. Geol.* **81**, 406–433, and *Trans. Instn. Min. Metall.* **87**, B79–86.

a major global late Proterozoic episode of copper deposition which included the Adelaide Geosyncline deposits in Australia, but due to lack of precise age control this remains highly dubious. One possibility is that intracontinental rifting was widespread in the Upper Proterozoic supercontinent, but that mineralization took place only in rift successions where palaeolatitudes were favourable for evaporite deposition and where the transgressive sequence resulted in particularly favourable lithological and geochemical environments during diagenesis; this could perhaps have resulted, as suggested for the younger Kupferschiefer deposits, from a sudden influx of sea water into a sub-sea-level basin.

(b) Stratabound silver–lead–zinc deposits

Deposits of Ag, Pb and Zn within very thick successions consisting largely of terrigeneous clastic sediments have been termed Sullivan-type deposits (Sawkins, 1976b) from the large Upper Proterozoic ore body in British Columbia (Fig. 38). The mineralization in deposits of this type is usually more or less stratiform within mudstones or siltstones, and associated volcanic rocks other than basaltic dykes or minor flows are absent. Sawkins (1976b) considered that Sullivan-type deposits formed in either intracontinental rifts which failed to develop into ocean basins or in aulacogens (Fig. 39).

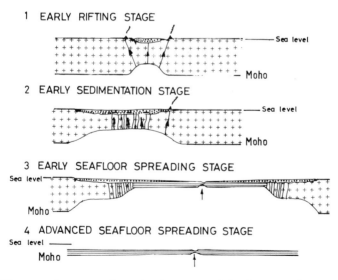

Fig. 39. Diagrammatic cross-sections showing stages of development of intracontinental rift and ocean basin with associated sulphide deposits. Stage 1 is analogous to the present day East African Rift System. Stage 2, a more advanced stage in which thinning and extension of the continental crust has produced an elongate sedimentary basin (Sullivan-type deposits). Stage 3, the rate of continental separation has increased and new oceanic crust is being formed (metalliferous sediments of the Red Sea brine pools). Stage 4, continental separation is complete and sea floor spreading is taking place (from Sawkins, 1976c). By permission of the author and publisher, *Spec. Pap. geol. Ass. Can.* **14, 221-240.**

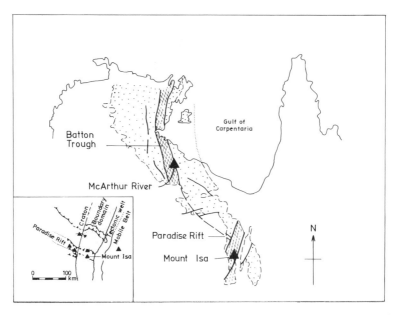

Fig. 40. McArthur Basin within mid-Proterozoic Batton Trough, northern Australia. Inset shows Mt Isa base metal deposits in relation to Paradise Rift – craton margin triple junction (after Dunnet, 1976, and Raybould, 1978). By permission of the authors and publishers, *Phil. Trans. R. Soc. Lond.* **280A**, 641–654 and *Trans. Instn Min. Metall.* **87**. B79–86.

Sawkins included in his Sullivan-type deposits the mid-Proterozoic Ag–Pb–Zn deposits of Mt Isa in Queensland, considered to have formed in a failed rift (Glikson *et al.*, 1974), and the Rammelsberg and Meggen lead–zinc deposits of Devonian age in Germany. The mid-Proterozoic stratiform lead–zinc sub-economic deposits of McArthur River, within the fault-bounded Batton Trough in the Northern Territory of Australia (Fig. 40), have also been interpreted as deposits of an intracratonic rift which extended southeastwards to include the Paradise Rift at Mt Isa (Dunnet, 1976; Raybould, 1978). However, in the case of these deposits the evidence for the presence of a syn-sedimentary and hence syn-mineralization rift is debatable. At McArthur River there is only limited evidence for the position of the inferred rift boundaries, and at Mt Isa the succession lies near the probable Proterozoic eastern margin of the continent. Nevertheless, in the absence of satisfactory alternative interpretations, formation of the mineralization in rift zones remain at least a possibility.

Within the late Proterozoic Namaqua Mobile Belt of amphibolite to granulite facies metamorphic grade in Cape Province, South Africa, ore bodies in the Gams Iron Formation have metal ratios suggesting analogy with the Sullivan-type deposits (Rozendaal, 1978). The ore deposits, from west to east with ore grades in parentheses are: Black Mountain (0.8% Cu, 2.9% Pb, 0.6% Zn), Broken Hill (0.36% Cu, 3.0% Pb, 2.2% Zn), Big Syn (1.23% Pb, 2.88% Zn) and the Gamsberg 95 million tonne deposit (0.5% Pb, 7.0% Zn).

At Gamsberg the upper mineralized zone is banded quartz–garnet–grunerite rock with 10–40% opaque minerals comprising Fe, Zn and Pb sulphides and accessory minerals. The lower horizon consists of mineralized cordierite rock overlying fine-grained quartzite and banded mineralized sillimanite schist with Fe, Zn and Cu sulphides; baryte locally forms intercalated lenticular bodies or layers in an upper horizon.

Despite the high grade of metamorphism in the area, the distinct metal zoning across three of the ore deposits has been interpreted (Rozendaal, 1978) as lateral zoning of Kupferschiefer-type (Wedepohl, 1971) with copper near the rim of the basin, lead intermediate and zinc near the centre. The vertical zoning also shows marked sedimentary facies changes. Within the Gams Iron Formation as a whole an oxide facies, a carbonate facies and sulphide facies can be recognized in a transgressive sedimentary basin. Electron probe analyses (Stumpfl *et al.*, 1976) showing that the pyrite lacks "magnetic" Ni or Cr and the magnetite is a non-magmatic type with no Ti, V or Cr, also support a sedimentary or diagenetic origin for the deposit.

While there is no evidence of syn-sedimentary rift or graben development at Gamsberg, some similarities to both the Sullivan and Kupferschiefer type of deposits necessitate inclusion of the Gamsberg ores with those having a more obvious relationship to intracontinental rifts.

(c) Stratabound sandstone-type uranium deposits

Uranium in sedimentary rocks occurs within the thick succession of mid-Proterozoic rocks of the "Athapuscow Aulacogen" (Hoffman, 1973), east of the Great Slave Lake in Canada. The uranium is mostly in the form of interstitial uraninite, replacing sulphides, and occurs in non-marine sandstones and conglomerates at several stratigraphic horizons in the 12-km thick succession of continental and marine sediments and volcanic rocks (Fig. 41). Morton (1974) has suggested that the uranium was precipitated by metamorphic hydrothermal fluids expelled either from the deepest levels of the succession or from the underlying Archaen basement. Hoffman (1973) has argued that the succession accumulated in the failed arm of a trilete rift system, the other two arms having opened and subsequently closed to form the Coronation Geosyncline, resulting in conversion of the failed arm to an aulacogen.

(d) Evaporites

With the recent reinterpretations of most stratabound copper and lead–zinc deposits within terrigeneous sedimentary rocks as diagenetic rather than sedimentary in origin, evaporites are the only type of rift-related economic deposit which are undoubtedly syngenetic with sedimentation.

Most evaporites associated with intracontinental rifts were deposited during the early stages of ocean floor spreading, and are hence considered with the passive continental margin stage of development (Ch. 3, I.B.1). However, in a few cases evaporite accumulation took place in rifts which failed to develop, one of the best-known examples being that of the North Sea.

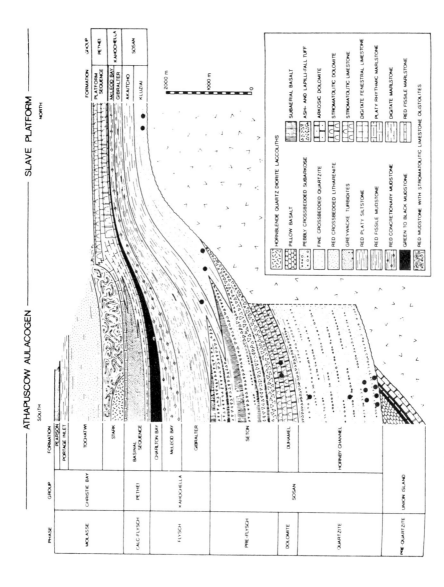

Fig. 41. Schematic cross-section of the Athapuscow aulocogen and adjacent platform showing stratigraphic positions (black dots) of known uranium mineralization (from Hoffman *et al.*, 1974, and Morton, 1974). By permission of the authors and publishers, *Spec. Publs. Soc. Econ. Palaeont. Miner.* **19** and "Formation of uranium ore deposits", 255–273, IAEA, Vienna.

The North Sea Zechstein evaporites, comprising halite and potash deposits, extend from Poland to eastern England, having a distribution similar to that of the underlying Lower Permian Kupferschiefer and Marl Slate described above. Evaporite deposition began with a marine transgression over the Rotliegende and deposition of the Marl Slate and Kupferschiefer, but unlike the rifted continental margin evaporites described later, there was no major marine transgression following deposition of the evaporites. In general the Zechstein shows a transition from clastic sediments at the basin margin through tidal carbonates and evaporitic shelf banks to thick salts in the basin centre (W. H. Ziegler, 1975). Smith and Crosby (1979) have described from the British side of the Southern North Sea the four main evaporite cycles, beginning with the Marl Slate-Lower Magnesian Limestone-Hayton Evaporite cycle, and ending with the Top Anhydrite. In the first cycle anhydrite predominates and is thickest at the basin margins, but in the second cycle thick sulphates border chlorides in basinal areas. In the upper two cycles chlorides predominate with minor sulphates.

(e) Recent deposits of sodium and potassium salts, magnesite and phosphate in the East African Rift

Lacustrine brines and evaporites in the Eastern Rift Lakes Magadi and Natron comprise huge reserves of sodium carbonate, chloride and fluoride (see Fig. 25). At Lake Magadi in Kenya annual production exceeds 100 000 tonnes of anhydrous sodium carbonate and 15 000 tonnes of salt; sodium fluoride (villiaumite) is also produced (Deans, 1966). Beds of trona up to 20 m thick are interbedded with clays, and carnallite (hydrated potassium–magnesium chloride) is also present. Precipitation of new trona in dredged areas approaches 7.5 mm per year (Heinrich, 1966).

The proportion of the vast salt and soda reserves derived from weathering of alkaline silicate lavas and tuffs, rather than from carbonatite magmas and hydrothermal solutions, is uncertain (Deans, 1966). The predominance of sodium carbonates in the anhydrous carbonatite lavas and tuffs erupted from Oldoinyo Lengai (Fig, 31A; Dawson, 1962) near Lake Natron suggests an important contribution from carbonatites. Heinrich (1966) claimed that hot springs emitting sodium salts are the main source of the soda deposits of Lakes Natron and Magadi, and also considered that carbonatite volcanism could explain Mg-rich bentonites at Gelai southeast of Lake Natron and adjacent magnesite deposits (Harris, 1961). In Lake Magadi and Natron extraordinary geyser-like spirals of sodium salts up to 1.5 km across (George, 1979) are believed to be the surface manifestation of alkaline geysers emitted from cracks produced during extension along the Eastern Rift.

Around the eastern and southeastern shores of Lake Manyara at Minjingu, south of Lake Natron, deposits of slightly radioactive sedimentary phospate (Harris, 1961) were derived either from guano deposits on former islands, or from the Oldoinyo Dili carbonatite vent.

3 Vein-type deposits

(a) Lead–zinc mineralization of the Benue Trough and Amazon Basin

Mostly sub-economic deposits of lead and zinc are known from several of the Cretaceous successions bordering the Atlantic coast of Africa, but few are of as much economic interest as the copper deposits. However, lead–zinc mineralization of economic significance has long been known from the Benue Trough failed rift in Nigeria (Fig. 42), and has been compared with Mississippi Valley carbonate-hosted deposits (Guild, 1972; Olabe, 1976). The Benue deposits are in the Abakaliki district where gently folded Albian sediments lie within the 5000 m thick Cretaceous succession. Galena and sphalerite with minor copper sulphides and pyrite are present in lodes and veins on steeply dipping fracture systems and sheeted zones mostly along the crests of anticlines, and are strata-bound in the sense that they occur within black carbonaceous shale host rocks (Ukpong and Olabe, 1979). The mineralization is considered to be related to connate waters, and while volcanic rocks occur in the Albian succession, there is no obvious spatial relationship of these to the sulphides.

Lead mineralization similar to that in the Benue Trough is associated with the Amazon fracture zone (N. Cameron, pers. comm.), which opened along the axis of the Palaeozoic intracratonic Amazon depression from a triple junction located west of Liberia and at the southern end of the North Atlantic Rift

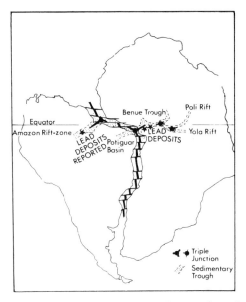

Fig. 42. Lead mineralization in Benue and Amazon rift zones formed at Cretaceous triple junctions during early separation of Africa from South America (after Burke and Dewey, 1974, modified by Mitchell and Garson, 1976). By permission of the authors and publishers, *Nature, Lond.* **249**, 313–316, Copyright © 1974, MacMillan Journals and *Minerals Sci. Engng* 8, 129–169.

(Fig. 42). Unlike the Benue Rift, no ocean floor spreading took place and the fracture zone, defined by a prominent linear gravity high, is occupied by a tholeiitic dyke and sill complex. Diorite, gabbro, and pyroxenite intrusives are also present.

The mineralization in both the Benue Trough and Amazon Basin may have been related to mixing of metal-bearing chloride brines, possibly partially magmatic, and basinal sulphate-bearing waters derived from evaporites and sour petroleum fluids. Hot (30° C) sulphurous springs are active at present on the limb of the Amazon depression, and Cameron suggests that the mineralization forms an intermediate type between continental Mississippi Valley orebodies unrelated to magmatic hydrothermal solutions and mineralization associated with spreading ridge geothermal brines.

(b) Fluorite deposits of western North America and the Illinois–Kentucky region

Economic deposits of fluorite or fluorspar have long been known from the East African Rift, important deposits occur in the Rhine Graben, and a major

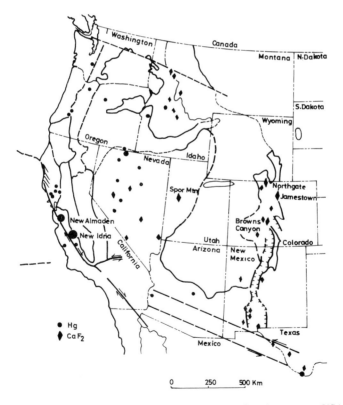

Fig. 43. Mercury in the Franciscan Complex and rift-related fluorite, western USA, mostly late Cenozoic (from Guild, 1978b). By permission of the author and publisher, *J. geol. Soc. Lond.* **135**, 355–376.

fluorspar province lies adjacent to Lake Baikal in USSR. The East African and
Rhine graben deposits, and possibly those near Lake Baikal, are associated with
alkaline rocks and carbonatites emplaced during the rifting.

Van Alstine (1976) has emphasized the close association of many fluorspar
deposits in North America with rift zones and lineaments, and suggested a
genetic relationship between rifting and the rise of fluorine from the lower crust
or upper mantle. In particular, he suggested that fluorspar deposits of western
North America, which are mostly epithermal and deposited from a mixture of
meteoric and magmatic hydrothermal water, were formed along a more or less
continuous system of rifts and lineaments extending from Mexico through the
Rocky Mountain Trench to Alaska (Fig. 43).

Many of the late Cenozoic deposits in North America, in particular those in
or adjacent to the Rio Grande graben, are undoubtedly related to back-arc
rifting associated with cessation of subduction beneath western North America
(Ch. 5, VI. B1). However, a rift association for some of the other deposits is
less certain, and a few, for example those in the Lost River area of Alaska, are
associated with tin and beryllium which show little evidence of a rift setting.

The Illinois–Kentucky fluorite district and baryte and base metal deposits in
the region provide an example of mineralization which, while its age relationship
to a rift is rather dubious, indicates the importance of regional tectonic controls
on mineralization adjacent to a failed rift landward from a rifted continental
margin.

The mineralization consists of fluorite veins and bedded deposits in Palaeozoic limestones lying north of the Mississippi Embayment (Fig. 21) with
base metals to the northwest and baryte lying further from the Embayment

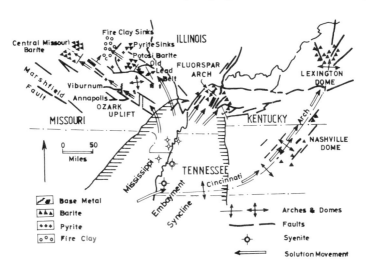

Fig. 44. Structural framework and mineral zoning, Illinois–Kentucky fluorite district and
adjacent areas (from Brecke, 1979). By permission of the author and publisher, *Econ. Geol.*
74, 1327–1335.

along the north margin of the Ozark Uplift. Brecke (1979) suggested that the baryte, fluorite and sulphides were precipitated in the Upper Cretaceous in structural traps around domes, and were carried by connate hydrothermal brines migrating up-dip from the thick Mesozoic sequence, including evaporites, of the Mississippi Embayment (Fig. 44). Migration of solutions, perhaps analogous to the metal-rich oilfield brines described from Central Mississippi, probably was aided by heating during late Cretaceous emplacement of syenite intrusions and numerous dykes into the Embayment succession.

Burke and Dewey (1973) considered that the Mississippi Embayment is the third arm of the postulated Jackson triple junction, the other arms having opened to form the Gulf of Mexico at the end of the Palaeozoic. However, the presence of Upper Cretaceous syenites and volcanic rocks within the Mississippi Embayment suggests a young episode of incipient rifting, presumably related to opening of the Atlantic Ocean. There is no obvious explanation for the late Cretaceous compressive event which according to Brecke (1979) resulted in expulsion of the hydrothermal solutions and their migration, although compression associated with the late Palaeozoic Appalachian orogeny may have resulted in earlier mineralization in the region (Ch. 6, IV. B.5).

(c) Vein-type deposits in the Oslo Graben and Keweenawan Rift

In addition to the molybdenum porphyry-type deposits at Glitrevann (Geyti and Schønwandt, 1979) and the apatite-rich jacupirangite with titaniferous magnetite at Kodal (Bergstøl, 1972), both described in previous sections, mineral occurrences in the Oslo Graben (see Fig. 33) include vein deposits carrying mainly sulphides of Mo, Cu, Pb and Zn within most rock types in the region. Contact-metasomatic deposits occur at or near the contact between granites and adjacent lower Palaeozoic to Permian sedimentary rocks downfaulted within the rift system, and there are vein-type deposits in the Precambrian basement and overlying Palaeozoic sediments (Vokes and Gale, 1976).

The most noteworthy vein deposits are molybdenite-bearing quartz in the biotite granite massif of the Drammen District. Contact-metasomatic deposits carry Zn, Pb, Cu and Fe with lesser Bi, Ni and Co and, recently, numerous W occurrences (scheelite) have been located within thermal aureoles. The vein-type deposits in the Precambrian basement and overlying Lower Palaeozoic sediments include the famous native silver deposits with Co-Ni arsenides in the Kongsberg area, potentially economic fluorite veins in the Kongsberg and Porsgrunn areas and some beryl and helvine occurrences.

The Keweenawan Rift (Fig. 32) includes, in addition to the Cu–Mo porphyry-type deposits, the Duluth Complex Cu–Ni mineralization, and the sediment-hosted stratabound copper deposits, described or referred to in earlier sections, niobium–uranium–rare earth deposits, native copper, and chalcocite vein and lava-hosted stratabound deposits (Norman, 1978).

The niobium–uranium–rare earth deposits formed in alkaline complexes in an early phase of rift magmatism; these alkaline complexes and the Cu–Mo porphyries tend to occur at the intersections of zones of rifting and offset or transform faults. Native silver, base metal sulphide deposits are mostly in tensional faults paralleling the rift axis, e.g. at Thunder Bay (see Fig. 32).

Deposits of native copper occur as vesicular fillings virtually through the basaltic volcanic pile with economic concentrations confined to flow-top breccias and interflow sediments. The main chalcocite vein deposits occur in the volcanic pile in association with felsic intrusive. The native copper deposits have been explained by a process of leaching of copper from the volcanics after deep burial followed by upward migration of liquids (White, 1968) while a possible origin for the chalcocite deposits is the mixing of sulphide-bearing hydrothermal fluids derived from felsic intrusions with copper-bearing waters similar to those postulated to have deposited the native copper deposit.

4 Preservation potential

The preservation potential of most mineral deposits in intracontinental rifts and aulacogens is largely dependent on whether the rifts developed into ocean basins or remained as failed rifts. Aulacogens and failed rifts are among the most commonly preserved tectonic settings as indicated by the abundance of Proterozoic and Palaeozoic examples, and as a direct result of their location within continental interiors remote from orogenies affecting the margins. While most mineral deposits are preserved and exposed in rifts which are not deeply eroded, exposure of metal ores associated with ultrabasic dykes can occur only in the most deeply dissected rift systems.

Rift or half-graben successions bordering ocean basins are initially preserved beneath continental shelf successions as the rifted continental margin subsides. Eventually, the margin may become active with the start of ocean floor subduction beneath it, and a magmatic arc may develop across the buried rift succession, with associated metamorphism, uplift at least locally, and probably destruction of recognizable rift features. Alternatively, the rift succession may remain on a passive continental margin until the margin collides with an arc system on the overriding plate following ocean closure. The rift succession, plutons and associated mineralization may then be eroded in the elevated foreland thrust belt, but will be preserved and exposed adjacent to the thrust belt beneath the overlying shelf succession. Examples are provided by the numerous Mesozoic rift facies, although largely unmineralized, described from both the overriding southern and subducting northern plate adjacent to the early Tertiary Alpine collision belt.

CHAPTER 3

Deposits Formed on Passive Continental Margins and in Interior Basins

Passive continental margins are those at which there is no relative movement between the continent and ocean floor. Most passive margins have developed as a result of ocean floor spreading in a developing intracontinental rift system, and are commonly referred to as Atlantic-type, rifted, trailing or divergent margins, although a few passive margins not considered here have formed following cessation of subduction or transform faulting and rarely following collision of an island arc with a continent. Although passive continental margins lie by definition within a single lithospheric plate, their initial development as rifted margins and their subsequent subsidence to form continental shelves and slopes is related to tectonic processes at and on the flanks of the developing spreading centre. Since intracontinental rifts are commonly underlain by basic intrusive rocks, the change in tectonic setting from an intracontinental rift to a developing ocean basin with passive continental margins is here arbitrarily defined to coincide with the eruption of tholeiitic lavas within the rift.

Some sedimentary basins within or on continents but of uncertain tectonic setting are included in this chapter because they are not obviously related to orogenic settings described later.

I PASSIVE CONTINENTAL MARGINS

A Tectonic Setting

The major morphological features of passive margins have long been known from oceanographic work in the Atlantic and Indian Oceans. The geological environments may include continental talus fans or epeiric seas and extend through the coastal plain and continental shelf to deep marine environments of the continental rise. In general these environments and associated sedimentary successions continue for thousands of kilometres along the continental margin, but

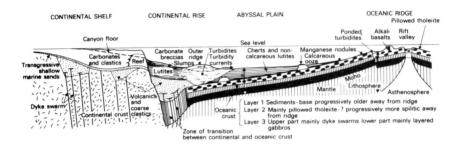

Fig. 45. Schematic cross-section, western Atlantic (from Dewey and Bird, 1970). By permission of the authors and publisher, *J. geophys. Res.* 75, 2625–2647.

are interrupted by major deltas commonly situated where failed rifts extend into the continent, and less distinctly by the non-transform extensions of oceanic ridge–ridge transform faults discussed below (Ch. 7, II.A).

The main process in the development of Atlantic-type margins is the accumulation of thick sedimentary successions as a result of post-rifting subsidence. Dietz and Holden (1966) indicated how the major features of the continental shelf, slope and rise bordering the Atlantic developed during subsidence, with a prism or miogeocline of shallow marine sediments beneath the continental shelf thickening seawards towards the shelf edge, and bordered by a thin slope succession passing into a time-equivalent prism of turbidites and mudstone forming the continental rise. Essential to Dietz and Holden's concept was the presence of a submarine barrier, considered in most cases to be a submerged reef, the growth of which kept pace with subsidence, situated at the seaward edge of the continental shelf, and of submarine canyons cutting the shelf and allowing detritus to reach the continental rise. Dewey and Bird (1970) described an idealized cross-section through the western Atlantic margin (Fig. 45) and subsequent sections

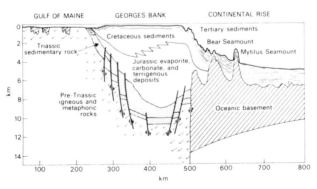

Fig. 46. Structural cross-section, Gulf of Mexico and Georges Bank, Atlantic margin, North America (after Sheridan, 1974, modified by Mitchell and Reading, 1978). By permission of the authors and publishers, "The Geology of Continental Margins", 391–407, Springer-Verlag, New York, and "Sedimentary Facies and Environments", 439–476, Blackwell Scientific, Oxford.

through specific areas have confirmed many of the postulated features (Fig. 46).

The subsidence so characteristic of passive margins results from cooling of the crust beneath the former intracontinental rift flanks, from tectonic thinning or plastic flow of the continental crust during rifting, from cooling of ocean floor beneath the continental rise as it moves away from the spreading centre, and from loading by sediment deposited on the continental rise and to a lesser extent on the shelf. Falvey (1974) has argued that crustal thinning is a result of erosion rather than stretching of the crust, and that subsequent subsidence reflects metamorphism deep in the crust; he related major pre-intracontinental rift and pre-shelf succession unconformities to these events.

During subsidence the continental shelf and rise successions may prograde seawards (Fig. 47), so that although the shelf sediments, which are up to 12 km thick, normally overlie the rift-related continental rocks, eventually the shelf may extend across continental rise turbidites deposited on the ocean floor.

The main features of continental margin sediments as described by Dietz and Holden and Dewey and Burke are common to most passive margins; ancient successions were referred to in geosynclinal terminology as the orthoquartzite-carbonate or miogeosynclinal facies and the flysch or eugeosynclinal facies, corresponding respectively to the continental shelf and rise successions of modern margins. Intensive exploration for petroleum on modern margins has resulted in recognition of a number of types of passive margin (e.g. Beck, 1972;

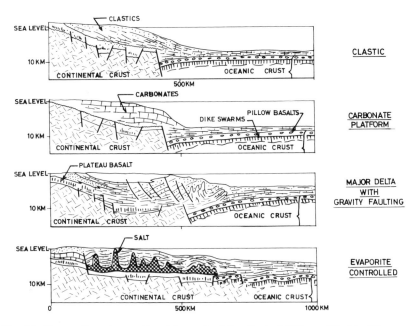

Fig. 47. Types of passive continental margin (after Beck, 1972, modified by Thompson, 1976). By permission of the authors and publishers, *Aust. Petroleum Explor. Ass. J.* **12,** 7–28, and *Bull. Am. Ass. Petroleum Geol.* **60,** 1463–1501.

TABLE VI

Mineral deposits characteristic of passive continental margins, and continental interiors.

Tectonic setting	Association	Genesis	Type of deposit/metals	Examples
Passive continental margins	Magnesian carbonates in transgressive sequence	Chemical sedimentary	Evaporites	S Atlantic (Aptian) Red Sea (Miocene)
	Black shale–chert–dolomite, commonly transgressive	Biochemical–chemical sedimentary	Phosphorites	Peru, W Africa (Recent); Florida (Miocene)
	Transgressive marine sequence	Pelagic sedimentary	Metal-rich black shales	Alum Shales, Sweden (Cambrian) Venetian Alps (Mesozoic)
	Shallow marine clastics in regressive sequence	Chemical sedimentary	Minette-type ironstones	W Europe (Jurassic); E USA (Silurian)
	Chert-shallow water clastics	Chemical/biochemical sedimentary	Banded iron formation	Labrador, S Africa (L Proterozoic)
	Mostly beach sands	Sedimentary	Placers: ilmenite rutile, zircon	S Africa, E coast Australia (Quaternary)

TABLE VI (Cont'd)

Tectonic setting	Association	Genesis	Type of deposit/metals	Examples
Passive continental margins	Shelf carbonates, mostly deeply buried	Epi (and syngenetic?) meteoric or connate hydrothermal	Carbonate-hosted Pb Zn	Mississippi Valley (Cambrian, Carboniferous); Eire (Carboniferous); S Alps (Triassic)
	Shelf carbonates	Epigenetic meteoric or connate	Carbonate-hosted Ba and F	Pakistan (Jurassic); Burma (Ordovician)
Uncertain intracontinental settings	Proterozoic terrigeneous sediments unconformable on L–M Proterozoic metasediments	Supergene diagenetic–epigenetic hydrothermal, metamorphic remobilization	Unconformity vein-type U	Athabasca basin, Canada; Alligator River, NT, Australia (L–early M Proterozoic)
	Orthoconglomerates, commonly above unconformity	Detrital sedimentary and partly epigenetic	Quartz-pebble conglomerate U, Au	Rand, S Africa; Elliott Lake, Canada; Jacobina, Brazil; Western Australia (L Proterozoic)

Thompson, 1976), defined by variations in the proportion of clastic and carbonate sediment on the shelf, by the presence or absence of a delta, and by the presence and significance of evaporites at the base of the shelf succession and in some cases postulated to underlie the continental rise turbidites (Fig. 47). Close similarities between the late Cenozoic of the East African rift system and the Mesozoic succession and structure of Western Australia, which separated from Greater India in the late Jurassic, have been described by Veevers and Cotterill (1976).

Continental margins bordering marginal basins, for example much of the Pacific margin of Asia, show some similarities to young Atlantic-type passive margins, with subsidence and sedimentation following back-arc rifting and oceanward migration of the arc system. However, it may be inferred that in continents bordering a marginal basin the shelf sediments commonly overlie part of the subsided magmatic arc which remained on the margin following rifting, rather than the rift or half-graben successions typical of Atlantic-type margins.

B Mineral Deposits

Although modern passive continental margins lack surface economic deposits other than heavy mineral placers and in a few localities evaporites, ancient shelf successions now exposed in or adjacent to orogenic belts contain very large sedimentary and stratabound deposits of a number of metals or minerals (Table VI). Most of these are largely restricted in age to the latest Proterozoic and Phanerozoic, e.g. evaporites, phosphorites, and carbonate-hosted lead–zinc deposits, but the early to mid-Proterozoic shelf successions are of extreme economic importance because of their enormous deposits of iron. Although not considered here, accumulation of some of the world's largest coal deposits was clearly associated with deltas formed on passive continental margins, e.g. the Upper Carboniferous deposits of England, Wales and eastern USA, deposited on the northwestern passive margin of the Rheic ocean immediately before its closure and consequent end-Carboniferous orogeny in southwest England and the Appalachians (see Fig. 35A).

In contrast to the considerable mineral potential of ancient continental shelf and deltaic rocks, the flysch-type sequences of ancient continental slopes and rises contain no economic deposits with the possible exception of petroleum and, very speculatively, Besshi-type stratiform sulphides referred to later (Ch. 4, III. B.2). Whether modern continental rises contain significant petroleum reserves has yet to be proved.

1 Evaporites

Salt deposits in continental margin successions are important not only for their economic value and because they commonly form diapirs with related petroleum traps, but also because they are essential to the formation of stratabound copper deposits in underlying rift successions, and probably form the source of mineralizing brines in overlying carbonate-hosted lead–zinc deposits. Moreover, and more speculatively, it has been suggested that massif-type anorthosites, restricted to an age range of 1000 to 1700 Ma, were derived metasomatically from ancient continental shelf evaporite sequences (Gresens, 1978). Although there are no present-day examples of extensive evaporite deposition on passive continental margins, modern evaporites being known in most detail from the sabkhas of the Trucial Coast and the salinas of Baja California, the close association between deposition of many evaporite successions and the early stages of ocean floor spreading has been appreciated since the discovery of the now well-known deposits on the margins of the Red Sea. We here consider evaporites which are considered to have accumulated on passive continental margins rather than in rift settings because of evidence that emplacement of oceanic crust preceded or accompanied their deposition.

Evaporites on the Ethiopian side of the Red Sea (Fig. 48) have been described by Hutchinson and Engels (1970). East of the Coastal Plain Mesozoic sediments are overlain in angular unconformity by Miocene evaporites, dominantly halite; these exceed 3500 m in thickness, thin seaward and pass landward into volcanic facies. The evaporites are considered to have developed during block faulting and subsidence of the coastal plain relative to the Danakil Alps, which accompanied an episode of sea-floor spreading along the Red Sea axial trough. Evaporites in the Danakil Depression to the east are mostly and possibly entirely younger than those bordering the Red Sea, but probably formed as a result of similar down-faulting along the margin of the Ethiopian Plateau.

Burke (1975) discussed the formation of *circum*-Atlantic salt deposits formed

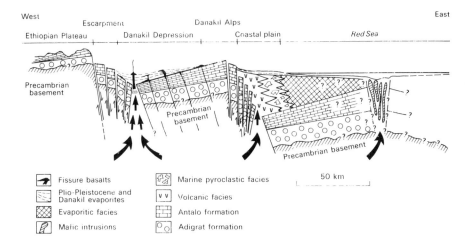

Fig. 48. Section across Red Sea and Danakil depression (from Hutchinson and Engels, 1970). By permission of the authors and publisher, *Phil. Trans. R. Soc. Lond.* **267A**, 313–329.

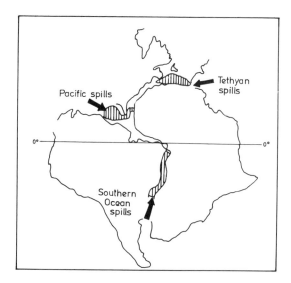

Fig. 49. *Circum*-Atlantic salt deposits, areas of massive salt deposition indicated by abundant diapirs (shaded areas); Africa, Americas and Iberia in their relative positions before opening of the Atlantic (after Burke, 1975). By permission of the author and publisher, *Geology* **3**, 613–616.

in the Mesozoic during the early stages of ocean opening. He described three provinces of salt diapirs, in Morocco, in the Gulf of Mexico and offshore Senegal, and in West Africa, Brazil and possibly the Benue Trough, and argued that formation of the evaporites within favourable latitudes was controlled largely by the distribution of hot spot related topographic highs which segmented the rift system. Similarly Leyden *et al.* (1976) suggested that evaporite deposition in the South Atlantic province was controlled by the hot spot related Walvis Ridge barrier as the Atlantic opened. Evaporites were deposited where ocean water, perhaps already highly saline, spilled over the domes or ridges into sub-sea-level sub-aerial graben, and overlay continental crust of the rift margin (Fig. 49). They possibly also extended onto newly emplaced oceanic crust at the rift axis, and have subsequently been buried by turbidites of the continental rise. The depositional environment was thus similar to that postulated for the Miocene salts of the Mediterranean, although the tectonic setting in at least most of the Mediterranean was undoubtedly different.

The on-land salt deposits of the Congo and Gabon Basins, forming part of the extensive South Atlantic evaporite province, have recently been described by de Ruiter (1979). The deposits are known in most detail from the Congo Basin where they occur in three main cycles each comprising a thin black shale overlain by halite passing up into a mixture of halite and carnallite and ending in the potassium chloride tachydrite. The overall characteristics are an abundance of potash and magnesium salts and scarcity of sulphates and carbonates indicating the presence of a brine enriched in highly soluble salts before entering the basin. Salt is mined only in the Congo Basin where economic deposits of sylvite are associated with the carnallite. The overall succession is transgressive, with fault-controlled rift sediments and the associated copper mineralization described above overlain by a transgressive non-marine to highly saline post-rift sequence passing up into the salt of Aptian age (see Fig. 34); this in turn is overlain by Albian–Cenomanian carbonates.

2 Phosphorites

Phosphorites, sedimentary rocks containing at least 10% phosphate, yield more than four-fifths of world phosphate production, dominantly from sources in Florida, Middle East and Morocco. Phosphorites are known from the Proterozoic, e.g. in the mid-Proterozoic stromatolitic beds of the Aravalli sequence in India, but are most abundant in the Cambrian, Permian and Jurassic to late Tertiary. They consist of phosphatic chemical or biochemical muds and detrital grains, mostly intraclasts, pellets, oolites and fossil skeletal material, with pelletal material becoming far more abundant in Phanerozoic than in Proterozoic deposits. Most phosphorites are associated with black shales, cherts and sometimes dolomites.

Limited evidence from present day areas of phosphorite deposition indicates that they occur largely in shallow water areas with little detrital sediment in low latitudes (Fig. 50). Recent phosphorite occurrences on the upper continental slope of Peru and Chile and off the coast of southern Africa have been interpreted as largely reworked older deposits (Fuller, 1979). Many ancient phosphorites can be shown from palaeogeographic reconstructions to have formed on continental margins within or near the tropics. Phosphorite deposition is largely restricted to areas of upwelling of deep cold ocean water rich in nutrients (i.e. phosphate, nitrate and bicarbonate) into a surface zone where organic activity is intense (e.g. McKelvey, 1967). Precipitation, probably in some cases as a result of accumulation of phosphate by phytoplankton, takes place either directly on the sea floor, or, less probably, during diagenesis through interstitial precipitation or phosphatisation. Formation of workable deposits from phosphate deposited on the sea floor probably involves a complex sequence of reworking and trapping of grains.

The generation of phosphorite in Florida has recently been described by Riggs (1979). Phosphorite deposition accompanied movement of cold upwelling water across the shallow platform into coastal environments, where benthic faunas produced faecal pellets, muds accumulated and were partly lithified and reworked as intraclasts, and local oolites were produced. The resulting detrital phosphorite grains and pebbles were transported along the shoals and accumulated in local "entrapment basins". Riggs (1979) discussed the possibility that subsidence of the continental margin related to ocean floor spreading was accompanied by discharge of phosphate-bearing hydrothermal solutions. However other deposits, e.g. the Sechura Desert deposits of Miocene age in Peru

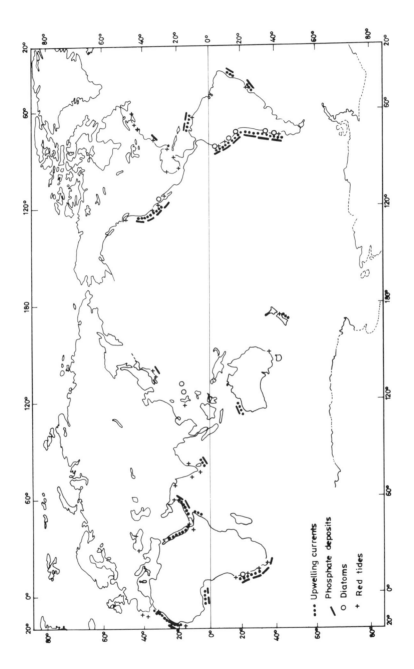

Fig. 50. Areas of upwelling currents and recent deposition of phosphorite (from Fauconnier and Slansky, 1979). By permission of the authors and publisher, Internat. Geol. Correl. Proj. 156, Canberra.

and the Miocene deposits of Baja California, accumulated respectively landward of a submarine trench and within a narrow marginal basin, indicating that phosphorite accumulation does not necessarily require either subsidence or the postulated hydrothermal solutions.

The association of upwelling with oceanic circulation indicates that phosphorite deposition is most common on west-facing continental margins, and on south-facing margins in the northern hemisphere and north-facing margins in the southern hemisphere (Sheldon, 1964), although topographically induced upwelling can also be important. Because western coastlines are preferred sites, passive continental margins, and continental margin arcs which tend to face westwards to open oceans (Ch. 5, II.A.4), are more likely to receive phosphate deposition than active margins bordered by marginal basins and island arcs, most of which face east. The location of the major Miocene deposits in Florida and the southeast Atlantic Coastal Plain provide perhaps the best evidence that passive continental margins are locally suitable for accumulation of economic deposits of phosphorite, while the distribution of upwelling currents and modern

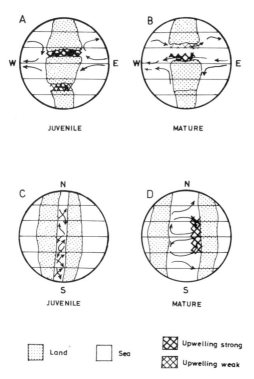

Fig. 51. Schematic current patterns in narrow oceans and areas of upwelling at continental margins; (A,B) latitudinal, (C,D) longitudinal orientations (from Cook and McElhinny, 1979). By permission of the authors and publisher, *Econ. Geol.* 74, 315–330.

phosphate occurrences (Fauconnier and Slansky, 1979) indicates that both passive and active west-facing continental margins are preferred settings.

Evidence that a major marine transgression preceded deposition of many phosphorite deposits (Eganov, 1979; Riggs, 1979) suggests a possible indirect plate tectonic control of some phosphorites, as many eustatic rises in sea level are probably related to development of ocean rises. The consequent flooding of epicontinental shelf areas may have resulted in higher organic productivity, with phosphate accumulation in mildly reducing environments produced by bacterial oxidation (Jenkyns, 1980). The related raising of the carbonate compensation depth in the oceans may also have aided phosphate formation in shallow water by removing a deep-sea phosphorous sink.

Cook and McElhinny (1979) have discussed the relative importance of continental rifting followed by ocean floor spreading along E-trending and N-trending spreading systems, and argued that deposition of phosphates takes place at an early stage of spreading in the case of E-trending ocean margins and later in the case of N-trending margins (Fig. 51). They also emphasized how phosphorite deposition commonly succeeds evaporite formation, in some cases resulting in phosphorite above evaporites, as an intracontinental rift opens into an ocean basin.

3 Metal-rich black shales

Black bituminous shales, and their metamorphic equivalents carbonaceous phyllites and graphite schists, commonly contain several hundred parts per million of certain metals, in particular Ag, Ni, Cr, V, Mo, Cu, Pb, Zn and U. Holland (1979) noted the similarity in enrichment factors for Ag, Cu, Ni, V and Zn relative to modern ocean waters, and suggested the enrichment was largely due to chemical precipitation. With a few exceptions, such as the Alum Shales of Cambrian age in Sweden, uraniferous black shales in the Venetian Alps, and certain Palaeozoic shales in USA, extraction of the metals from bituminous shales is uneconomic; nevertheless the shales form potential future ore bodies in some countries, dependent partly on improvement in extraction methods.

Black bituminous shales in continental settings are largely pelagic and perhaps occur most commonly on, although they are by no means restricted to, the deeper parts of passive continental shelves, the main environmental requirement being anoxic conditions at the sediment–water interface. The relationship of black mud deposition to marine transgression has been discussed by Schlanger and Jenkyns (1978) and Jenkyns (1980), who emphasized how abundant organic material produced during flooding of epeiric and shelf seas yields oxygen-depleted waters via bacterial oxidation. In general it can be seen that a major marine transgression will tend to result in deposition of phosphates in suitable shallow marine areas, and black metal-rich muds in topographic lows, in place of hydrogeneous manganese nodules and crusts formed in oxidising environments (Fig. 52).

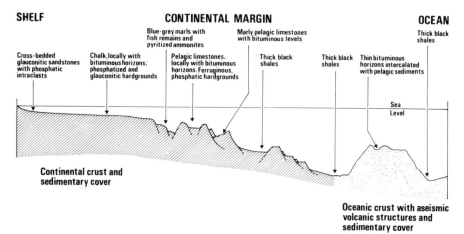

SHELF · CONTINENTAL MARGIN · OCEAN

Blue-grey marls with fish remains and pyritized ammonites

Marly pelagic limestones with bituminous levels

Thick black shales

Cross-bedded glauconitic sandstones with phosphatic intraclasts

Chalk, locally with bituminous horizons; phosphatized and glauconitic hardgrounds

Pelagic limestones, locally with bituminous horizons. Ferruginous, phosphatic hardgrounds

Thick black shales

Thick black shales

Thin bituminous horizons intercalated with pelagic sediments

Sea Level

Continental crust and sedimentary cover

Oceanic crust with aseismic volcanic structures and sedimentary cover

Fig. 52. Suggested facies distribution across a continental margin of a restricted ocean during a Cretaceous oceanic anoxic event (after Jenkyns, 1980). By permission of the author and publisher, *J. geol. Soc. Lond.* **137**, 171–188.

4 Minette-type ironstones

Ironstones of minette-type, of which the type-deposit is that in the middle Jurassic succession of Lorraine, are sedimentary iron ores consisting largely of chamosite oolites in a chamositic or sideritic mud matrix. Other examples are the Northampton Sand Ironstones of central England, of the same age as the Lorraine deposits, the Silurian "Clinton" ironstones of the eastern USA, and Ordovician deposits of Europe. *2/76/3*

All ironstones of this type show clear evidence from underlying and overlying sedimentary facies of deposition in a shallow marine environment, and Stanton (1972) has emphasized the significance of a continental source for the iron, which is derived either directly from the adjacent continent in solution and as fine particles, or from iron dissolved during breakdown of continental detritus in deeper water as visualized by Borchert (1960).

The minette-type ores of the English and French Jurassic successions undoubtedly accumulated in a more or less stable environment adjacent to a land mass of low relief, although subject to fluctuating sea levels, and it seems reasonable to suppose a similar environment for comparable deposits elsewhere. Hallam and Bradshaw (1979) have suggested on the basis of sedimentary facies that oolitic ironstones of Northampton and various other ironstones in the British Jurassic each accumulated towards the end of a marine regression, with repeated sea level changes resulting from eustatic fluctuations, a situation which could result in separation of the iron oolites from clastic material due to slight irregularities in the sea floor. The palaeogeography of Western Europe in the middle Jurassic consisted of a number of shallow marine basins separated by low-lying land areas but did not form a continental shelf as the North Atlantic had not yet opened and the central Atlantic Ocean lay far to the south. How-ever, it seems probably that the basins were related to opening of the central Atlantic rather than to rift-related processes in the North Sea to the east, and so the minette-type deposits can be considered in a very general way as typical of continental shelves or epeiric seas landward of passive continental margins.

5 Banded iron formations

The largest post-Archaen banded iron formations, e.g. those of Labrador, South Africa and western Australia, are restricted to the Lower and earliest Middle Proterozoic, within the age range of about 2600 to 1800 Ma. As in the case of the Lower Proterozoic Witwatersand type uranium–gold deposits, deposition of the detrital siderite found in some deposits is considered to have required an oxygen-free atmosphere (Garrels *et al.*, 1973). However, Klein and Bricker (1977) have suggested that the physicochemical environments of some modern sedimentary basins are similar to those in which the Sokoman Iron Formation of Labrador and probably other Proterozoic banded iron formations accumulated, with a sub-sediment surface reducing environment resulting from the presence of organic matter oxidized by bacteria.

In the absence of recognized Cenozoic analogues, suggested sedimentary environments include intracontinental lakes, and either marine or barred lagoons with wide supratidal flats (Eugster and Chou, 1973), in all cases in basins only

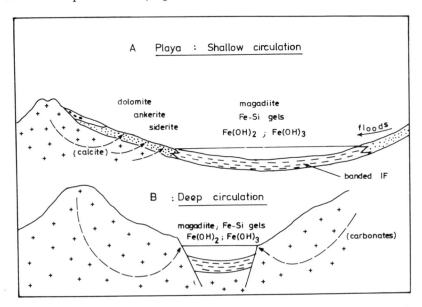

Fig. 53. Playa-lake model for banded iron formation. (A) Shallow circulation with playa lake. (B) Deep circulation, no playa fringes. Carbonates form mainly in playa mud flats. Magadiite, hydroxide and silicates form in the lake (from Eugster and Chou, 1973). By permission of the authors and publisher, *Econ. Geol.* **68**, 1144–1168.

HYPOTHETICAL STRUCTURAL SECTION THROUGH THE LABRADOR TROUGH

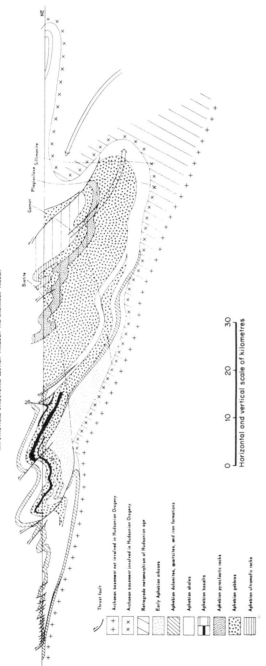

Fig. 54. Schematic cross-section through the Labrador geosyncline showing location of banded iron formation near margin of Superior Province continental foreland (from Dimroth, 1972). Note similarity to Himalayan collisional orogen (Fig. 139). By permission of the author and publisher, *Am. J. Sci.* **272**, 487–506.

partially open to the ocean (Fig. 53). Button (1976) has suggested partially barred basins of the Persian Gulf type to explain deposition of banded iron formations in the Lower Proterozoic Transvaal Supergroup of South Africa, the iron being a Lower Proterozoic analogue of younger evaporites. Microfossils in these banded iron formations have recently been described and interpreted by Klemm (1979) as oxygen-producing silica algae, which in the presence of anaerobic sea water in a semi-enclosed basin resulted in alternating deposition of chert and haematite. Drever (1974) stressed the importance of high biological activity in marginal marine environments during deposition of banded iron formations and suggested that they formed in similar environments to modern shallow water carbonates; his emphasis on the significance of upwelling of anaerobic water onto shallow platforms suggests an analogy with Phanerozoic phosphorite depositional environments.

Common to almost all interpretations is the presence of a broad gently subsiding but otherwise stable shallow-water depositional area open at least partially to the ocean, an environment which could occur within continents as well as on their margins. However, deformation of the iron formations of the Superior Province foreland in the Labrador Geosyncline (Fig. 54) during the mid-Proterozoic Hudsonian orogeny (Dimroth, 1972), interpreted as the result of collision with the Churchill Province (Kearey, 1976), suggests that the iron deposits lay near the margin rather than interior of the continent, and hence for this deposit at least a continental shelf seems the most probable tectonic setting.

6 Placer deposits

That economic concentrations of heavy minerals can occur in the coastal areas of passive continental margins is indicated by the rather numerous Quaternary deposits of zircon and the titanium minerals rutile and ilmenite in beach sands, and by the presence of off-shore and on-shore diamonds.

The abundance of beach placers on the fragmented margins of Gondwanaland has been emphasized by Guild (1974a), and major economic deposits of ilmenite and rutile are known, for example, from Richards Bay in South Africa, the southern coast of India and the east coasts of Sri Lanka, Australia and New Zealand. While a suitable source, probably metamorphic rocks of granulite facies, is clearly essential for formation of these deposits, perhaps of equal importance is the coastal morphology and off-shore current distribution. Formation of workable concentrations of diamonds to form the on-shore and off-shore deposits of Namibia is similarly closely related to the local conditions, and although the source is clearly of even more importance, despite intensive investigation it has not yet been located.

There are no economic fossil placer deposits of ilmenite, rutile or zircon, and little evidence that any of the fossil placer deposits of other minerals, which include economic deposits of gold and tin and sub-economic fossil diamond placers, accumulated preferentially on passive margins.

7   Carbonate-hosted lead-zinc deposits

Deposits of this type are sometimes termed Mississippi Valley or Alpine-type lead–zinc deposits from the very large ore bodies of Cambrian and Carboniferous age in the USA and the smaller but long-known deposits of Triassic age in the Alps (Fig. 55). The main source of lead and zinc production in USA and Europe is from deposits of this type, and the Navan deposit in Ireland with at least 40 million tonnes of ore averaging more than 13% combined metal is the largest single lead-zinc deposit in Europe.

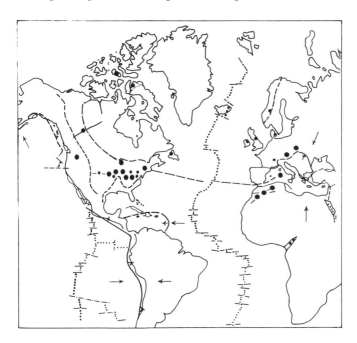

Fig. 55. Some major carbonate-hosted lead–zinc deposits of the Western World (from Guild, 1974a). By permission of the author and publisher "Metallogenetische und geochemische Provinsen", 10–24, Springer-Verlag, Vienna.

The carbonate-hosted ores consist largely of sphalerite and galena in some cases with economic quantities of silver, and are associated with baryte or fluorite. Mineralization is strata-bound within limestones and dolomites and rarely in calcareous terrigeneous sediments. In some deposits the ore is stratiform, forming layers parallel to bedding, but in most, while confined to a particular carbonate member or formation, it forms replacement or cavity-fill deposits cross-cutting the bedding. In deposits of this type, stratabound but not stratiform, the mineralization is commonly related to facies changes associated with carbonate reef margins, to solution structures in "platform" carbonates beneath an intraformational unconformity (Callahan, 1967), e.g. in the Cornwallis Lead–Zinc District of the Canadian Arctic (Fig. 56) described by Kerr (1977), and less commonly to faults; preferred sites for mineralization are where these features coincide with the axes or flanks of anticlines.

(a) Regional settings

Almost all carbonate-hosted lead–zinc deposits are in sedimentary successions accumulated in stable environments lacking evidence of intrusive rocks. The host rocks often lie not far above an angular unconformity, in some cases on the flanks of basement highs (e.g. Fig. 56). While there are no known examples of stratiform lead–zinc deposits forming today in sea-floor carbonate environments, the regional depositional environment of ancient mineralized carbonates is sufficiently understood for it to be evident that the host rocks for nearly all carbonate-hosted lead–zinc deposits accumulated in either shelf areas or on epeiric seas landward of passive continental margins. Upper Palaeozoic examples are the Tri-State deposits of the Mississippi Valley province, mineralization at Grays River in Nova Scotia and the major deposits of Tynagh, Silvermines and Navan in Eire: all these occur in Lower Carboniferous carbonate host rocks deposited on or landward of the continental shelf of the northwestern margin of the Rheic ocean prior to end-Carboniferous closure following eastward subduction. The Carboniferous mineralized host rocks are relatively undeformed in the Mississippi Valley area and the British Isles, but tightly folded where involved in the Appalachian orogeny.

Older examples are the major deposits in the Upper Cambrian Bonneterre Formation of Southeast Missouri, which also lies in the Mississippi Valley lead–zinc province, zinc deposits in Lower Ordovician dolomite in the Appalachians of east and central Tenessee and in the Upper Cambrian to middle Ordovician St George Group in Newfoundland, and galena mineralization in the Durness Limestone of Scotland; the Laisvall lead-zinc ore body in Cambrian sandstones in Sweden is possibly the northwestern continuation of the same province (Scott, 1976). These Cambrian-Ordovician deposits all occur in shelf sediments deposited on the northeastern margin of an ocean which finally closed in a late Lower Palaeozoic collision.

The Alpine lead–zinc deposits similarly formed in or on a continental carbonate shelf, in this case of Triassic age, but unlike the deposits referred to above which remained in a passive continental margin setting until involved in continental collision, the Alpine shelf margin host rocks became an overriding plate margin before their involvement in late Cretaceous or earliest Tertiary collision and northward thrusting to form the Calcareous Alps.

(b) Genesis

An epigenetic origin for most stratabound carbonate-hosted deposits is widely accepted, but many authors consider the locally stratiform Alpine-type deposits to be syngenetic with sedimentation, and possibly related to volcanism and submarine–exhalative in origin. Sangster (1976a) has emphasized the

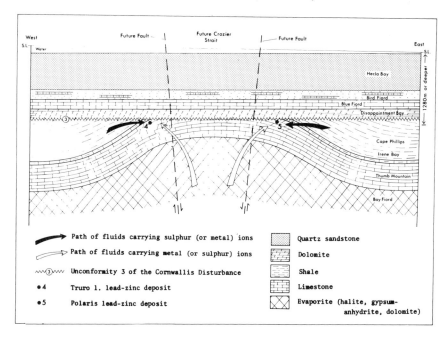

Fig. 56. Carbonate-hosted lead-zinc mineralization of the Cornwallis District, central Canadian Arctic, during late Devonian. Epigenetic mineralization occurs beneath unconformity in pores and caverns; note postulated metal and sulphur-bearing connate brines expelled from shales and evaporites within thick shelf succession (after Kerr, 1977). By permission of the author and publisher, *Can. J. Earth Sci.* **14**, 1402–1426.

contrast between the stratabound epigenetic and stratiform syngenetic deposits. The epigenetic hypothesis is supported by similarities between fluid inclusions in ores and those in oil field brines (Carpenter *et al.*, 1974), and most geologists now favour warm metal chloride-bearing connate brines as the mineralizing fluids. Some authors emphasize the role in ore genesis of hydrocarbons which result in either bacterial or inorganic reduction of sulphate to hydrogen sulphide, allowing precipitation of the metals. The observation that petroleum and carbonate-hosted sulphides are often mutually exclusive, the latter occuring in carbonates from which oil and gas may have been expelled, suggests that mineralization is perhaps related to brine circulation following heating and cracking of oil previously present in or beneath the carbonates (D. Taylor, pers. comm., 1980). However, some deposits, e.g. the Middle Devonian carbonate-hosted Pine Point in Canada, occur in the same formation as, and in this case up-dip from, oil and gas fields (Dunsmore, 1973).

There are several explanations for the source and cause of circulation of the hot brines. Expulsion of saline connate water from sediments during compaction has been suggested as the source of some mineralizing brines (Fig. 56), e.g. those which precipitated the Cornwallis District deposits in the Lower Devonian

(Kerr, 1977), but in a number of deposits it can be demonstrated that mineralization was younger than compaction of sediments. Other authors have favoured either later spreading of brines from evaporitic deposits, or gravity flow of dense brines down-dip followed by upward movement along the temperature gradient. Lange and Murray (1977) suggested that dense brines from evaporites overlying the ore host rocks moved downwards under gravity and displaced older hot metal-bearing brines formed from solution of older evaporites at depth; the displaced hot brines then migrated upward into permeable host rocks where the metals were deposited. This brine "reflux" model would require evaporite deposition both before and after accumulation of the carbonate host rocks; the pre-carbonate evaporites may be expected as early passive continental margin facies but there is less reason to expect the common occurrence of evaporites above the carbonates.

In almost all hypotheses the metals are considered to be leached either from terrigeneous sedimentary rocks lying topographically or stratigraphically beneath the ore host rocks, or from "basement". While few deposits are associated with intrusive rocks, it is possible that granitic plutons distant from the mineralization, and intruding either the mineralized stratigraphic unit or older rocks, could have caused convective circulation of the brines as a result of radiogenic heating.

Ohle (1980) has recently emphasized that while most deposits are now commonly interpreted as the result of circulation of warm brines, many aspects such as the source of the metals, origin of the brines and causes and location of ore deposition have yet to be explained satisfactorily.

The lead-zinc deposits of Ireland (Fig. 57A, B) are mostly stratiform and often considered to be exhalative; Russell (1978) has suggested that they formed where convecting sea water, which penetrated Lower Palaeozoic rocks beneath the carbonate host rocks, reached the sea floor along faults. The underlying thick Palaeozoic geosynclinal succession postulated by Russell as the source of

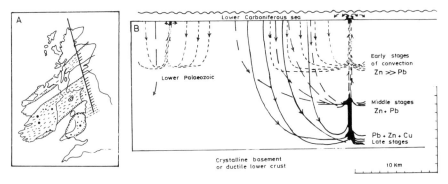

Fig. 57. (A) Location of- carbonate-hosted lead–zinc deposits in the Lower Carboniferous of Ireland and Great Britain; (B) Schematic cross-section through deposits of (A) showing downward-excavating ocean-water hydrothermal convective system; sea water penetrated syn-sedimentary faults into thick hot geosynclinal prism of Iapetus Suture (after Russell, 1978). By permission of the author and publisher, *Trans. Instn Min. Metall,* 87, B168–171.

the metals possibly consisted of imbricate flysch wedges continuous with that in the Southern Uplands of Scotland, indicating the significance of outer arcs as sources of ore-forming fluids (Ch. 5, I. B. 5).

The age of mineralization relative to the age of the host rocks is of interest with regard to tectonic setting, as if sulphide emplacement is much younger than the host rock, it would have taken place when the latter were in a tectonic setting different from that in which they were deposited, e.g. either during continental collision, or still later during post-collision rifting. Radiometric dates on the ores themselves are characteristically difficult to interpret and in some cases yield anomalously young ages. Stratigraphic evidence also rarely fixes the age of mineralization within an interval of less than 100 Ma. Sangster (1970) has emphasized how in the East Tenessee district sphalerite and dolomite sand in cavities in Lower Ordovician carbonates is parallel to the host-rock bedding, and thus mineralization pre-dated orogeny in the Carboniferous. This example is of value as it demonstrates that while deposition of some carbonate-hosted lead–zinc deposits may have resulted from brine circulation consequent on folding and compaction of adjacent rocks following closure of the ocean basin, in at least some cases mineralization preceded the collisional orogeny.

There is no obvious explanation for the scarcity of Precambrian carbonate-hosted deposits, as it may reasonably be supposed that accumulation of evaporites as well as of the preserved extensive carbonate deposits took place at least as early as the late Proterozoic. Possibly the restriction of major deposits to the Phanerozoic is related to the absence of significant Precambrian petroleum source rocks.

8 Carbonate-hosted baryte and fluorite deposits

Economic deposits of baryte or fluorite are present with several carbonate-hosted lead-zinc deposits. Less commonly, the minerals occur in economic amounts as stratabound deposits in carbonates lacking significant lead and zinc mineralization. They commonly form veins or irregular bodies within a preferred horizon but generally markedly transgressive to the bedding. Examples are the baryte deposits worked on a small scale and widespread in Middle Ordovician limestones which extend from northern Burma to Malaysia, and baryte in Jurassic rocks of Pakistan. In both cases there is little doubt that the carbonate host rocks accumulated on the shelf of a passive continental margin. The fluorite deposits of the Illinois-Kentucky district provide an example of mineralization emplaced in shelf carbonates long after the carbonates were involved in continental collision and probably during incipient rifting (Ch. 2, II. B. 3. (b)).

9 Preservation potential

With the exception of marine placer deposits, commonly destroyed before burial, mineral deposits within continental shelf successions have a high chance of being preserved. This is due partly to their forming in a tectonic setting which is characterized by subsidence throughout the spreading history of the adjacent ocean floor. During the subsequent closure of the ocean basin, the formerly passive continental margin may develop into a magmatic arc with consequent uplift and erosion as a result of subduction beneath it. However, as most oceans close by subduction beneath only one margin, it is almost equally probable that the passive margin will remain as such during ocean closure. Under these circumstances it will continue to subside during ocean floor subduction, and eventually, during collision, will become sutured with and possibly under-thrust the overriding plate on the opposite side of the former ocean (see Ch. 6). Part of the shelf succession may be either lost beneath thrusts, or buried beneath foreland basin sediments, but part will be elevated and metamorphosed in the immediately post-collision foreland thrust belt, while the shelf lying furthest from the suture may escape major deformation. Consequently the mineral deposits of the shelf will be preserved, sometimes after metamorphism, both within thrust slices bordering the suture and within gently folded successions on the continental side of the foreland thrust belt.

II INTRACONTINENTAL BASINS OF UNCERTAIN ORIGIN

A Tectonic Setting

A number of thick sedimentary successions of Proterozoic age are known from basins or broad synclinal structures within intracontinental settings which lack features typical of modern basins within continents. It is probable that many of the basins formed in rift-related settings; others are possibly remnants of successions deposited in depressions either between or on the flanks of intracontinental swells and their associated graben, analogous to the African depressions west of the broad uplifts which include the East African Rifts. Whether some apparently intra-continental basins represent preserved segments of continental shelf successions, as suggested by Dietz and Holden (1966) to explain the early Proterozoic Witwatersrand Basin succession in South Africa, now seems very doubtful. A more probable explanation for this and some other Proterozoic predominantly terrigeneous successions lacking evidence of syndepositional rifting, e.g. the Upper Proterozoic–Lower Palaeozoic Amadeus Basin in Australia, is that they accumulated in back-arc cratonic basins (Ch. 5, V. A), far from but related to an active continental margin. As many of these basins of uncertain tectonic setting either contain or are associated with important ore bodies they cannot be neglected, and we briefly discuss some examples here while recognizing that they may have diverse origins.

As might be expected the basin successions are highly variable, consisting of different proportions of continental clastic and shallow marine sediments including carbonates. Volcanic rocks are scarce although in the Witwatersrand

Basin a thick succession of lavas, the Ventersdorp, occurs high in the sequence. The basement usually consists of metamorphic rocks while the basin successions themselves are commonly but not invariably only weakly metamorphosed. Major syn-sedimentary thrust belts adjacent to the basins and characteristic of orogenic settings are scarce, as are linear graben within or confining the basins and indicative of intracontinental rifts.

Most of the basins considered are of Proterozoic age. This, together with the absence of recognized Cenozoic analogues, could be explained by sedimentation in a type of tectonic setting which existed only in Proterozoic time, implying that tectonic regimes were significantly different from those in the Phanerozoic. However, it seems more probably that the basins were characteristic of large continental areas which existed during at least part of the Proterozoic, and that in the late Phanerozoic continents were mostly too small for development of comparable basins. It is also possible that preservation and exposure of only remnants of the Proterozoic basins has obscured their relationship to underlying rifts or adjacent thrust belts which provide the criteria for identification of the tectonic setting for many of the Phanerozoic basin successions.

An example of a modern intracontinental basin related neither to thrusting nor contemporaneous rifting is the Chad Basin of Pliocene to Quaternary age which developed over an extension of the Cretaceous Benue Rift System. Burke (1976) has shown that about 0.5 km of subaerial sandstones and diatomaceous clays overlain by late Quaternary lacustrine deposits accumulated in a depression surrounded by uplifts and volcanic areas, which include Tibesti, Ahaggar, Air and Jos; these source areas are believed to be related to hot spots developed in the continental crust in the last 30 Ma. Burke (1977) has suggested that the early Palaeozoic Michigan Basin developed over the Keweenawan Rift is an ancient analogue of Chad, and it is possible that some of the mineralized Proterozoic basin successions referred to below formed in similar settings.

B Mineral Deposits

Only two types of deposit are considered here, both of major economic importance but known only from, and almost certainly restricted to, Proterozoic successions (Table VI).

1 Unconformity vein-type uranium deposits

Uranium deposits associated with a major unconformity beneath thick mid to late Proterozoic non-marine sediments have in the last decade attracted widespread interest. The "unconformity vein" –type uranium ores of pitchblende and uraninite occur around the margins of the Athabasca Basin in Saskatchewan (Dahlkamp, 1978) (Fig. 58A), in the Alligator River area of the Northern Territory in Australia adjacent to the overlying Kombolgie sandstones (Hegge and Rowntree, 1978), and in the Rocky Downs area of South Australia; the Alligator River deposits comprise a large proportion of Australia's huge uranium

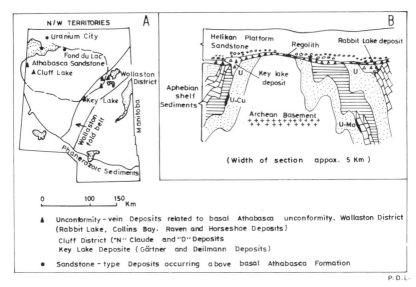

Fig. 58. Uranium deposits in the Athabasca Sandstone and Wollaston Fold Belt, Saskatchewan (From McMillan 1978). (A) Location of deposits. (B) Epigenetic or secondary concentration of uranium into unconformity veins in carbonate and metapelitic rocks in Wollaston and Key Lake Districts (width of section approx. 5 km). By permission of the author and publisher, *Min. Ass. Can. 3.*

reserves and include important gold values, and the Rocky Downs deposits with 550 000 tonnes of ore grading 0.1 to 0.2% U_3O_8 include more than 10 million tonnes of copper ore. Recently, unconformity vein-type deposits have also been reported from the Lianshanguan region of northeastern China (Finch, 1980).

In both the Athabasca and Alligator River areas, lower to middle Proterozoic metasediments overlie Archaen "basement" and are in turn overlain in angular unconformity by thick relatively undeformed Proterozoic successions of terrigeneous sediments. Uranium occurs mostly beneath but within a few hundred metres of the angular unconformity, although locally also in the overlying sandstones. In the Athabasca Basin (Fig. 58B) the mineralization is commonly associated with graphitic Aphebian rocks. Langford (1978a) has stressed that in the Australian deposits the ore mineral texture indicates space filling, and that dilatant structures such as faults and breccia zones tended to localize the ore deposits (Fig. 59). The primary mineralization is characteristically simple consisting mostly of pitchblende, and uraninite and minor sulphides with S, As and Se commonly combined with Fe, Cu, Pb, Co, Ni, Bi and Mo. Gold and silver may also be present.

Interpretations of the origin of the mineralization range from near-surface supergene through, in the case of the Athabasca uranium, hydrothermal, to diagenetic hydrothermal activity coeval with the Hudsonian and Grenville orogenies (Morton and Beck, 1978). Another recently suggested origin involves metamorphic enrichment processes which preceded deposition of the uncon-

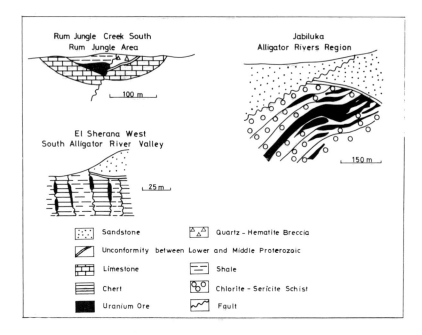

Fig. 59. Sections through unconformity-vein uranium deposits in the Darwin area, Australia showing the diversity of host rocks and the relationship of the deposits to the unconformity (from Langford, 1978a). By permission of the author and publisher, *Min Ass. Can.* **3.**

formably overlying sandstones, mineralization at the unconformity and radiometric ages on pitchblende being explained by remobilization of primary ores (Finch, 1980). Langford (1978b) however favoured formation of pitchblende vein deposits during early deposition of the cover rocks (Fig. 58B), due to groundwater circulation under the streams which deposited the sandstones. Extensive envelopes of chlorite and sericite were produced during uranium deposition and formed effective seals to initial movement of solutions. The deposits were subsequently deeply buried and subjected to diagenetic recrystallization. This hypothesis may explain many features of the deposits such as the temperature of fluid inclusions, penetration of the chlorite–sericite rocks by the fluids, the presence of a few U–Pb dates of similar age to the cover rocks, the abundance of later dates and the presence of limited amounts of pitchblende in the overlying sandstones, and above all why the deposits only occur at unconformities overlain by fluvial sandstones.

The size of the sedimentary basins above the unconformity suggests that they are related in some way to tectonic processes at plate boundaries; possibly they may have formed either on overriding continental plates landward of an arc system or within continents undergoing incipient rifting.

2   Quartz-pebble conglomerate uranium and gold-uranium deposits

Uranium deposits in quartz-pebble conglomerates of earliest Proterozoic (2700 or 2800 to 2300 Ma) age are known from Canada, (e.g. Elliott Lake), South Africa (Witwatersrand), Brazil (Jacobina) and western Australia, and comprise more than a quarter of the world's reasonably assured uranium reserves (Dahlkamp, 1980). The uranium, commonly associated with pyrite, occurs above an angular unconformity at the base of a thick succession of shallow marine or continental sedimentary rocks.

In the Elliott Lake deposits of Canada, the uranium has been interpreted as largely detrital, derived from pegmatite and gneisses (Robertson, 1974), and transported and deposited as uraninite in an oxygen-free atmosphere, although an epigenetic origin for some of the metal has also been suggested. In the Witwatersrand deposits of South Africa, uranium and gold were partly detrital and partly in solution, forming both fluvial fossil "placers" and biochemical deposits by interaction with carbon-forming algae mats (Pretorius, 1975; Minter, 1976).

Most detailed sedimentological work has been done on the deposits of the earliest Proterozoic Witwatersrand Basin within the Archaen Kaapvaal Craton, and suggests an intracontinental basin up to 350 km long by 200 km wide, fed by streams draining Archaen rocks to the northwest. The goldfields were formed as fluvial fans that built up at several points along the periphery of the basin. There is some evidence for syn-depositional faulting along the northwestern rim of the basin (Pretorius, 1975) but there are few obvious structural controls or features typical of an aulacogen, although the tectonic environment for deposition of the overlying Ventersdorp lavas is uncertain. A possible rift-related environment is suggested by the presence to the northeast of the Great Dyke of Zimbabwe, of similar age to the Witwatersrand succession, interpreted by some authors as the axis of a failed rift as discussed above.

The following geological and mineralogical aspects are summarized from Feather and Koen (1975). The Witwatersrand System is up to 8000 m thick with the most important mineralized conglomeratic reefs in the Upper Division. These reefs consist largely of well-rounded pebbles mainly of quartz but including some chert, jasper, quartzite, quartz porphyry and metamorphic rocks. The matrix is mainly secondary quartz and phyllosilicates. The most abundant or important minerals identified are gold, pyrite, uraninite (in rounded detrital and euhedral secondary form), carbon, brannerite, arsenopyrite, cobaltite, galena, pyrrhotite, gersdorffite, chromite and zircon. Numerous detrital diamonds from less than a carat up to 8 carats were found in the early stages of mining before grinding methods were employed (Young, 1917), indicating that kimberlites were probably emplaced in the Kaapvaal Craton in Archaen times.

The temperature of metamorphism in the Witwatersrand System is estimated at about 600°C and caused redistribution of several minerals. Gold mostly occurs in the fine-grained matrix between pebbles, in veinlets and in association with carbon, pyrite and other sulphides. Gold values in the Vaal Reef and

Ventersdorp Contact Reef are 50 ppm and 44 ppm respectively with corresponding U_3O_8 values of 870 ppm and 290 ppm.

3 Preservation potential

In view of the uncertain extent of Proterozoic basin successions which hosted the unconformity-vein type uranium and quartz-pebble conglomerate uranium–gold deposits, discussion of their preservation potential is largely futile. The known deposits have clearly been preserved largely because of their location within Archaen or early Proterozoic cratons which have not suffered subsequent rifting and fragmentation; the mineralized successions have hence escaped the collisional orogenies and consequent uplift and erosion which have affected the craton margins.

CHAPTER 4

Deposits Formed in Oceanic Settings

The world's oceans are characterized by rises or ridges sometimes termed Mid-Ocean Ridges from the medial position of the Atlantic Ridge. These are flanked by deep basins which, although interrupted by features such as extinct ridges and plateaux, extend from the base of the ridge to the continental rise. Superimposed on the ocean ridges and basins are oceanic linear island and seamount chains, commonly trending at a high angle to the ridge and either rising from the abyssal plain or extending across the plain to the ridge axis, and transform faults, mostly forming offsets in the ocean ridge axis and in some cases extending across the deep basins. Of these four major tectonic settings, only the ocean ridges and transform faults are plate boundaries. Nevertheless the intraplate linear island chains and deep basins form sufficiently distinct and extensive morphological and crustal features to be considered as tectonic settings on a global scale.

Development of ocean basins by spreading in former rift zones corresponds to the late generative and development stage of the Aubouin geosyncline, and to the initial stage of the Russian geosynclinal model (Table II).

I OCEAN RIDGES AND BASINS

A Tectonic Setting

Next to the fundamental division of the earth's surface into continent and oceans, ocean ridges are the largest morphological features on earth. They are known in most detail from the Atlantic and Pacific Oceans, where they have been investigated both from the surface and using submersibles. The axes of active ridges, characterized by shallow seismic activity, volcanism and high heat flow, are bordered by more or less symmetric patterns of magnetic anomalies parallel to the ridge axis and extending down its flanks, giving rise to

114

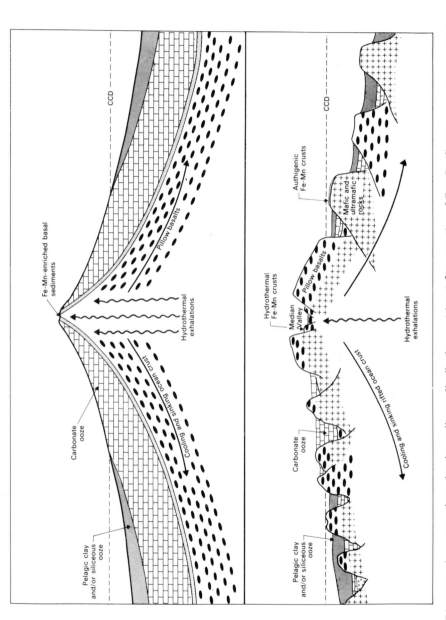

Fig. 60. Diagrammatic cross-sections showing sediment distribution on (above) fast-spreading of East Pacific Rise-type ridge, and (below) slow-spreading ridges of Atlantic-type, CCD-carbonate compensation depth (after Jenkyns, 1978). By permission of the author and publisher, "Sedimentary environments and facies" 313–371, Blackwell Scientific, Oxford.

the concept that the anomalies are related to the creation of ocean floor at the ridge axis and cooling of basaltic rocks in the magnetic field prevailing at the time.

Comparison of the Mid-Atlantic Ridge and East Pacific Rise indicates significant morphological differences, the former having a more rugged topography indicating relatively extensive block faulting (Fig. 60); this difference is widely attributed to the slower spreading rate in the Atlantic than Pacific.

The Mid-Atlantic Ridge has been studied in most detail in the FAMOUS area, around Lat 36°N, where the spreading rate is estimated at around 1.1 cm/yr, and where N-trending ridge segments around 30 km long are offset by E-trending transforms. In the absence of an articulate indigenous population to name the ridge features, a rather tedious physiographic terminology has resulted. In general the Ridge is characterized by a series of blocklike linear steps, comprising a central ridge within a narrow inner rift valley floor at least 2400 km deep, and several kilometres wide, bordered by inner fault scarps above which are terraces up to 15 km wide; the terraces lie at the foot of outer fault scarps which form the inside of the rift shoulders and are usually around 30 km apart (Ramberg, *et al.*, 1977). The central ridge of the inner floor of the rift (Fig. 61), described by Ballard and Van Andel (1977) (see Fig. 71A) is a basaltic edifice produced by axial ridge volcanism. The floor on each side of the ridge consists of collapsed axial volcanics cut by vertically dipping tension fractures; at the outer margins of the rift the floor is uplifted to form the inner wall or scarp. Predominant lithologies are pahoehoe pillow lavas, massive basalt and fault breccia; the crustal thickness is estimated at 2-3 km, with a thermal gradient of 400-600°C/km. The model of crustal generation derived from the detailed work in the FAMOUS area may be compared with an earlier model for the same area (Moore, *et al.,* 1974). Other explanations for

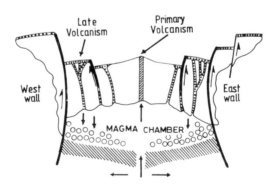

Fig. 61. Diagrammatic cross-section of inner rift valley crust over magma chamber in which lateral magmatic differentiation and deposition of cumulates is taking place, Mid-Atlantic Ridge, Lat 36° 50' N. (from Ballard and Van Andel, 1977). By permission of the authors and publisher, *Bull. geol. Soc. Am.* **88**, 507–530.

generation of the median rift at slow-spreading ridge include caldera collapse, analogous to that of some on-land volcanoes (Francis, 1974).

Rather different features to those observed in the FAMOUS area have been described from the East Pacific Rise at the mouth of the Gulf of California, where the spreading rate is around 3 cm/yr. Here zones are similar to but wider than those on the Mid-Atlantic Ridge and offsets on the fault scarps are relatively subdued, suggesting similar tectonic and volcanic processes despite differences in morphology (Normark, 1976). Elsewhere on the East Pacific Rise, where spreading rates are faster, an axial block or ridge up to 15 km wide occupies the inner rift–scarp–terrace system of the ocean ridge (Rea, 1975).

The elevation of ocean ridges above the surrounding sea floor can be explained by thermal expansion with or without associated phase changes. As the oceanic crustal rocks emplaced at the ridge move outwards away from the rift shoulders they cool, resulting in the regular decrease in elevation of the sea floor on the ridge flanks, with the ocean depth proportional to the square root of the age of the crust. There is also an increase in thickness of both lithosphere and ocean crust with age, although the actual thickness of the oceanic lithosphere remains uncertain.

The deep ocean basins consist of broad abyssal plains at depths of at least 4000 m, interrupted by topographic irregularities carried in from the spreading ridge, by hot spot related linear island chains or ridges, by extinct spreading ridges and by plateaux. Adjacent to continents the abyssal plains have a characteristically rather smooth topography due to submarine fans deposited from turbidity currents which are mostly fed by submarine canyons extending down the continental slope; elsewhere there may be a thin sedimentary layer, with in places basaltic crust exposed due to erosion by abyssal currents.

As the ocean floor moves away from the ridge axis it becomes mantled in sediment (Fig. 60). Work on a fast-spreading segment of the East Pacific Rise has shown that the first sediments deposited are chemical precipitates of iron-manganese oxides and hydroxides found on the ridge crest and described below. With continued movement down the ridge flanks and onto the abyssal plains sediments are derived from three main sources. Oceanic islands supply volcanogenic or calcareous sediment to form submarine aprons of turbidite and other mass flow deposits, and near continental margins not bordered by a trench extensive submarine fans may build out over the abyssal plain. A second major source of sediment is provided by pelagic organisms, their remains forming either calcareous or siliceous oozes depending on the latitude and whether the sea floor is above or below the carbonate compensation depth. The third type of sediment is pelagic red clay. Turbidites may be interbedded with this sequence. The sediments may largely smooth out ocean floor topographic irregularities derived from the ridge, although sedimentation is not continuous and unconformities due to submarine erosion by abyssal currents are commonly present.

Data on the composition and structure of the oceanic crust is based on seismic refraction data, on dredge and to a lesser extent on deep sea drilling core samples, and on comparison of the rocks in on-land ophiolites with oceanic seismic layers. Because dredge hauls rarely give a clear picture of stratigraphy in the upper oceanic crust as exposed in submarine fault scarps, the most powerful tool in determining the nature of the crust is that of ophiolite-seismic layer comparison.

Ophiolites, interpreted as oceanic crust and uppermost mantle rocks tectonically emplaced on land, have been defined (e.g. Penrose Conference, 1972; Coleman, 1977) as comprising five layers (Fig. 62): an uppermost pillow lava layer passes transitionally downwards through either sills or massive basalts into a complex of sheeted dykes; this is underlain by a gabbro layer which shows cumulate textures in its lower part and is succeeded downwards by cumulate pyroxenite and dunite; the lowermost layer consists of tectonized harzburgite, with minor dunite. The cumulate and tectonized ultramafic layers are commonly largely serpentinized, while overlying layers are sometimes metamorphosed to greenschists and amphibolites facies, as a result of circulation of hot sea water.

Seismic models of the oceanic crust identify an uppermost sedimentary layer 1, underlain by 2, 3 or 4 crustal layers above the mantle–crust boundary or Moho. In the more widely accepted 4-layer models (Table VII), layer 2A beneath supra-crustal layer 1 sediments is equated with pillow lavas, layer 2B with sheeted dykes, which probably give rise to the ocean floor magnetic anomalies, and layer 3A with altered gabbro or metagabbro. There is less agreement on the position of the Moho, at the base of layer 3B, within the ophiolite sequence. Early interpretations had suggested that the Moho was the boundary between fresh ultramafic rock below and partially or wholly serpentinized ultramafic

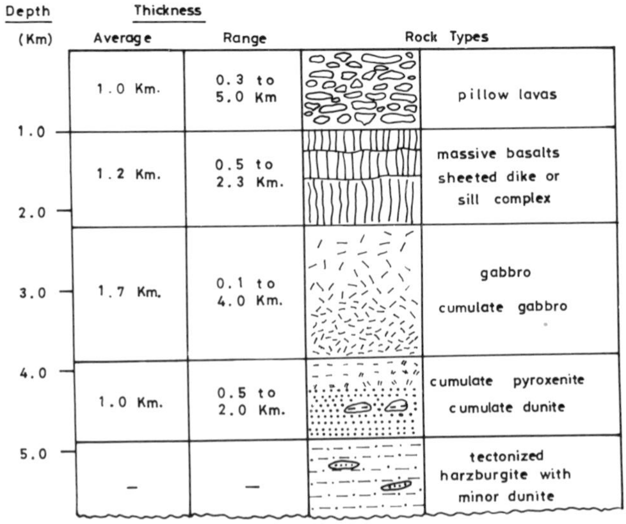

Fig. 62. Igneous stratigraphy of typical ophiolite, metamorphism omitted (from Clague and Straley, 1977). By permission of the authors and publisher, *Geology* 5, 133–136.

TABLE VII

Models of oceanic Moho, oceanic layer determined by seismic refraction in parentheses; compressional wave velocities in km/s (after Clague and Straley, 1977). By permission of the authors and publisher, Geology **5**, *133–136.*

2-layer model of Shor and others	3-layer model of Woollard	4-layer model of Woolard	Model 1 — Moho=contact between gabbro and cumulate ultramafic rocks	Model 2 — Moho=contact between cumulate and tectonized ultramafic rocks	Model 3 — Moho=contact between serpentinized and unserpentinized tectonized ultramafic rocks	Depth (km)
						0
1.5	1.5	1.5	Water	Water	Water	−2
2.1	1.7	1.7	Sediments	Sediments	Sediments	−4
	2.75	4.12	Pillow lava	Pillow lava	Pillow lava	
5.15	4.33	5.76	Sheeted dikes	Sheeted dikes	Sheeted dikes	−6
	6.55 (3A)	6.77 (3A)	Altered gabbro	Altered gabbro or metagabbro	Metagabbro	−8
6.82 (3)	7.32 (3B)	7.42 (3B)	Fresh cumulate gabbro	Fresh gabbro — Partly serpent. cumulate dunite and pyroxenite	Partly serpent. cumulate dunite and pyroxenite — Partly serpent. tectonized harzburgite	−10 / Moho
8.15	8.2	8.2	Cumulate dunite and pyroxenite — Fresh tectonized harzburgite	Fresh tectonized harzburgite	Fresh tectonized harzburgite	−12

TABLE VIII

Mineral deposits characteristic of oceanic settings.

Tectonic setting	Association	Genesis	Type of deposit/metals	Examples
Mid-ocean ridges and basins	Pelagic red clays and basalts	Hydrogeneous (authigenic) sedimentary	Oxide and hydroxide nodules and encrustations Mn Ni Co Cu	Atlantic, Pacific, Indian oceans (Recent)
	Ocean ridge basalts	Hydrothermal exhalative sedimentary	Mn Fe oxides and hydroxide nodules and encrustations	Mid-Atlantic Ridge; E Pacific Rise (Recent)
	Ocean ridge basalts	Sea water hydrothermal exhalative sedimentary	Cu Fe Zn sulphides	E Pacific Rise; Red Sea deeps (Recent); Troodos, Cyprus (Cretaceous)
	Pelagic carbonaceous sediments	Sedimentary	Metal-rich shales	S Uplands, Scotland (L Palaeozoic)
	Dunite within harzburgite of uppermost mantle	Magmatic	Podiform chromite	Cyprus, Cuba (Mesozoic); Philippines (Tertiary)
	Peridotites and serpentinites of uppermost mantle	Magmatic and metasomatic or epigenetic hydrothermal	Ni, Fe, Ti, Au, Pt and asbestos, talc, magnesite	Philippines, Italy, Greece (U Mesozoic–L Tertiary)

TABLE VIII (Cont'd)

Tectonic setting	Association	Genesis	Type of deposit/metals	Examples
Oceanic transform faults	Fan sediments, high Ba basalts, Mn	Sea water hydrothermal exhalative sedimentary	Ba	San Clemente Fault Zone (Recent)
	Ocean crust igneous rocks	Sea water hydrothermal sedimentary	Fe Mn oxides and hydroxides	Romanche Fracture zone (Recent)
Oceanic linear island and seamount chains	Undersaturated alkaline intrusions	Magmatic	Carbonatite?	Canary and Cape Verde Is., Tahiti?

rock above, and Clague and Straley (1977) suggested that this serpentinization boundary lay within the tectonized harzburgite zone. This could explain the increase in thickness of layer 3 with distance from the ridge axis due to downward progression of serpentinization as the crust cooled.

Various hypotheses of magmatic processes at ocean ridges have been advanced to explain the ophiolite sequences, e.g. that of Dewey and Kidd (1977) based largely on western Newfoundland, Lister (1977) in connection with hyrdothermal flow at spreading centres, and Duncan and Green (1980).

B Mineral Deposits

These can conveniently be divided into occurrences of hydrogeneous and hydrothermal metalliferous sediments known from present day ocean basins and active ridges, and stratiform sulphides, podiform chromite, and rare Ni, Pt, Fe, Ti and Au deposits found in on-land ophiolites and considered to have formed on and beneath the ocean ridge surface. Asbestos, magnesite and talc are also found in ultrabasic rocks of ophiolites (Table VIII).

1 Mineralization on ocean ridges and in ocean basins

(a) Hydrogenous manganese nodules of the ocean floor

Ferromanganese nodules and encrustations occur both overlying basalt lava at spreading ocean ridges where they are considered to be hydrothermal deposits, and overlying sediments away from the ridge crests, where they are hydrogenous or authigenic in origin.

The hydrogenous deposits can be distinguished by their Fe/Mn ratios of around 1, high trace metal contents (Ni, Co, Cu) (Fig. 63A), and high Th values and low U/Th ratios (Fig. 63B), considered to reflect their relatively slow rate of accumulation (Bonatti *et al.*, 1976b). They also differ from hydrothermal deposits in their strong positive Ce anomalies in the chrondrite-normalized distribution of rare-earth elements. The deposits, either in the form of crusts or more commonly of nodules and hence long-known as manganese nodules, (Fig. 64A), occur in a variety of settings on the ocean floor surface. They are most abundant, locally exceeding 5 kg/m^2, in areas where strong ocean bottom currents prevent accumulation of non-chemical sediments and may have resulted in an erosion surface. Thus they have been reported from the summits and flanks of seamounts and from oceanic plateaux, but are particularly abundant in some areas of the deep ocean basins. Here they mostly overlie red clays, although some rest on carbonates, and can form an extensive "pavement".

The nodules are of economic interest because of their high content of Cu, Ni and Co, these metal abundances being particularly high where nodules are widely dispersed on the sea floor, and there is some evidence that seamount nodules are richer in Co and possibly Ba, Pb and V than those of deep basins, which are richer in Ni and Cu (Cronan and Tooms, 1969). Because of the possibly economic grades of Cu, Ni and Co, currently calculated as at least 2%

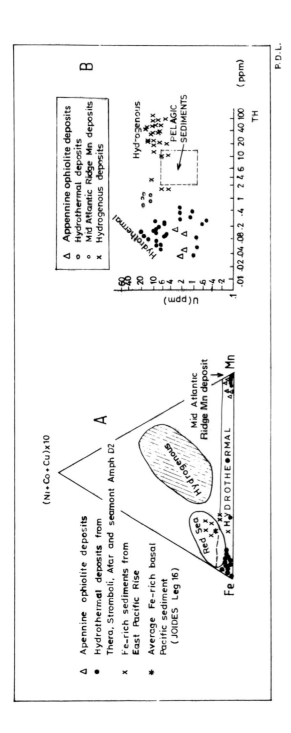

Fig. 63. (A) Fe/Mn/(Ni+Cu+Co) x 10 plot, illustrating the different composition of "hydrogenous" and "hydrothermal" deposits from the ocean floor (from Bonatti *et al.*, 1976b). (B) U/Th ratio in metalliferous deposits from the ocean floor (from Bonatti *et al.*, 1972). By permission of the authors and publishers, *Bull. geol. Soc. Am.* 87, 83–94, and *Econ. Geol.* 67, 717–730.

Fig. 64. Hydrogenous manganese nodules from the ocean floor at about 5000 m depth in the Madagascar Basin, collected during the cruise VA 07 of the motor vessel Valdivia (Bäcker and Schoell, 1974). Photo by Professor Dr H Bäcker. The average metallic composition of similar manganese nodules with Ni + Cu $>$ 1.5% from the north equatorial Pacific is 28.1% Mn, 6.22% Fe, 0.22% Co, 1.49% Ni, 1.17% Cu and 0.133% Zn. (B) Bottom photograph (field of view approximately 3 m by 4 m) from the TAG Hydrothermal Field showing hydrothermal manganese oxide as a crust on (upper left) and matrix in breccia of basalt fragments (centre) at 2600 m below sea level on the east wall of the rift valley of the Mid-Atlantic Ridge (Rona, 1978). Photo by Dr Peter A. Rona.

combined metals in the nodules (Piper, 1980), and less probably economic manganese, numerous feasibility studies are being conducted on the possibility of mining the deep sea deposits, mostly using dredging techniques.

Absence of manganese nodules from deep sea cores indicates that they do not normally survive burial by younger sediments. However, a number of ancient on-land occurrences, interpreted as analogous to those of modern oceans, are known, and accounts of some of these have been summarized by Jenkyns (1978). In Timor, Cretaceous red clays in Miocene olistostromes contain scattered concentrically laminated ferromanganese nodules interpreted as ancient abyssal plain deposits. Possible examples of ancient seamount manganese nodules are those of Eocene age overlying tholeiitic lavas in the Olympic Peninsula, Washington. The lavas have been interpreted as ancient seamount-like features tectonically accreted to North America during eastward subduction. While it is unlikely that these and other on-land occurrences will be economic, they do demonstrate that preservation of nodules can occur.

(b) Hydrothermal metalliferous sediments of oceanic spreading systems

In the last 15 years metal-rich chemical sediments have been described from young spreading centres in the Red Sea, the Afar rift and Salton Sea, and later from the Pacific, Atlantic and Indian ocean ridges. Most of the deposits are iron-rich but at least one iron-poor, manganese-rich sediment has been located in the Mid-Atlantic Ridge at Lat 26°N (Scott *et al.,* 1974). The Red sea and Salton Sea deposits are sulphide-rich muds associated with metalliferous brines, and most of the ocean ridge deposits are oxides and hydroxides of iron and manganese, although recently sulphide deposits have been found on the East Pacific Rise. The recently formed Afar deposit is rich in Fe, Mn and Ba (Bonatti *et al.,* 1972). While all these types of deposit probably have a similar origin, they differ in their significances as analogies of ancient mineral deposits in ophiolites and we describe them separately below.

(i) Iron- and manganese-rich sediments of the ocean ridges. The presence of iron and manganese oxides and hydroxides overlying the basaltic rocks at ocean ridges was referred to above (see I. A.). Deposits of this type have also been recovered from drill cores penetrating the base of the sediment layer at various off-ridge localities. Examples of on-ridge deposits are those of the TAG hydrothermal field on the Mid-Atlantic Ridge (Fig. 64B), deposits on the East Pacific Rise crest around 13°S, and on the Indian Ocean Ridge south of Lat 16°S; examples of deep sea basin deposits are those in the Bauer Deep and in the Central North Pacific.

Rona (1978) has provided a useful and readily accessible summary of these deposits, which occur as metalliferous sediments and encrustations, and near the Galapagos Islands as cappings to submarine mounds. The metalliferous sediments, up to a few metres thick, consist largely of goethite, iron-rich montmorillonite and manganese hydroxides, considered to accumulate at least an order of magnitude faster than pelagic sediments. Metalliferous encrustations, forming layers up to 2m thick overlying basalt, are more local in occurrence than the metalliferous sediments. The encrustations have either a very low or very high Fe/Mn ratio, which distinguishes them from hydrogenous deposits described above, and results in manganese-rich deposits or umbers and iron-rich deposits or ochres. In manganese-rich encrustations birnessite, todorokite and pyrolusite are the major minerals, while in iron-rich encrustations goethite and iron-smectite predominate. Trace metal contents are low, although baryte may occur.

(ii) Metal-rich muds and brines of the Red Sea. Metal-bearing muds of the Atlantis II deep in the median valley of the Red Sea were the first metalliferous sediments reported from an oceanic spreading centre (Degens and Ross, 1969). The deposits consist of multi-coloured sediments with concentrations of fine-grained Fe-montmorillonite, goethite, manganite, manganosiderite and sulphides; in general iron-rich sediments overly the sulphides (Figs. 65 and 66A). Sulphide minerals comprise pyrite, sphalerite and chalcopyrite with up to 20% Zn and form beds a few kilometres long and a few metres thick. Metal-rich brine occurs interstitially within the muds and overlies the deposits. The probable tonnage of dry salt-free mud is estimated at 100–200 million tonnes with up to 5% Zn, 1% Cu and some Ag, forming potentially minable bodies.

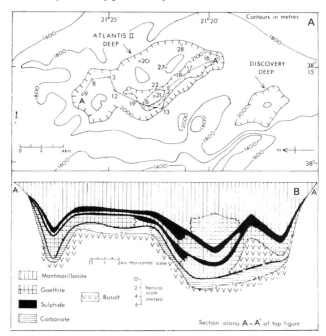

Fig. 65. Bathymetric map of the main hot brine region of the Red Sea showing core locations (A) and a generalized section through Atlantis II Deep showing mineral facies relations (B) (from Hackett and Bischoff, 1973). By permission of the authors and publisher, *Econ. Geol.* **68**, 533–564.

The Red Sea deposits have been widely interpreted as the result of circulation of heated sea water through the volcanic rocks, leaching of metals and deposition as sulphides on the sea floor, and similarities between the Red Sea axial zone and ocean ridges led to suggestions that similar sulphide deposits could be present at spreading centres within major oceans. However, the extensive Miocene evaporites (Ch. 3, I. B. 1) flanking the Red Sea provided a possible alternative source of brines, perhaps transported in connate or meteoric water, and hence the concept of the Red Sea as a model for spreading ocean ridge metallogenesis for long remained in question.

(iii) "Massive" sulphides on the East Pacific Rise. In contrast to the numerous accounts of oxide-hydroxide minerals at ocean ridges, "massive" sulphides have been reported only from one locality, at around 21°N on the East Pacific Rise, west of Mexico. The sampled deposits, described by Francheteau *et al.* (1979), were from two sites on the flanks of a structural depression within a fissured or faulted zone about 600 m west of the youngest axial lavas at a depth of around 2600 m. At both sites the hydrothermal material is in the form of vertical columnar mounds overlying pillow lava.

The deposits, mostly sulphides but in some cases partly oxidized, are multi-coloured porous and friable compounds, comprising ochre-coloured iron oxides, and sulphides of sphalerite and pyrite with minor chalcopyrite and marcasite forming agglomerates of crystalline phases and amorphous material. Two main types of deposit are present: Zn-rich ore with around 25% Zn and an Fe-rich ore with 20 to 40% Fe (these samples contained 2 - 6% Cu). Minor Co, Pb, Ag and Cd are present, but no manganese. Pure sulphur is widespread.

(iv) Origin of the hydrothermal metalliferous sediments. A hydrothermal origin for the ferromanganese sediments and encrustations of the spreading ocean ridges is now generally accepted, and supported by evidence that holocrystalline basalts from the Mid-Atlantic Ridge are depleted in Fe, Mn, Co, Cu and Pb relative to nearby glassy basalts (Corliss, 1971), and by bottom water thermal anomalies on ridge crests (Rona, 1978). It is probable that circulating sea water penetrates through cracks to a depth of several kilometres, at least to the base of the cumulate gabbro layer and possibly to the top of the ultramafics (Lister, 1977), and leaches metals before rising to the sea floor and precipitating them in the discharge zones, adding H_2O, K, Na, Ca and CO_2 to the basalts during alteration and metamorphism (Fig. 67A).

The preferred origin of the sulphides of the East Pacific Rise is analogous to that of the Red Sea metalliferous muds and ferromanganese deposits of other ocean ridge deposits. The metals, including Au, Ag, Cu, Zn, Pb, Ni, Ba, Mn and Fe, are leached by circulating heated seawater from underlying oceanic crust, transported as chlorides, and deposited in the relatively narrow discharge zone of a hydrothermal seawater convecting system at the basalt–seawater boundary, commonly on topographic highs. Preservation of sulphides requires either the presence of anoxic bottom waters, or rapid burial of sulphides beneath either

Fig. 66 (A) Interbedded normal and metalliferous sediments in core VA03-334K of the Suakin Deep in the Red Sea spreading centre. The width of the core is 0.15 m. The light-coloured sediment is largely marl with about 50% fine-grained terrigenous silicates, while the darker intercalations indicate hydrothermally precipitated components, mostly iron and manganese hydroxides but also iron or heavy mineral sulphides. Photo by Professor Dr H. Bäcker. (B) Photograph taken by Professor Peter Lonsdale from the observation port of the submersible DSV-4 Seacliff of baryte mounds at a deep sea hydrothermal site on the San Clemente transform fault zone. 70km southwest of San Diego, California (Lonsdale, 1979; see Fig. 78). One of the baryte mounds, about 1 m high is white an irregular with a large stalkless crinoid while the other is darker and veneered with sediment; the mound with the fresh white exterior is believed to be slowly leaking barium-rich water. The baryte deposits are interpreted as recent precipitates from hydrothermal springs and by analogy with Kuroko-type ore bodies in which baryte overlies sulphides, Professor Lonsdale has suggested that sulphides could be forming beneath the baryte columns.

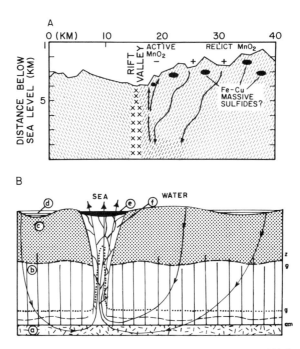

Fig. 67. Postulated sub-sea floor hydrothermal convection systems. (A) based on TAG Hydrothermal Field: +, zone of recharge; -, zone of discharge; x, zone of igneous intrusion (from Rona *et al.,* 1976); (B) Hydrothermal convection, metamorphism and mass transfer within the ophiolitic rocks of the Troodos Massif, Cyprus. General metamorphism occurred in zones of recharge flow, whereas formation of cupriferous pyrite orebodies occurred in local discharge zones. Precipitation of Fe/Mn enriched sediments (umbers) was also probably a consequence of discharge of hot metalliferous solutions. a, gabbros, metagabbro and variable altered trondhjemite; b, sheeted dyke complex; c, pillow lavas; d, umbers; e, massive sulphide ore; f, mineralized stockwork. Dashed lines approximate to metamorphic boundaries between zeolite (z), greenschist (g) and amphibolite facies (am). Continuous lines are schematic flow lines for hydrothermal convection; dotted line is isotherm produced by convection (after Spooner and Bray, 1977). By permission of the authors and publishers, *Bull. geol. Soc. Am.* **87**, 661-674, and *Nature, Lond.* **266**, 808-812. Copyright © 1977, Macmillan Journals.

detrital sediments or umbers. Ochres, such as those found elsewhere on the East Pacific Rise, probably result from oxidation of sulphides, while umbers found in the FAMOUS area, in the Galapagos and in the Gulf of Aden, are probably deposited from a relatively weakly convecting hydrothermal fluid, and it is possible that minor associated sub-surface sulphides could be present. The discovery of massive sulphides only on the relatively fast-spreading segment of the East Pacific Rise suggests that high spreading rates may favour rapid hydro-thermal flow and sulphide accumulation.

Sub-surface sulphides have been found at a few ocean ridge localities as disseminations and stockwork Cu-Fe sulphide occurrences within basalts, but in concentrations far lower than in the "massive" sulphide reported by Francheteau *et al.* (1979).

(c) Metal-rich pelagic shales of the ocean floor

The occurrence of metal-rich bituminous shales related to marine transgressions on continental margins has been referred to above (Ch. 3, I. B. 3), and several authors have shown how similar metal-rich muds can form in topographic lows on the ocean floor, largely as a result of worldwide oceanic anoxic events also related to transgressions (e.g. Jenkyns, 1978). While none of the Mesozoic and Cenozoic pelagic muds in the oceans is likely ever to be economic, a few examples are known of sub-economic metal-rich pelagic shales deposited in ancient oceans and subsequently elevated, for example the Glenkiln Shales of Ordovician age in the Southern Uplands of Scotland, preserved in the Lower Palaeozoic accretionary prism or outer arc described later (Ch. 5, I. B. 2. (b)).

(d) Precious metals in ocean ridge basalts

In pillow basalt dredged from the Mid-Atlantic Ridge, Keays and Scott (1976) have demonstrated the depletion in gold and silver of pillow interiors relative to their glassy rims, and attributed it to high-temperature interaction of gold in early crystalline phases of the interiors with sea water. However, it is considered that most of the precious metal released from the pillow interiors remains within the lava pile, only a small proportion being transported to and deposited on the sea floor by convecting hydrothermal fluids. Thus while some gold and silver can be deposited with sulphides on the sea floor, much of it remains available in the basalt for possible concentration into economic deposits by hydrothermal fluids either before or after the basalts have been emplaced on land as part of ophiolite sequences.

2 Mineral deposits of ophiolite sequences

The tectonic settings in which these deposits and associated ophiolitic rocks are emplaced are discussed in the relevant sections on outer arcs and collision belts in Chs 5 and 6 below.

(a) Stratiform pyritic copper sulphides

The basaltic pillow lava layer of several ophiolite bodies either is overlain by or contains in its upper part interbedded lenticular stratiform sulphide ore bodies of pyrite and copper with or without zinc, lead and minor gold and silver. Examples are the Lower Palaeozoic deposits of Newfoundland and Scandinavia, Upper Cretaceous deposits of the Semail nappe in Oman, deposits in the Mesozoic ophiolites of the Appennines and the kies-ore deposits of the Eastern Alps. The best known deposits of this type, which are sometimes termed Cyprus-type deposits, are those from the Troodos Massif, Cyprus.

Discovery of the metalliferous sediments of the Red Sea led numerous authors to suggest that Cyprus-type deposits formed at ocean ridges and were subsequently tectonically emplaced with the ophiolite on land (Pereira and Dixon, 1971; Sillitoe, 1972a). Comparison of hydrothermal mineralizing processes inferred for the on-land deposits with those postulated at ocean ridges continued throughout the 1970s as increasing numbers of oxide deposits were found on the ridges. However, the discovery of "massive" sulphides on the East Pacific Rise, referred to above, was of major significance in indicating that hydrothermal sulphides could form in the ocean and not only in narrow oceanic rifts as in the Red Sea.

Pearce and Gale (1977) pointed out that if most massive sulphide deposits result from hydrothermal circulation of sea water through oceanic crust, they could form at almost any site of submarine igneous activity with a permeable crust and sufficiently high geothermal gradient. They defined tectonic environ-

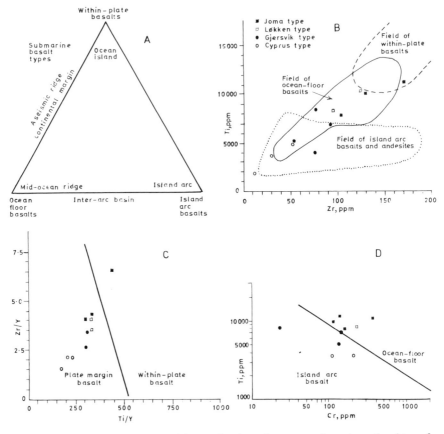

Fig. 68. Identification of ore-deposition environment from trace-element geochemistry of associated igneous host rocks (from Pearce and Gale, 1977). (A) Distribution of three main magma types within ocean basins. (B,C,D) Mean analyses of metabasalts from various massive sulphide deposits plotted on: (B) Ti–Zr discrimination diagram; on (C) Zr/Y – Ti/y discrimination diagram, and on (D) log Ti – log Cr discrimination diagram. By permission of the authors and publishers, *Instn Min. Metall.* and *Geol. Soc. Lond. Spec. Publs. No. 7.*

ments for volcanogenic massive sulphide deposition based on the geochemical characteristics of the associated igneous rocks, in particular the "stable" trace element contents (Ti, Zr, Y, Nb, Cr and trace elements) of the metabasalts, and recognized three distinct classes of deposit: (1) Cyprus-type possibly formed during early stages of back-arc basin development, (2) Løkken-type possibly formed at back-arc spreading centres and (3) Joma-type possibly formed in a small ocean of Red Sea type. There is also a distinct class related to island arc settings termed the Gjersvik-type (Fig. 68).

We use this terminology for convenience although the precise setting in which most stratiform pyritic copper sulphide deposits were generated is still uncertain.

(*i*) *Cyprus-type sulphides and associated ochres and umbers.* In *Cyprus* about 90 massive sulphide deposits and prospects, genetically related to the Troodos ophiolitic complex (see Fig. 88a) of Upper Cretaceous age, have become known as the type-examples of stratiform massive sulphides associated with basaltic pillow lavas and cherts (Constantinou and Govett, 1972; Searle, 1972; Sillitoe, 1972a). Iron-rich sedimentary formations, termed ochres, directly overlie the sulphide deposits while manganese- and iron-rich sediments devoid of sulphides, termed umbers, were the first sediments deposited after volcanic activity terminated (Constantinou and Govett, 1972). The underlying basaltic rocks and upper part of the gabbroic layer have undergone low to medium-grade metamorphism considered to be due to circulation of hot sea water (Spooner *et al.*,

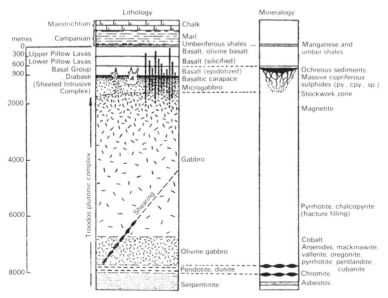

Fig. 69. Diagrammatic sections (from Searle, 1972) through Troodos igneous complex, Cyprus showing distribution of rock types and associated mineralization (A, acidic intrusions). By permission of the author and publisher, *Trans. Instn Min. Metall.* 81, B189-197.

1977b) analogous to hydrothermal metamorphic processes postulated at oceanic ridges.

The massive sulphide deposits occur as irregular elongate bodies in basin-like depressions, largely at the contact of the lower and upper pillow lavas, but also more rarely throughout the pillowed lava formation (Fig. 69). Searle (1972) showed that the deposits form three distinct zones (Fig. 70), and considered that the rapid decrease in sulphur content from about 40% in the quartz-bearing pyrite zone 2A, to 30% at the top of the stockwork zone 3, occurred at the original lava-sea floor interface on which the massive sulphides accumulated.

Ore bodies range in size from about 15 000 tonnes to 15 million tonnes with from 0.2% Cu (e.g. at Mathiati mine where pyrite only is mined) to 4% Cu with 8.5% Zn at the rich Agrokipia B mine (Searle, 1972). Gold and silver are also present in small amounts in the massive sulphides.

Most authors have favoured a syngenetic volcanic-exhalative origin for the massive sulphides, considering them to have formed by fumarolic activity on the sea floor (Searle, 1972) in an analogous way to the Red Sea metal-rich brines and muds (Constantinou and Govett, 1972). Smitheringale (1972) suggested an alternative origin for the Cyprus deposits and similar sulphides at Lush's Bight in Newfoundland, requiring metal-rich hydrothermal solutions expelled from deeper crustal layers and circulated through clastic volcanic rocks. Recently however Sr isotope geochemistry has demonstrated that sea water was the source of the hydrothermal ore-forming fluid which formed the Cyprus deposits (Spooner *et al.*, 1977b; Spooner, 1977; Heaton and Sheppard, 1977), the metals being scavenged from the volcanic rocks by the hydrothermal convection system, confirming previous suggestions based on hydrogen and oxygen (Heaton and Sheppard, 1977) and sulphur (Spooner, 1977) isotope geochemistry. The model by Spooner and Bray (1977) (Fig. 67B) shows a pattern of metamorphism in zones of recharge flow and of formation of massive sulphides in discharge zones of stockwork type.

The ochres are intercalated with chert, tuffaceous material and occasional limestone, and conformably overlie the sulphide ore (Constantinou and Govett, 1972). They characteristically contain bands or fragments of sulphides and are enriched in copper and zinc. Constantinou and Govett regard the ochres as products of submarine oxidative leaching of the sulphide ore exposed on the sea floor.

The umbers (Fig. 71B) lie unconformably above the lavas in association with jaspers and other sedimentary rocks. Deposits are mostly from 1 m to 10 m thick infilling irregularities on the volcanic surface, and consist largely of poorly crystalline manganiferous goethite together with maghemite (Constantinou and Govett, 1972). Robertson (1975) has described an almost invariable upward gradation in umbers from orange ferruginous through dark brown manganiferous to grey clay-rich umber succeeded by pelagic sediments, and interpreted them as chemical precipitates rapidly deposited from submarine thermal springs, analogous to the umbers of the spreading ocean ridges.

Fig. 70. Vertical zonation of Cyprus sulphide deposits (from Searle, 1972). By permission of the author and publisher, *Trans. Instn Min. Metall.* 81, B189–197.

Genesis	Mode of formation	Mode of deposition	Ore type	Geol. column	Ore zones	Mineralogical description	Type of mineralization	Minerals other than pyrite	Average ore grade %S	%Cu
Magmatic	Oxidation / Sedimentary	Exhalative-sedimentary	Sulphides and iron mudstones		Post-mineralization lava					
				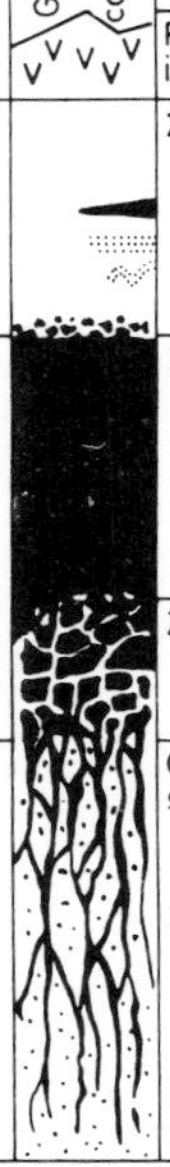	Zone 1	Iron-rich and siliceous mudstones; oxidation of sulphides; lenses of friable black ore. Graded bedding, with grains of pyrite, and slumping. Boulders of colloform sulphide	Type 1	Sediment: goethite, silica, montmorillonite. Sulphides: covellite, sphalerite, chalcopyrite, marcasite	45–50	0·5–1·5
	Fumarolic		Massive sulphide		Zone 2	Fragmentary ore consisting of angular and hard blocks of yellow ore in matrix of black sandy ore	Type 2	Marcasite, sphalerite, po (v.r.), bn (v.r.), galena (v.r.), ∗ chalcopyrite	40–45	1·0–4·0
					Zone 2A	Fragmentary sulphide, blocks of pyrite in siliceous matrix with decreasing amount of black, triable sulphide	Type 3	Marcasite (r), chalcopyrite, sphalerite (r), po (v.r.) quartz, jasper, chalcedony	30–40	0·5–1·5
		Hydrothermal	Stockwork, cavity, fracture-filling, disseminations		Original lava surface	More coarsely crystalline splendent pyrite as cavity-filling with much quartz and jasper	Type 4	Chalcopyrite, sphalerite, quartz, illite, jasper, chlorite	15–30	0·4–1·2
					Zone 3	Discrete fracture fillings with pyrite and quartz and jasper. Disseminations of pyrite in vesicles and cracks	Type 5	Chalcopyrite (r) sphalerite (v.r.), rutile (r), illite, quartz, chlorite, jasper	5–15	0·2–0·5

∗ From Agrokipia *B* only. r, rare ; v.r, very rare.

Fig. 71. (A) Bottom photograph of pillow basalts elongated in the downslope direction in a central zone of extension about 0.5–1.0 km wide within the rift valley of the Mid-Atlantic Ridge at 36°N (Ballard *et al.*, 1979). This central zone is composed of elongated hills about 100-250 m high built up of lavas which reached the sea floor through fissures resulting from plate separation and which quenched quickly, flowing only a short distance over the ocean floor. Photo taken from the DSRV Alvin by courtesy of Dr Robert D. Ballard of Woods Hole Oceanographic Institution. (B) Troulli umber pit, east Troodos Massif, Cyprus (see Figs 69, 70 and 88). The leached and brecciated Upper Pillow Lavas (pale-coloured) are overlain by up to 6 m of well-bedded umber with strong parallel lineation, grading and local slumping. The umber has been folded post-depositionally into a syncline probably due to basement faulting soon after the final stages of lava extrusion. The umbers, pillowed lavas, associated stratiform sulphide deposits and underlying dyke complex form part of the upper layer of the ophiolitic suite known as the Troodos igneous complex, interpreted as a tectonically emplaced slice of oceanic crust formed in a late Cretaceous back-arc or outer arc setting. Photo by Dr A. H. F. Robertson.

In the *Semail ophiolite of Oman*, the Upper Cretaceous volcanic layers comprise three distinct units which in order of eruption are termed the Geotimes, Lasail and Alley Units (Alabaster *et al.*, 1980). The Geotimes Unit directly overlies the sheeted dyke swarm throughout the ophiolite and typically consists of non-vesicular aphyric basaltic pillow lavas forming sections up to 1.5 km thick. The metamorphic grade is low with smectite as the important alteration mineral; greenschist facies alteration is confined to discharge zones of later hydrothermal convective systems. This Unit is considered on stratigraphic grounds to have been erupted at a spreading axis. The Lasail Unit, a kilometre thick, consists of a fractionation system from pillow basalt to andesite to felsite. The pillow lavas are associated with swarms of inclined sheets of basaltic andesite and andesite which take the form of cone-sheets focussed on acidic centres; these occur as high-level trondhjemite plugs with peripheral felsite sheets or massive flows. Intense hydrothermal alteration is exhibited throughout the sequence

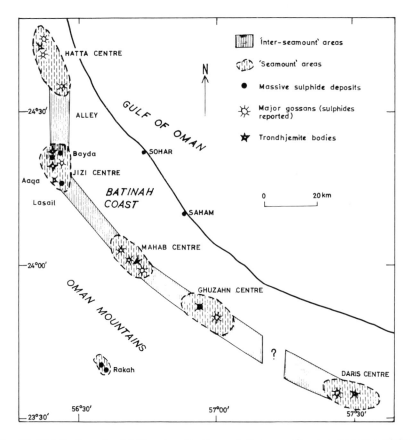

Fig. 72. Distribution of sulphide mineralization and "seamount" areas, Semail ophiolite, Oman (from Alabaster *et al.*, 1980). By permission of the authors and publisher, Geol. Surv. Dept. Nicosia.

especially in the greenschist-facies pillow lavas and in the actinolite (after pyroxene)-bearing andesite and epidote-rich felsites. The Lasail Unit, which is restricted to centres spaced at 25-30 km intervals along the N-S strike of the ophiolite complex, is considered to have formed seamounts erupted on the oceanic crust.

The Alley Unit represents a second basalt-rhyolite fractionation sequence up to 0.5 km thick with discrete acidic centres with occasional acidic cone-sheets. These centres are smaller but more numerous than the Lasail Unit seamounts, and are particularly well developed in the faulted depressions that lie between the seamount areas. They are considered to have erupted in a post-seamount rift setting.

All the massive sulphide deposits occur within the Lasail seamounts areas (Fig. 72), the largest within the Jizi centre at Lasail consisting of 12 m tonnes of ore with 2.4% Cu. They are characteristically similar to the Cyprus-type massive

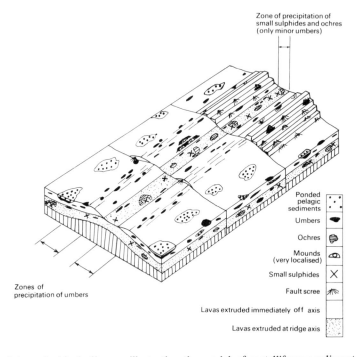

Fig. 73. Schematic block diagram illustrating the model of metalliferous sediment deposition at an oceanic spreading axis which can be derived from the sequence of sediments occurring within and above the extrusive rocks of the Semail Nappe. The diagram shows where different types of metalliferous and pelagic sediments might be expected to form in relation to the extrusives erupted both on a ridge axis and immediately "off-axis". The diagram is *not* intended to be to scale and it is *not* an actualistic representation of the ridge at which the Semail ophiolite formed: the effects of the volcanic centre in the area studied are ignored in this model (after Fleet and Robinson, 1980). By permission of the authors and publisher, *J. geol. Soc. Lond.* **137**, 403-422.

sulphide deposits with an underlying quartz-chlorite–sulphide stockwork. Alabaster *et al.*, (1980) have determined that the Lasail Unit seamounts with associated sulphide mineralization, and the later Alley Unit centres, are geochemically similar to present day island arc lavas but consider that they probably formed in a back-arc setting.

The upper levels of the basaltic sequence of the Semail Nappe also include a varied assemblage of ferromanganiferous umbers and ferruginous ochres (Fleet and Robinson, 1978, 1980). In structurally simple areas the ochres are related to the major sulphide deposits and are confined to the stratigraphically lower lavas of greenschist facies while the umbers are largely located within and above the overlying lavas of zeolite-facies. In contrast, in a major N-trending depression termed the Alley, umbers, ochres and sulphides are reported by Fleet and Robinson from throughout the higher levels of the volcanic pile. The schematic relationship of these deposits to an oceanic spreading axis is shown in Fig. 73.

The Lower Palaeozoic rocks of the *Central Mobile Belt of Newfoundland* include ophiolites (Upadhyay *et al.*, 1971; Smitheringale, 1972; Strong, 1972)

Fig. 74. Distribution of ophiolitic rocks in Newfoundland with detail of Betts Cove area (from Upadhyay and Strong, 1973). By permission of the authors and publisher, *Econ. Geol.* **68**, 161–167.

with typical Cyprus-type pyrite-chalcopyrite deposits such as Betts Cove, Tilt Cove, Whalesback and Little Bay (Upadhyay and Strong, 1973). Throughout the Newfoundland ophiolites, the Cyprus-type mineralization appears at the contact between sheeted dykes and pillow lavas. While there is some evidence that the host rocks suffered deformation in an oceanic transform fault zone (Karson and Dewey, 1978) it is probable that the sulphide deposits formed earlier on a spreading ridge.

The Betts Cove area (Fig. 74) is underlain by the Lower Ordovician Snooks Arm Group, a belt of ophiolitic rocks stretching over nearly 15 km from Tilt Cove in the northeast to Betts Cove in the Southwest. The Group comprises from bottom to top: pyroxenite, harzburgite and dunite with cumulate features; a complex transition zone of interlayered pyroxenite and gabbro grading upwards through gabbro cut by basic dykes into a sheeted dyke complex; a sequence of pillow lavas with tuffs and abundant interstitial red chert, overlain by sedimentary rocks.

The remarkable copper-rich Betts Cove deposit produced nearly 130 000 tonnes of hand-picked ore averaging 10% copper, and several million tonnes of cupriferous sulphides were worked at Tilt Cove. The sulphides range from interpillow banded massive lenses to disseminated zones occurring through the pillows. Massive and banded deposits show interbedded sedimentary layers rich in pyrite, chalcopyrite or sphalerite indicating repeated precipitation of each phase.

Similar mineralization to that at Betts Cove occurs at Pilleys Island to the south (1.7 million tonnes with 2% Cu), and at the Whalesback–Little Deer mine (4 million tonnes with 0.8% Cu; Upadhyay and Strong, 1973). The occurrences contain mainly pyrite and chalcopyrite as massive to disseminated tabular or elongate zones in chlorite schists within mafic volcanic rocks of the Lush's Bight Group. The schists probably formed from glass-rich aquagene tuffs and pillow breccias during sea water hydrothermal alteration.

The lavas associated with the Cyprus-type deposits are characterized by low concentrations of the small ion lithophile elements, including chromium, and correspond to island arc tholeiites in Pearce and Gale's classification (Fig. 68 B, C, D). However, as the lavas are part of a complete ophiolite sequence and hence formed at a constructive margin, the geological and geochemical data are apparently contradictory. Pearce and Gale (1977) proposed that the deposits formed in a back-arc basin where oceanic crust was generated above a subduction zone, on the basis of the overall low concentrations of Ti, Zr and Y, the presence in the upper pillow lavas of komatiitic rocks and the depletion of the underlying " upper mantle" ultramafics in their basaltic component (Menzies and Allen, 1974). These features can be explained in terms of partial melting under hydrous conditions, characteristic of island arc development.

(ii) Løkken-type sulphides. The Løkken deposit in Norway (Vokes and Gale, 1976 and the York Harbour deposit in the Bay of Islands ophiolite complex in Newfoundland (Duke and Hutchinson, 1974) were termed evolved back-arc

deposits by Pearce and Gale (1977). Compared to the Cyprus-type lavas, the Løkken and York Harbour lavas contain higher concentrations of all the small ion lithophile elements, especially in the lower units. The lavas are therefore classified as ocean floor basalts (Fig. 68 B, C, D) but Pearce and Gale concluded from their geological setting and high concentration of large ion lithophile elements that a back-arc setting is most likely.

The Løkken deposit is one of an economically important group of massive sulphide deposits within Lower Palaeozoic volcanic rocks which extend along the Caledonide orogen for about 1400 km. Massive sulphides occur near the top of a thick inverted sequence of submarine volcanics resembling both geo-chemically and petrographically the Lower Pillow lavas of the Troodos Complex (Vokes and Gale, 1976). The overlying pillow lavas are similarly low-K island arc tholeiites overlain by a thick sedimentary pile. Sulphide ore totals about 20 million tonnes with 2% Cu, 1.8% Zn and a trace of lead. The main sulphide minerals are pyrite, chalcopyrite, sphalerite and galena; pyrrhotite is virtually absent in contrast to other deposits in Norway such as Røros. There is evidence for feeder channels and sedimentary reworking, and some metamorphic recrystallization occurred during the Caledonian orogeny (Vokes, 1968).

The York Harbour deposit lies within the belt of allochthonous Ordovician volcanic rocks extending through the Appalachians from Tennessee through Quebec, New Brunswick and Nova Scotia to the Bay of Islands area of western Newfoundland. The ore-bodies, which occur at the contact between upper and lower volcanic units (Duke and Hutchinson, 1974), total about 200 000 tonnes with 2.3% Cu and 8.4% Zn. They consist of pyrite, sphalerite, chalcopyrite and pyrrhotite in veins and stringers along the top of pillow lava flows and as massive bodies between pillows; adjacent lavas have suffered strong chlorite alteration.

(*iii*) *Joma-type sulphides.* Lavas at the Joma-type deposits contain high Ti/Y and Zn/Y ratios (Fig. 68 B, C, D), and light rare earth element-enriched rare earth patterns, both indicative of a within-plate origin according to Pearce and Gale (1977), and a small ocean setting of Red Sea type has been suggested. Examples include the Palaeozoic deposits at Joma and Røros in central Norway and the Proterozoic deposit at Bidjovagge in Finnmark, Norway.

The deposits at Joma are typically conformable lenses in horizons of graphite-pyrrhotite schist overlying metabasic lavas. The sulphides consist mainly of allotriomorphic pyrite with only minor interstitial gangue and non-ferrous sulphides (Waltham, 1968). Irregular cross-cutting bodies consist of pyrrhotite-chalcopyrite (Baugsberg-type) ore which is also found at several other localities in the Caledonides. Vokes (1968) suggested that these resulted from mobiliza-tion of a pyrrhotite-chalcopyrite fraction during regional metamorphism.

At Røros the orebody is crudely lenticular in form but lacks sedimentary banding: the dominant texture is that of fracture infilling or veining as in a stockwork. The sulphides are entirely of Baugsberg-type, lacking pyrite; copper and cobalt values are higher than in the Løkken-type deposits.

(iv) Other examples of stratiform pyritic copper sulphides in ophiolite sequences.
Metal sulphides of probable Cyprus-type have been worked from ophiolitic
bodies of the northern Appennines of Italy, where they commonly occur close
to the manganiferous sediments described below (Bonatti *et al.*, 1976b). Sulphide
concentrations occur as three main types: massive-type deposits form lenticular
bodies from less than one to a few thousand cubic metres in volume, consisting
primarily of pyrite and chalcopyrite; disseminated deposits consist of small
pockets of pyrite and chalcopyrite aggregates within the basalt flows; stockwork-
type deposits consist mainly of pyrite, chalcopyrite and quartz together with
small amounts of sphalerite, marcasite, pyrrhotite, mackinawite and magnetite.
The Appennine deposits are considered to have originated at a spreading centre
from sub-sea floor hydrothermal systems similar to those which gave rise to the
metalliferous sediments described earlier (Spooner *et al.*, 1977a).

The stratabound kies-ore deposits in ophiolites of Italy and Austria are
distinctive. The Alpine ophiolites are more highly metamorphosed than their
Appennine equivalents and in place of the normal ophiolitic sequence and a
part of the sedimentary cover there is a volcanic-sedimentary sequence termed
the "calc schists with greenstone series" (equivalent to the schists lustrés or
Grunschiefer). These consist of repeated layers of metavolcanics, metatuffites,
serpentinite, calcareous mica schists, phyllites and occasional marbles (Zuffardi,
1977; Derkmann and Klemm, 1977). The kies deposits, which occur within
finely-banded greenschists, are invariably overlain by calcareous mica schists
with thin intercalations of black phyllites. The mineralization comprises up to
80 cm of quartzite–pyrite (locally altered to magnetite)-rich rock overlain by
0.5–2.0 m of kies ore (pyrite, chalcopyrite, sphalerite, and Au, Mo, Pb) in a
quartzitic–chloritic matrix; this sequence can be repeated up to three times with
barren greenstone intercalations. High chrome values in the ores (>300 ppm) are
considered to indicate ascendant synsedimentary mineralization, and the lack
of manganese and silica-enriched layers is believed to exclude post-volcanic ore
deposition (Derkmann and Klemm, 1977).

(b) Manganese-rich hydrothermal sediments

A few manganese-rich metalliferous deposits of hydrothermal type, other than
those associated with sulphides in Cyprus and Oman and described above, have
been identified in ophiolite complexes on or within continents, e.g. within the
northern Apennine belt of Italy (Bonatti *et al.*, 1976b), and in chert-greenstone
complexes of the western USA (Snyder, 1978).

In the Apennines, banded and massive manganese deposits have been exploited from cherts at the base of the sedimentary sequence above non-vesicular
basalts (Fig. 75), the banded deposits consisting of layers up to a metre thick
extending laterally for hundreds of metres while massive deposits occur as
lenses from a few metres up to 20 m thick with a lateral extent of 10 – 200 m.
Most deposits consist of mixtures of quartz and braunite with small amounts of
other minerals including manganite, pyrolusite, calcite, rhodochrosite, baryte and

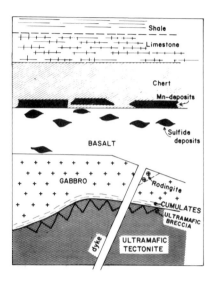

Fig. 75. Schematic and simplified diagram showing stratigraphic column of the northern Apennine ophiolite complexes with hydrothermal manganese deposits (from Bonatti *et al.* 1976b). By permission of the authors and publisher, *Bull. geol. Soc. Am.* 87, 83-94.

piedmontite. Braunite, not found in manganese deposits from the present ocean floor, probably formed during mild metamorphism. The Fe/Mn and U/Th ratios, concentration of minor metals (Ni, Co, Cu), and chondrite-normalized distribution of rare earth elements are similar to those of metalliferous sediments of hydrothermal origin forming at spreading centres (Bonatti *et al.*, 1976b). The lack of iron probably reflects extensive metal fractionation during sub-bottom circulation, especially during formation of sulphide ores when manganese is kept preferentially in solution.

In the western USA, Palaeozoic and Mesozoic chert–greenstone complexes include manganiferous sediments interbedded with radiolarian chert; the deposits are manganese-bearing carbonate and opal accompanied by braunite, bementite and occasionally by hausmannite and minor baryte (Snyder, 1978). The original minerals were probably syngenetic layers of manganese-bearing carbonate, amorphous silicates and manganese oxides, later affected by hydrothermal alteration. The lenses, up to 250 m long and 25 m thick, are considered to have formed from hot springs on the ocean floor.

Other examples of probably hydrothermal manganiferous deposits include the Proterozoic Långban deposit in Sweden (Boström *et al.,* 1979) consisting largely of hausmannite, braunite and hematite ores, and the Franklin Furnace and Sterling Hill deposits (Frondel and Baum, 1974).

(c) Podiform chromite deposits in ultrabasic rocks

Podiform chromite or chrome-spinel deposits are known from many Alpine-type ultrabasic bodies now interpreted as on-land oceanic lithosphere. The chromite, mostly forming lenticular bodies each rarely exceeding several thousand tonnes of ore, but exceptionally up to 14 million tonnes as in Coto in the Philippines (Dickey, 1975), occurs either in dunite near the harzburgite contact or in the harzburgite within dunite pods and lenses. Upper Mesozoic and Cenozoic examples include those in Cyprus, Greece, Turkey, Oman, Cuba and the Philippines, while Lower Palaeozoic deposits are known in west Newfoundland and in Unst in the Shetlands, and Proterozoic deposits in the Eastern Desert of Egypt and Saudi Arabia. Deposits generally comprising zones of several podiform bodies typically range in size from a few thousand tonnes up to several million tonnes with over 45% Cr_2O_3 and Cr/Fe ratios exceeding 2.0.

Interpretation of the ophiolites as ocean floor rocks, and the chromite-bearing rocks as uppermost mantle or lower crust, implies that the chromite concentrations formed beneath either ocean ridge or marginal basin spreading centres. The harzburgite unit forms the basement to the cumulate rocks. It is generally tectonized, lacks recognizable cumulus magmatic textures, and is considered to be a depleted mantle residue after partial fusion and extraction of basalt magma (Thayer, 1969). In the Cyprus deposits (Fig. 76A) the presence of abundant chromite at the base of the cumulate succession, its upward decrease and its absence in the late mafic cumulates suggest fractionation involving early chromium depletion in the melt (Greenbaum, 1977), in accord with experimental evidence (Ulmer, 1969). This implies evolution of a pseudostratiform body of igneous cumulates produced by gravity-aided fractional crystallization.

Chromite in ophiolites contrasts with that in continental stratiform complexes (Jackson and Thayer, 1972) described above, in forming isolated deposits of limited lateral extent and of variable physical form and chemical composition, probably due to episodic crystallization at localized centres (Greenbaum, 1977). It shows a wider range of Cr/Al and Mg/Ro ratios, and has a lower and more uniform iron oxidation ratio (Irvine, 1965; Thayer, 1970); titanium levels are low and there is a lack of significant correlation between titanium and major elements (Dickey, 1975).

In the Oman chomitites, Brown (1980) has determined that there is a general increase in Cr/Fe ratios of the chromites with depth over the first 5.5 km away and perhaps up to 12 km away from the overlying cumulate sequence, and also that the Cr/Fe ratios of disseminated chromites are much less than those in chromitites. This shows that the magma which deposited the chromitites could not have been in equilibrium with the harzburgite, and therefore the harzburgite is not a residue from an equilibrium partial melting episode which produced that magma.

The chromite occurs not only as podiform deposits but as schlieren, as disseminated grains and as dyke-like deposits possibly due to later mobilization (Peters and Kramers, 1974). Some dunites exhibit either occluded silicate texture, with olivines surrounded by a cloud of smaller chromites, or chromite net texture, with adjacent olivines in partial contact (Thayer, 1969). Massive chromitite layers at Troodos in Cyprus are composed of a granular aggregate of coarse-

Fig. 76. (A) Podiform chromite in dunite close to contact with harzburgite, Chrome Mine, Mount Olympus, Cyprus. The ultrabasic rocks form part of the lower layer of the óphiolitic suite emplaced tectonically in the Upper Cretaceous. Photo by Dr. D. Greenbaum. (B) Angled aerial photograph taken over Peru looking eastwards across the Altiplano basin to the tin-bearing granite mountains (snowclad) of the Cordillera Real of Bolivia. The hills in the foreground are volcanics of the Andean Cordillera in a calc-alkaline section (with porphyry copper mineralization) of the *circum*-Pacific belt (see Figs 109 and 110). This part of the belt is associated with eastward subduction of the Nazca Plate under the South American continental plate. The tin-bearing granites of early Mesozoic age in the La Paz area, and in the Tertiary magmatic arc to the south are considered to have been intruded in a back-arc thrust belt. Photo by Professor Peter Laznicka.

grained (up to 10 mm) interlocking chromite anhedra and minor serpentinized olivine; increase of serpentine gangue leads to streaky and banded olivine-chromitite (Greenbaum, 1977). Cumulus textures, although rarely preserved, include layers of settled chromite crystals, e.g. in the Canyon Mountain complex, Oregon, and cross-bedded layers of chromite crystals in an ophiolite complex south of Istanbul, Turkey (Thayer, 1970).

Podiform deposits comprise ellipsoidal, disc- or sack-like bodies from a few centimetres up to several hundred metres long, consisting largely of nodular chromite in a dunitic matrix. Pull-apart texture is commonly well developed in coarse-grained nodules, and torpedo- or pencil-shaped masses and schlieren are formed by strong deformation (Thayer, 1969). In several deposits, for example, in northern Oman (Peters and Kramers, 1974), the South Eastern Desert of Egypt (Ivanov *et al.*, 1973) and the Guleman-Soridag district of Turkey (Thayer, 1969) the podiform masses show a strong lineation.

Dickey (1975) noted that the silicate matrices of podiform chromites are more varied than the surrounding rocks. Although olivine and serpentine are most common at the Celebration Mine in Oregon (Thayer, 1969) the chromite has a clinopyroxene matrix, and at the Dubostica deposit in Yugoslavia (Pamic, 1970) the matrix contains amphibole. In Cuba some deposits have a troctolite matrix (Thayer, 1969), and at Troodos some nodules consist of a core of dendritic chromite with interstitial olivine, clinopyroxene and feldspar, and isolated dendrites and skeletal crystals are locally present between the nodules (Greenbaum, 1977). The variety of matrices and the wide chemical variation are considered by Thayer (1969) to result from gravity differentiation of hot crystals mushes and later mixing during magmatic re-emplacement.

The coarse grain size of chromite and presence of thick nearly monomineralic rock units suggest much slower accumulation rates than in stratiform complexes. In the Philippines, lobate chromite structures projecting downwards from overlying cumulate (?) dunite may have formed by gravitational sinking (Dickey, 1975). At Vourinos in Greece, podiform chromite concentrated in fold noses (Moores, 1969) perhaps resulted from slumping of chromite layers. George (1975) and Thayer (1969) both concluded that deformation is characteristic of magmatic sedimentation in ophiolite complexes, suggesting that the podiform nature of the chromite deposits is syn-depositional. Mechanisms proposed to explain the origin of podiform structures include overgrowth from dendrites (Greenbaum, 1977), pelletization resulting from chromite crystals rolling down earlier consolidated silicate banks (Borchert, 1964), snowballing within a turbulent zone of magma segregation (Dickey, 1975), abrasion of consolidated chromite ore during rock flowage (van der Kaaden, 1970) and crystal accumulation *in situ* from magma rising through the harzburgite sequence with later deformation causing deformation into pods (Brown, 1980). In this last proposed mechanism, which we consider the most realistic, Brown has suggested that the podiform chromitites and lensoid dunite bodies represent "mini" chambers within a residual mantle host where crystal settling from a rising melt took place (Neary and Brown, 1978). Chromitites as cumulates were then obscured by later deformation causing pull-apart textures and boudinage of lenses into pods with surrounding shear zones filled with serpentinite.

(d) Ni, Fe, Ti, Au, Pt and asbestos, talc and magnesite deposits in basic and ultrabasic rocks

Compared to the well-known occurrences of podiform chromite and stratiform Cyprus-type cupriferous pyrite deposits, other types of economic mineralization are relatively uncommon in ophiolitic rocks, perhaps largely due to a real scarcity of mineralization but also to a lack of properly-directed exploration. Examples of economically interesting mineralization in the gabbroic layers include copper sulphides in the Oman Mountains (Hassan and Al-Sulaimi, 1979) and disseminations of ilmenite and rutile in cumulate gabbros in western Liguria, Italy (Zuffardi, 1977). In the ultrabasic rocks, economic mineralization includes: nickel sulphides associated with minor platinum in the Luzon ultrabasic complex in the Philippines (Bryner, 1969); asbestos in peridotites and serpentinites in Italy (Zuffardi, 1977), Cyprus (Ingham, 1959), the South-Eastern Desert of Egypt and in the Sudan (Garson and Shalaby, 1976); magnesite and talc in serpentinized peridotite layers in Greece, Italy (Zuffardi, 1977) and the Sudan (Whiteman, 1971); magnetite deposits in the Mesozoic serpentinites of Italy; gold-bearing veinlets in serpentinites at Ovada in western Liguria (Zuffardi, 1977).

The copper mineralization in the northern part of the Oman Mountains occurs in layered gabbros within linear zones of fracturing and kaolinization up to 300 m wide and over a kilometre long (Hassan and Al-Sulaimi, 1979); anomalous concentrations of As, Se, U, Sb and Ag are also present. The fracture zones are interpreted as channel-ways for mineralization indicating that the copper sulphide may represent the lower parts of stockwork systems of Cyprus-type massive sulphide deposits, analogous to those located elsewhere in Oman.

Disseminations of ilmenite ± magnetite or rutile ± magnetite, of uncertain economic potential, have been reported in the cumulate gabbros and eclogites derived from the gabbro layers of the ophiolites at Val di Vara in western Liguria, (Zuffardi, 1977).

Nickel sulphides together with minor platinum are worked in the lower ultrabasic portion of the Luzon ophiolite complex in the Philippines (Bryner, 1969). In the Mesozoic ophiolites of Italy, Zuffardi (1977) has reported averages of from 0.18 – 0.20% Ni and the presence of nickel-cobalt arsenides in the basal serpentinites. In the South-Eastern Desert of Egypt, serpentinite layers in the Proterozoic ophiolites at El Geneina carry sub-economic disseminated copper and nickel sulphides locally concentrated in thin massive layers (Garson and Shalaby, 1976). While Ni, Cu and Pt in the basal sections of ophiolitic complexes are rarely economic, further exploration for these deposits is clearly warranted.

Economic veinlets and minor disseminations of native gold occur in association with pyrite, pyrrhotite, sphalerite, graphite, marcasite and magnetite in a gangue of dolomite and chalcedonic quartz in serpentinized ultrabasic rocks at Ovado in western Liguria, Italy (Zuffardi, 1977). The origin of these minor deposits is uncertain and may be related to remobilization of underlying basement.

Iron, in the form of magnetite or haematite, has been located in economic massive concentrations in association with minor iron, copper and nickel sulphides, carbonates, olivine and brucite in serpentinized ultramafic parts of the ophiolitic complex at Cogne in the Aosta Valley of Italy (Zuffardi, 1977). Other occurrences of this type are rare.

Deposits of chrysotile asbestos and/or talc and magnesite are worked in serpentinized layers of ophiolitic belts in Italy, Greece, Cyprus and the South-Eastern Desert of Egypt, and huge relatively undeveloped deposits occur in an ophiolitic complex in the Sudan (Whiteman, 1971). The talc–carbonate rocks in the Sudan form bodies up to 3 km × 1 km with deposits containing 33–49% talc and 42–55% magnesite; chrysotile occurs in fractures in serpentinite and several hundred thousand tonnes of ore with a grade of +5% asbestos have been located.

3 Preservation potential

The crust and upper mantle of ocean basins are undoubtedly normally lost by subduction. However, oceanic lithosphere adjacent to a continental margin, and hence formed at an early stage in the spreading history of the ocean basin, can be elevated and exposed in an outer arc by the development of a subduction zone seaward of the continental-oceanic crust boundary (Ch. 5, I. A). On-land emplacement of the more extensive sheets of ophiolite probably takes place only where an oceanic ridge or transform system enters a subduction zone shortly before collision and is thrust onto the continental margin of the subducting plate (Ch. 6, II. A. 2). It is possible that in rare cases oceanic crust could remain beneath the thick sedimentary fill of residual basins formed as a result of collision of irregular continental margins, for example in the Caspian Sea (Ch. 6, I. A. 2), but crust in this setting is not normally exposed.

Most large ophiolite bodies are no older than Triassic and the scarcity of Proterozoic ophiolites suggests that they are normally destroyed by erosion within a period of several hundred million years. This presumably reflects the erosion rate of high-level nappes, flanking the core of orogenic belts, in which the larger ophiolite bodies mostly occur.

II OCEANIC TRANSFORM FAULTS

A Tectonic Setting

Oceanic transform faults were first described by Wilson (1965a) to explain why the sense of offset in spreading ocean ridge segments is opposite to that expected from a transcurrent fault. A similar sense of movement applies to transform faults joining two trenches, and oceanic transform faults may also link a trench and a spreading ridge. Gilliland and Meyer (1976) proposed that oceanic ridge–ridge transforms are a second-order feature resulting from the spreading mechanism, rather than related to original offsets in the margins of rifted continents as suggested by Wilson, and that other oceanic transforms, which form plate boundaries and which they hence termed boundary transform faults, are original or fundamental breaks in the lithosphere. We are here concerned largely with ridge–ridge transforms.

Ridge–ridge transforms are commonly a few tens to hundreds of kilometres apart and often form major submarine scarps. Atwater and Macdonald (1977) showed that the trends of many transforms at a slow-spreading ridge were oblique to the ridge, while those at fast-spreading ridges are perpendicular to the ridge. Ridge transforms comprise an actual transform segment, between the offset ends of the ridge, and traces or non-transform segments, extending beyond the offset ridges and with a length depending on the rate of spreading and the age of the transform fault. The transform segments theoretically lie on small circles, but in practice deviate from this pattern, and it has been suggested that the ridge transforms are initiated as thermal contraction joints in the cooling oceanic lithosphere (Collette, 1974).

The dominant motion at transform faults is by definition strike-slip, with rocks sliding past one another, although Thompson and Melson (1972) argued that in some Atlantic fracture zones there was evidence of intrusion of ultrabasic and basic plutonic rocks including undersaturated types and suggested that a type of sea floor distinct from that emplaced at ocean ridges could be created in fracture zones. In general it is now recognized that crust within transform zones is anomalous, and can include gabbroic and ultramafic rocks at the surface.

A schematic model of fracture zone evolution has been presented by Karson and Dewey (1978). They emphasize the difference between the transform segment between the ridges, with a broad zone of displacement, and the non-transform segment beyond the ridge, in which oceanic crust generated at a ridge axis termination and deformed in the transform segment is subsequently accreted to or intruded by younger crust generated at the other ridge (Fig. 77). Cooling of the lithosphere, also discussed by De Long *et al.* (1977), is accompanied by subsidence proportional to the square root of its age, resulting in varying rates

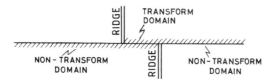

Fig. 77. Plan view of off-set spreading ridges showing areas (hachured) which are either in or have passed through transform segment. Fracture zone continuations beyond the spreading ridges are not fossil transforms but intrusive igneous contacts (from De Long *et al.*, 1977). By permission of the authors and publisher, *Geology* 5, 199–202.

of subsidence of the two sides of the transform with distance from the spreading axis. In general, dip-slip and strike-slip motion combine to give oblique-slip displacements within the transform segment, while pure strike-slip motion predominates in the non-transform segment: subsequently with movement away from the ridge axis the sense of dip-slip displacément on the fracture zone reverses. The age difference in lithosphere each side of a fracture zone far from the axis thus largely explains the major scarp features. Karson and Dewer (1978) point out that basic magma may be injected into the fracture zone opposite a ridge termination, and that serpentinite diapirs could be emplaced in both the transform and non-transform segments; these processes could explain some of the features described by Thompson and Melson (1972).

B Mineral Deposits

There is a considerable literature on the importance of transform faults as sites for formation of copper sulphide deposits of Cyprus type. The arguments in favour of transform fault sulphide formation are partly observational: iron and manganese hydroxide and oxide-rich sediments and encrustations similar to those at ocean ridges have been reported from the Romanche and Vema fracture zones on the Mid-Atlantic Ridge (Bonatti *et al.*, 1976a; Rona, 1978); the metal-rich Salton Sea active geothermal field (White, 1968) is associated with a transform zone in the San Andreas system; metalliferous sediments in the Red Sea deep are concentrated at intersections of transform faults with the axial zone (Garson and Krs, 1976). It is also argued that transform faults are favourable for mineral-ization because they, and their intersections with ridge axes, are highly fractured and hence permeable, allowing high rates of circulation of hot sea water. More recently, the hypothesis that many on-land ophiolites can be explained as obducted rocks which had passed through a former transform domain has led to suggestions that some Cyprus-type deposits within the ophiolite were similarly generated in transform fault zones.

1 Hydrothermal baryte on the San Clemente Fault Zone

The recent discovery (Lonsdale, 1979) of major deposits of baryte at a depth of 1800 m on the San Clemente fault zone off San Diego has increased the potential of oceanic transforms as sites of hydrothermal mineralization (see Fig. 66B). Investigation by manned submersible of the San Clemente fault, one of a number of major shear zones which occur on-shore and offshore in the California Border-land and form the Pacific–America plate boundary, revealed columns and cores of friable baryte beneath the SW-facing fault scarp (Fig. 78). The columns are up to 10 m high and consist of open meshworks of small tabular crystals, with a local thin manganese oxide crust; near some of the baryte deposits the late Quaternary submarine fan sediments are coated with sooty material interpreted as probable manganese.

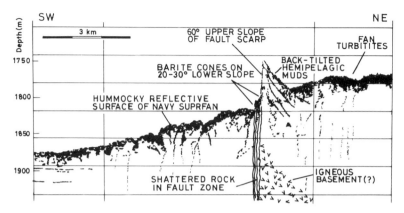

Fig. 78. 3.5-kHz acoustic profile across the San Clemente fault. At the scarp, where the steep slopes returned no echo on the original record, features have been sketched on from visual observations; structures sketched beneath the fault scarp are inferred from a nearby reflection profile (after Lonsdale, 1979). By permission of the author and publisher, *Nature, Lond.* **281**, 531–534. Copyright ©1979, MacMillan Journals.

Lonsdale (1979) has interpreted the baryte deposits as recent precipitates from hydrothermal springs, resulting from penetration of sea water into fault-shattered bedrock, reduction of sulphate by reaction with lavas, and removal of barium by the chloride solution. Barium, present in anomalously high amounts in the volcanic rocks of the region, is precipitated when the hydrothermal fluids mix with sulphate-rich bottom waters, the fault zone providing a channel through the relatively impermeable sediments. By analogy with Kuroko-type ore bodies of volcanic arcs (Ch. 5, II. B. 2), in which baryte overlies sulphides, Lonsdale suggested that sulphides could be forming beneath the baryte columns, although there is a rather surprising lack of metal concentrations other than Ba, Fe, Al and Mn with the baryte.

2 Preservation potential

The preservation potential of oceanic transform faults is greater than that of ocean basins because the transform zones form major positive features, and are hence more likely to be tectonically accreted to the overriding plate in a sub-duction zone and in some cases subsequently thrust or obducted onto the foreland during continental collission (Ch. 6, II. A. 2). As discussed later (Ch. 6. II. A. 2) we consider that there is little evidence to support the suggestion that transform fault zones are preferentially obducted because they are transforms, rather than because they are topographic highs.

III OCEANIC LINEAR ISLAND AND SEAMOUNT CHAINS

A Tectonic Setting

In the Atlantic, Pacific and Indian oceans there are aseismic linear submarine chains of seamounts locally forming volcanic islands and in some cases submarine ridges, up to several thousand kilometres in length. Examples are the Hawaiian chain in the Pacific, Walvis Ridge in the Atlantic and the Ninety East Ridge in the Indian Ocean. These chains and ridges are unrelated to present and ancient spreading ocean ridges, to subduction zones, and in most cases to transform faults, and are characterized by a distinctive type of volcanism.

In the North Pacific, the Hawiian chain extends from the active volcanoes of Kilauea and Mauna Loa northwestward through Midway Atoll, with Lower Miocene volcanic rocks, and joins the southern end of the N-trending Emperor Seamount chain. The Seamounts increase in age northwards and geophysical evidence for the existence of ancient coral reefs suggests that they formed in the early Tertiary at least $10°$ south of their present position (Greene *et al.*, 1978). In the South Pacific, the Tuamoto and Austral island chains each trend north-westward, the former including the active volcano of Pitcairn at its eastern end; there is some evidence that these chains also continue northward as a line of seamounts (Fig. 79).

In the South Atlantic (Burke and Wilson, 1976), the Rio Grande and Walvis Ridges extend respectively from the South American and African coasts to within a few hundred kilometres of the Mid-Atlantic Ridge. Tristan da Cunha Island, lying on 30 Ma old oceanic crust, east of the Mid-Atlantic Ridge and at the western end of the Walvis Ridge, includes Miocene and Quaternary volcanic rocks. In the North Atlantic submarine ridges have been identified east and west of Iceland, and near the Faroes.

Only one major linear chain or ridge is known from the Indian Ocean, the Ninety East Ridge, extending northwards through nearly 40 degrees of latitude, west of the Sunda Arc. Unlike the other chains referred to the Ninety East Ridge is seismically active and there is some possibility that it could be an active transform fault.

On most island or seamount chains the lavas are predominantly basaltic, but differ from those of ocean rises in the predominance of alkaline rocks. In a few islands at which volcanism has continued since at least the mid-Tertiary, alkaline intrusions are also present. The Hawaiian Islands show a characteristic change from early tholeiitic to later alkaline basalt before volcanism ceased. Island chain volcanic rocks are generally richer in large ionic lithophile elements (e.g. K, U, Th, Rb, Sr, Ba, Nd, La) than those of spreading ocean ridges, suggesting to some geochemists a shallower depth of origin in the upper mantle.

Hypotheses of origin of the chains have to explain their linearity, progressive increase in age away from an active volcano, and in the case of the Pacific chains,

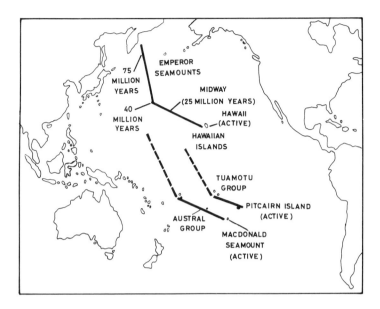

Fig. 79. Pacific island and seamount chains interpreted as tracks formed by northwestward movement of oceanic lithosphere over mantle plumes more or less stationary relative to each other; broken lines are seamount chains of uncertain age (from Burke and Wilson, 1976). By permission of the authors and publisher, *Sci. Am.* **238**, 46-57.

their parallelism and abrupt change in trend. Two main explanations have been put forward, the hot spot and the splitting hypotheses. According to the more widely accepted hot spot concept (Wilson, 1963, 1965b; Morgan, 1972), the chains are caused by movement of the lithosphere over a population of hot spots related to plumes rising from either the asthenosphere or core–mantle boundary and inferred, as discussed in Ch. 2, I. A, to be more or less fixed relative to each other. Thus the Hawaii chain is interpreted as a hot spot track formed during the northwestward movement of the Pacific plate in the last 40 Ma, and both the Hawaii and Tuamotu chains can be explained by rotation of the Pacific about the same pole. The northerly trend of the older Emperor Seamount chain, west of Hawaii, could similarly be explained by a pre-40 Ma northward movement of the Pacific plate, corresponding to a different pole of rotation. Similarly the origin of the Ninety East Ridge in the Indian Ocean is considered to be the result of northward movement of the Indian plate over a hot spot now lying beneath the Kerguelen Plateau (Curray *et al.*, 1980). The formation of discrete islands within the Hawaiian chain has been explained in terms of the interaction of a mantle plume with a low viscosity shear zone beneath the plate (Skilbeck and Whitehead, 1978).

Explanation of the South Atlantic ridges in terms of hot spots (Burke and Wilson, 1976) is dependent on the inference, based on the abundance of hot spots in the African plate, that hot spot related magmas can penetrate continents only where the lithosphere is more or less stationary relative to the underlying

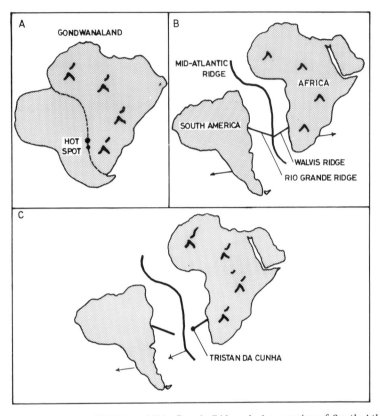

Fig. 80. Development of Walvis and Rio Grande Ridges during opening of South Atlantic and spreading in Mid-Atlantic Ridge. Tristan de Cunha hot spot is now on African plate, due to westward movement of Mid-Atlantic Ridge since Africa became stationary 30 Ma ago. (A) pre-120 Ma ago; (B) 120-30 Ma ago; (C) 30-0 Ma (after Burke and Wilson, 1976). By permission of the authors and publisher, *Sci. Am.* **238**, 46-57.

asthenosphere, but that they can penetrate oceanic lithosphere whether moving or stationary. Consequently hot spots initially situated beneath an intracontinental rift subsequently tend to underlie the crests of spreading ocean ridges, forming major islands, e.g. Iceland, with characteristic rare earth patterns and large ionic lithophile trace element abundances (e.g. Schilling, 1973). Burke and Wilson (1976) suggested that when the South Atlantic opened Tristan da Cunha was a volcano in the initial intracontinental rift, and subsequently remained as a volcanic centre on the developing Mid-Atlantic Ridge (Fig. 80); erupted rocks formed a submarine chain on the growing oceanic crust each side of the Ridge, the age of these rocks increasing with distance from the volcano. When Africa ceased moving eastwards about 30 Ma ago, the Mid-Atlantic Ridge was forced to move west; the Tristan da Cunha hot spot was stranded east of the Ridge, and has since provided volcanic debris to the E-moving African plate only.

Burke and Wilson have identified more than 50 hot spots in ocean basins, with 14 related chains in the Atlantic and three in the Pacific. More hot spots and oceanic chains probably remain to be identified, but as yet none have been found within marginal basins.

Other hypotheses for explaining oceanic linear chains postulate cracking of the ocean floor as a result either of membrane stresses arising during movement of the oceanic lithosphere over the earth's elliptical surface (Turcotte and Oxburgh, 1973), or of thermal contraction due to cooling as the crust moves away from a spreading ridge (Collette, 1974). In both cases the cracks propagate across the moving plates, and control the site of volcanism (Turcotte and Oxburgh, 1978). However, there is evidence that the forces of contraction due to both variable curvature of the earth's surface and cooling are too small, and that the age distribution of volcanoes on the island chains is not in accord with that predicted by the postulated cracking. A further hypothesis for the origin of the chains involves postulated hot lines in the earth's mantle, with roll-like curvature cells aligned parallel to the direction of plate motion (Bonatti and Harrison, 1976). These explanations have all met with less success than that based on hot spots.

B Mineral Deposits

1 Possible carbonatite occurrences

No economic metallic deposits are known from ocean ridges and islands, and their mineral potential is largely restricted to possible deposits associated with basic alkaline magmatic rocks, in particular carbonatites. While alkaline magmatism is found at several oceanic chain eruptive centres, instances of carbonatite activity are infrequent or uncertain. This may indicate, as in the case of tin mineralization, that carbonatites are largely restricted to areas of continental crust. Although carbonatite occurs at volcanic centres in the Canary (Dietz and Sproll, 1970; Stillman *et al.*, 1975) and Cape Verde Islands (Burri, 1960) in the Atlantic, in an apparent oceanic environment, continental crust detached from the African continent may be present. However at Tahiti in the central Pacific, where a central intrusion of nepheline syenite has been reported, carbonatite would not be unexpected, although its presence in this setting would be unique.

2 Besshi-type stratiform copper–zinc sulphides

In Japan (see Fig. 95) cupriferous sulphide ores of similar composition to those of Cyprus-type are associated with mafic volcanics in structurally complex settings characterized by thick sequences of greywackes, together with carbonaceous mudstones, clastic carbonates or quartzite mostly of fairly deepwater facies (Kanehira and Tatsumi, 1970; Mitchell and Bell, 1973). The deposits have high Co/Ni ratios, similar to those of Cyprus-type but distinct from Kuroko deposits, and have been termed Besshi-type deposits from the type-locality on Honshu Island. At Besshi an elongate tectonically flattened ore body consists mainly of massive pyrite and subordinate chalcopyrite with mafic schists in an Upper Palaeozoic succession of pelitic schists or meta greywackes; minor quartzose schists are possibly metamorphosed cherts (Sawkins, 1976c).

Sawkins (1976c) has listed other deposits with broadly similar host rocks, metal content, and deformation style; these include deposits in the Prince William Sound area in Alaska, copper-bearing massive sulphides at Matahambre in Cuba, and the Lower Palaeozoic Birtavarre, Folldal, Rødhammer, Killingdal and Kvikne deposits in Norway. A possible Besshi-type stratiform deposit of Dalradian age at Vidlin in the Shetland Islands, Scotland, consists of massive pyrrhotite and some chalcopyrite closely associated with amphibolites, probably representing tholeiitic flows and basic tuffs, in a metamorphic assemblage including mica-schists with thin intercalated quartzites (Garson and Mitchell. 1977).

The setting in which Besshi-type deposits were generated is uncertain (e.g. Sawkins, 1976c), and could be interpreted as an intracontinental rift, magmatic arc, outer arc trough or oceanic environment. However, the presence of thick greywacke successions with either tectonic or stratigraphic intercalations of the basalt and chert host rocks to the ore suggests that the sulphides could have accumulated where basaltic volcanism took place on a continental rise or sea floor fan, perhaps during drift of the rise across an oceanic hot spot. The subsequent deformation of the ore body and host rocks would result from accretion of the rise succession in an outer arc prior to continental collision (Ch. 6, I. A. 1).

3 Preservation potential

Oceanic linear island and seamount chains have some potential for preservation by tectonic accretion in subduction zones because they form topographic highs (Ch. 6, II. A. 2). The Coast Range seamount complex of western USA is a possible example of a linear oceanic chain tectonically accreted to the western margin of North America during the Eocene (Dickinson, 1976).

CHAPTER 5

Deposits of Subduction-Related Settings

DEPOSITS OF SUBDUCTION-RELATED SETTINGS

We describe in this chapter the various major tectonic settings and examples of associated mineral deposits and occurrences within arc systems beneath which ocean floor is consumed along a seismic Benioff zone. The settings thus include the essential elements of both continental margin arcs and oceanic island arcs, comprising magmatic arcs in which the formation of magma is directly related to descent of oceanic lithosphere, submarine trenches and outer arcs where ocean floor consumption results in dominantly tectonic processes, and outer arc troughs in which subsidence and sedimentation are related to subduction indirectly through the influence of the adjacent magmatic and outer arc source areas. We also include as subduction-related settings the back-arc magmatic belts and associated thrust zones formed during subduction but situated many hundreds of kilometres landward of some continental margin magmatic arcs, and sedimentary basins on the landward side of these belts (Fig. 81A). Marginal basins and interarc troughs (Fig. 81B), situated behind oceanic island arc systems and generated as a result of back-arc spreading of oceanic-type crust indirectly related to the Benioff zone, also form important settings associated with subduction and are described below.

The cause of subduction and the location in which it starts are still not understood. Postulated causes include gravitational movement away from a spreading ocean ridge, sinking of cold and hence dense lithosphere, viscous drag due to convection cells in the underlying mantle and a curious process known as lithospheric suction (Shoemaker, 1978). The distribution of geologically young subduction zones suggests that subduction commonly starts at the boundary between oceanic and thicker crust (e.g. Karig, 1980), the latter consisting in some cases of continental crust and in others of abnormally thick oceanic crust, and it is generally accepted that older and hence more dense lithosphere is more likely to subduct than young hot lithosphere. Once subduction has started, it

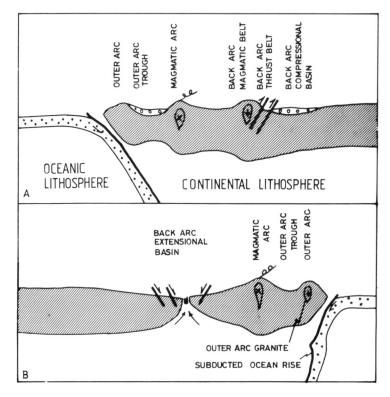

Fig. 81. Schematic cross-sections through (A) west-facing continental margin arc and (B) east-facing incipient island arc system. Relationship to lithosphere shown in Fig. 91.

normally continues until terminated by either a collision or change in relative plate motion resulting in a transform fault.

The start of subduction of ocean floor is equivalent to the beginning of the flysch stage of development of Aubouin's (1965) geosynclinal model; with continued subduction, elevation and oceanward movement of flysch nappes takes place in the outer arc bordering the outer arc trough of the overriding plate, while the miogeosynclinal shelf succession on the opposite side of the ocean begins to approach the subduction zone (Table IX). These processes are described in more detail in the section on outer arcs which follows, and in Chapter 6.

In the following account the position of arc systems with respect to continental margins and ocean basins, and the eastward or westward facing direction of the arc, are considered together with discussion of magmatic arcs; this is because the magma composition and type of mineralization in magmatic arcs are to some extent controlled by the arc's position and orientation, while processes in outer arcs are largely independent of the position or facing direction of the arc.

Subduction-related settings are important for mineralization largely because

TABLE IX

Plate or global tectonic terms with equivalent geosynclinal terminology; subduction, or early orogenic stage of development of Aubouin (1965).

Tectonic setting	Trench	Outer arc	Outer arc trough	Magmatic arc	Back-arc magmatic and thrust belt	Back-arc compressive basin
Alternative "plate tectonic" terms		Fore-arc, Trench-slope break, Mid-slope basement high — Accretionary prism Arc–trench gap	Fore-arc basin	Volcanic arc	Foreland thrust belt	Retro-arc basin. Foreland basin
Equivalent geosynclinal terms — Van Bemmelen 1949	Fore-deep	Outer arc	Inter deep	Inter arc, Volcanic arc		Back deep, Idiogeosyncline
Aubouin 1965	Eugeosynclinal furrow — Flysch and ophiolites — Eugeosyncline		"Meso-Hellenic Furrow", Molasse intra-deep	Eugeanticlinal Ridge		Epieugeosyncline

they include the commonly highly mineralized magmatic arc and back-arc magmatic belts, with abundant syngenetic and epigenetic magma-related hydrothermal deposits, as well as tectonically emplaced mineralized ophiolitic rocks in the outer arc. Back-arc marginal basins are probably also important sites of mineralization, although whether the economic potential of crustal and uppermost mantle rocks formed in these settings differs from that of oceanic lithosphere created at ocean ridges has yet to be determined.

I SUBMARINE TRENCHES AND OUTER ARCS

A Tectonic Setting

Submarine trenches border arcuate chains of active volcanoes either situated on continental margins or forming island arcs within the oceans; the total length of present trenches exceeds 50 000 km. They commonly have a steep "wall" or slope on the inner or volcanic arc side, and a relatively gentle outer slope extending upwards to a low swell on the ocean floor. The inner slope of the trench commonly coincides with a negative gravity anomaly, and an inclined seismic Benioff zone extends to depths of up to 700 km beneath the landward side of the trench.

The deepest trenches, including those bordering island arcs of the Western Pacific, contain less than 500 m of latest Mesozoic to Quaternary mostly pelagic or hemipelagic sediments (Scholl and Marlow, 1974). Some other trenches are partly filled with up to 2000 m of sediments. In the Peru–Chile Trench, landward-dipping pelagic and hemipelagic Tertiary sediments form the lowest unit, and in the Aleutian Trench pelitic turbidites and pelagic beds occur at the base. In both trenches an upper unit of Quaternary terrigeneous turbidites forms a wedge-shaped body thinning towards the ocean.

The landward slope of the trench in some arc systems, e.g. in the southern part of the New Hebrides island arc, is continuous with the seaward slope of the volcanic arc. However in many arcs the trench slope rises to a submarine or subaerial ridge, forming a major topographic high up to 100 km wide separated from the volcanic arc by an outer arc trough. The ridge in the Sunda Arc (Fig. 82) forms the outer arc of van Bemmelen (1949) and Hamilton (1973) while the ridge crest in these and other arc systems has been termed the trench-slope break (Dickinson, 1971, 1976); the outer arc is also known as the fore-arc (e.g. Dickinson, 1976). In arc systems such as the southern part of the Lesser Antilles and the northern part of the Sunda Arc in Burma, the outer arc is the major topographic feature seaward of the volcanic arc, and the trench may be poorly defined or absent. In other arcs, e.g. the Aleutians and the Shikkoku arc of Southwest Japan, the outer arc forms a subaerial or submarine shelf. Well-developed outer arcs coincide with the axis of the negative gravity anomaly normally associated with the landward side of submarine trenches. Heat flow beneath outer arcs is invariably low.

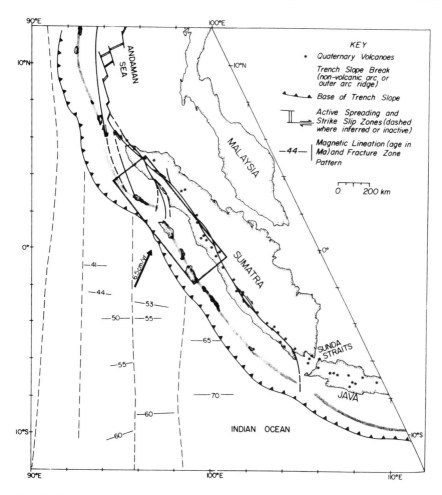

Fig. 82. Western Sunda Arc showing regional tectonic setting (from Karig *et al.,* 1980). By permission of the authors and publisher, *J. geol. Soc. Lond.* 137, 77–91.

All outer arcs consist predominantly of deformed flysch-type sediments, but in many, e.g. the Franciscan Complex in California and the Indoburman Ranges in western Burma, two major terrains can be recognized. A landward belt, on the volcanic arc side of the outer arc, usually consists of a tectonic complex of ultra-basic rocks, basic lavas, and metamorphic rocks which may include glaucophane together with flysch, and a seaward belt consists almost entirely of flysch with minor igneous rocks, mostly younger than the inner belt rocks. A few modern outer arcs, e.g. Crete in the Hellenides arc, are highly complex and the landward belt, in this case including carbonates, has largely overridden the younger flysch of the seaward belt.

The seaward belt is usually the most extensive of the two and there is con-siderable variation among different arc systems: in some outer arcs it is described

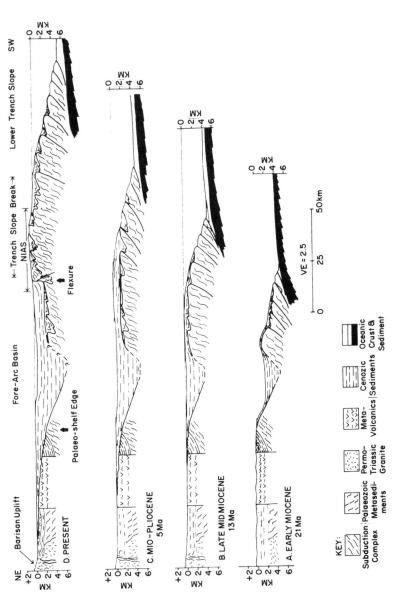

Fig. 83. Interpretative geological sections through Nias island and Sumatran continental margin showing Neogene evolution (from Karig *et al.*, 1980). By permission of the authors and publisher, *J. geol. Soc. Lond.* 137, 77–91.

as a tectonic melange, and in others as an imbricate stack of turbidites and mudstones deposited either on the ocean floor or in the trench and deformed and tectonically accreted in or above the subduction zone. Pelagic sediments and pillow lavas at the base of some sedimentary tectonic units are interpreted as slices of underlying oceanic crust. In the Indoburman Ranges the seaward or western belt consists of folded turbidites and mudstones with relatively few thrusts, evidently deposited on the landward side of the trench, and lying stratigraphically on late Cretaceous olistostromes, turbidites and mudstones, possibly representing deformed trench deposits.

Karig and Sharman (1975) have referred to outer arcs as forming accretionary prisms. The concept of tectonic accretion is based on seismic evidence that sediments in the trench, particularly if thick, are not subducted but are scraped off above the thrust plane between the subducting and overriding plates (Silver, 1971). Continued attempted subduction of trench floor sediments results in rotation of the thrust with elevation of the overriding plate margin above it; a second thrust plane then develops in younger sediments deposited in or carried into the trench, on the seaward side of the first thrust, and is in turn rotated. With continued suduction an imbricate stack of thrust sheets is produced (Fig. 83); within each tectonic unit the sediments young towards the magmatic arc (Fig. 84), but over all the age of the sediments decreases towards the trench (Mitchell, 1974a). Uplift of the prism of imbricate thrust slices usually accompanies subduction, but may accelerate as a result of isostatic adjustment when subduction ceases. Subduction or thrusting can occur not only on the youngest thrust plane, but also along older rotated thrust planes within the outer arc, although movement decreases rapidly with distance of the thrusts from the trench (Fig. 83). Consequently blocks and olistostromes, derived from nappes or thrust slices of landward and seaward belt rocks including ophiolite blocks, can form slope deposits and subsequently be incorporated into the outer arc by thrust movements, as suggested for the Franciscan Complex of California.

The scarcity of deep-sea pelagic sediments and associated hydrogeneous and hydrothermal metalliferous sedimentary deposits in most on-land outer arcs suggests that these sediments are partly subducted with the underlying basaltic layer of the ocean floor. Sediment balance calculations for the Middle America Trench and Japan Trench, and geophysical data from the eastern Aleutian Trench (Plafker and Brune, 1980) suggest that part of the sediment deposited in the trench is also subducted. The imbricate wedges of outer arcs thus probably consist very largely of deep-sea fan deposits carried into the trench, together with overlying trench sediments, off-scraped in thrust slices and overlain by deposits of the trench slope.

In the landward belt of outer arcs basaltic rocks are mostly described as ocean floor or island arc tholeiites, but the ultrabasic rocks show features typical of oceanic upper mantle, and the basic and ultrabasic rocks in general are interpreted as slices of oceanic lithosphere tectonically emplaced or obducted during ocean floor subduction. These ophiolitic rocks are variously explained as remnants of an inter-arc marginal basin beneath what subsequently becomes the outer arc trough, as the original leading oceanic edge of the overriding plate, or as part of a topographic high on the subducting ocean floor tectonically accreted to the overriding plate at an early stage of subduction. We consider that the third, and possibly in some cases the second, hypothesis can explain the presence of most ophiolites in outer arcs, and that elevation and oceanward thrusting of the ophiolites is a result of arrival of thick sedimentary successions in the trench, and their subsequent underthrusting beneath ophiolites previously accreted to the overriding oceanic plate margin (Fig. 85). The emplacement of outer arc ophiolites is thus analogous to that of ophiolite emplacement in collision belts described later (Ch. 6, II. A.2), with the important difference that in outer arcs the place of the continental shelf on the subducting plate is taken by flysch-type sediments deposited either as submarine fans or in the trench.

In contrast to the tectonic accretion of sedimentary rocks to form outer arcs,

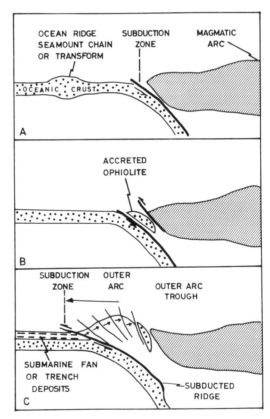

Fig. 84. Schematic cross-sections, developing outer arc. (A) early stage of subduction; (B) accretion of oceanic topographic high to overriding plate; (C) tectonic accretion of submarine fan or trench deposits to overriding plate, beneath ophiolite. Short arrows—younging direction of sediment in individual tectonic packets, long arrow – overall younging direction. Note in (C) outer arc flysch inhibits accretion of further oceanic topographic highs.

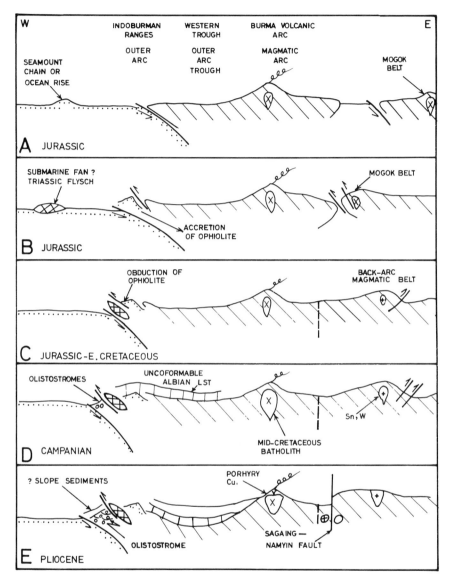

Fig. 85. Schematic cross-sections showing evolution of Western Burma arc system with emplacement of ophiolite in Indoburman Ranges outer arc. Western Burma–Asia collision belt bordering Mogok Belt and back-arc tin belt also shown (after Mitchell, 1981a). By permission of the publisher, *J. geol. Soc. Lond.* **138**, 109–122.

it has been suggested that landward of some trenches, e.g. the Middle America Trench and Japan Trench, removal of the continental crust of the overriding plate margin by underthrusting or subcrustal tectonic erosion (Fig. 86) has taken place (Scholl *et al.*, 1980; Scholl and Vallier, 1980). However, landward of the

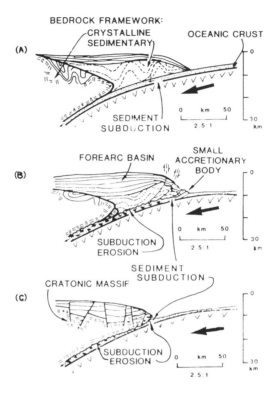

Fig. 86. Conceptual models of sediment subduction and subduction erosion. (A) Tectonic consumption of oceanic deposits beneath bedrock framework of an ocean margin; (B) subduction of oceanic deposits and subduction erosion of margin's bedrock framework, and *temporary* outgrowth of small wedge of accreted oceanic beds at base of margin; (C) advanced stage of subduction erosion, which has exposed igneous and metamorphic framework of a cratonic massif at inner wall of a deep sea trench (from Scholl *et al.*, 1980). By permission of the authors and publisher, *Geology* 8, 564-568.

Middle America Trench it is possible that the rocks of the overriding plate were moved laterally by a transform fault rather than subducted.

Many eugeosynclinal successions of ancient orogens include or comprise a wide belt of isoclinally folded greywackes which can now be interpreted as consisting at least partly of either continental rise or ocean floor submarine fans and to a lesser extent trench sediments tectonically accreted above a subduction zone to form an outer arc. The belt may include on the outer or former seaward side thick sequences of flysch or shallow marine sediments deposited in a remnant basin during the early stages of continental collision, and thrust onto the continent on the subducting plate, as discussed later. Examples of outer arcs within Palaeozoic orogens include the Southern Uplands of Scotland, the Ouachita flysch of the Appalachians, and the western part of the New England fold belt in Australia.

Ancient outer arcs are recognized chiefly by the presence of thick, poorly

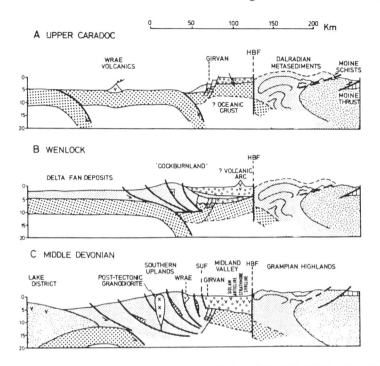

Fig. 87. Diagrammatic cross-sections showing evolution of Scottish Caledonides. HBF ⁻ Highland Boundary Fault; SUF - Southern Uplands Fault; tectonically emplaced ophiolite at Girvan omitted (after Mitchell and McKerrow, 1975). Location of outer arc granitic intrusions with uranium and porphyry copper mineralization shown in Fig. 89. By permission of the authors and publisher, *Bull. geol. Soc. Am.* **86**, 305–315.

fossiliferous, tightly folded and commonly imbricate thrust belts of turbidites and pelagic sediments, with olistostromes and in some cases minor ophiolitic rocks. There is commonly a predominant stratigraphic dip towards and structural vergence away from an adjacent ancient magmatic arc of similar age (Fig. 87). The outer arc succession on the magmatic arc side may be overthrust by slices of ophiolite, metamorphic or sedimentary rocks other than flysch. Deeper structural levels of ancient outer arcs are characterized by glaucophane schists, jadeite and nephrite, probably elevated above thrusts, and sheared serpentinites and meta-melanges. Melanges and olistostromes with blocks of ophiolitic rocks, although not always present, can be considered diagnostic of outer arcs.

A few modern outer arcs, e.g. that of southern Alaska, Southwest Japan, and the islands west of Sumatra, include granitic plutons intrusive into the flysch, while in some other outer arcs gabbroic intrusions are present. These plutons lie seaward of the volcanic arc and hence differ in origin from the volcanic arc magma which is more directly related to subduction. Marshak and Karig (1977) have suggested that the granitic bodies of the Southern Alaska and western Sumatra outer arcs were generated above a trench–ridge–trench triple junction, i.e. during subduction of the ridge, and that those of Southwest Japan were

TABLE X

Mineral deposits characteristic of outer arcs.

Tectonic setting	Association	Genesis	Type of deposit/metals	Examples
	Obducted ophiolite	Submarine exhalative sedimentary sulphide	Cyprus-type stratiform Cu Fe	Troodos, Cyprus (Cretaceous)
	Obducted ophiolite	Magmatic	Podiform Cr	Indoburman Ranges (Jurassic)
	Granite		Sn W	SW Japan (Miocene)
Outer arcs	Granite	Magmatic-meteoric hydrothermal	U	Southern Uplands Scotland (L Devonian)
	Granodiorite		Porphyry Cu	Southern Uplands Scotland (L Devonian)
	Flysch	Epigenetic connate hydrothermal	Au-quartz	Taiwan (Pliocene)
	Magnesian carbonate	Epigenetic hydrothermal	Hg	Coast Ranges, California (Tertiary); Tyan Shan, USSR (Palaeozoic)
	Flysch–quartz–carbonate		Sb	Pakistan (Tertiary)

related to a trench–trench–trench triple junction; De Long and Schwartz (1979) have argued that magmatism in outer arcs is compatible with the thermal effects of ridge subduction in a trench–ridge–trench triple junction. However, Hudson *et al.* (1979) recently described several plutons of biotite granodiorite and granite from the South Alaska outer arc, and considered that they resulted from anatexis of the accretionary prism greywackes either during a decrease in rate, or following cessation, of subduction.

Granitic plutons in ancient outer arcs have mostly been interpreted as magmatic arc rocks. However, by analogy with the Cenozoic plutons in modern outer arcs, there is evidence that they are not directly related to the "steady-state" subduction of normal ocean floor.

B Mineral Deposits

There are no reports of metallic mineral occurrences in the sedimentary layer of submarine trenches, and outer arc flysch belts interpreted as accretionary prisms of trench, ocean floor and continental rise rocks form virtually barren provinces with regard to mineralization, although pelagic sediments at the base of individual flysch units may contain hydrogeneous sedimentary manganese as in the Franciscan of California.

While in the flysch itself mineral deposits are virtually absent, tectonically emplaced ophiolitic rocks and blocks of ophiolite derived from them and deposited in olistostromes can contain major deposits of chromite and Cyprus-type sulphides (Dewey and Bird, 1971). The generation of these types of deposit has been described above (Ch. 4, I. B. 2), and we discuss below the tectonic setting of a few of those which are preserved in outer arcs.

Analogy with granitic plutons in modern outer arcs, some of which form major batholiths, has recently led to suggestions that a number of mineralized granites were emplaced in outer arc settings rather than either in magmatic arcs, or as proposed more recently, in foreland thrust belts of continental collision zones. We describe here three examples of mineralization in granites considered to have been emplaced in outer arcs during subduction, but it is possible that silicic plutons within many ancient orogenic or eugeosynclinal flysch belts may have been intruded into outer arcs (Table X).

Other minor deposits in outer arcs include mercury in the Franciscan Complex and antimony in outer arc rocks of Pakistan. It has also been suggested that outer arc sediments undergoing compaction and deformation could be important sources of ore-forming fluids.

1 Tectonically emplaced deposits of outer arcs

(a) Podiform chromite, Indoburman Ranges

Deposits of podiform chromite, broadly stratabound within dunites and serpentinized dunite and interpreted as having formed in upper mantle rocks of the ocean floor, are known from tectonically emplaced rocks of the outer arc in

Burma. The outer arc, the northward continuation of the Outer Sunda Arc of van Bemmelen (1949), comprises well-defined eastern or landward and western or seaward belts.

The eastern belt (see Fig. 85E) consists of recumbently folded Upper Triassic turbidites, in places metamorphosed in the greenschist facies, overlain by pillow lavas and locally overthrust by sheet-like bodies of serpentinite up to 2 km thick; tectonic slices of quartz–mica schist and amphibolite are also present locally. Albian limestones, intruded by thin sill-like bodies of serpentinite, lie at the base of the outer arc trough succession and rest unconformably on the Upper Triassic turbidites and pillow lavas. The eastern belt rocks are thrust westwards over the more extensive western belt, which in the north comprises a broadly folded succession up to 10 km thick of Campanian olistostromes with interbedded micritic limestones overlain by mudstones and turbidites and capped by Eocene cross-bedded sandstones preserved in synclines at elevations up to 3000 m.

The serpentinite sheets and pillow lavas are interpreted as part of an oceanic topographic high, tectonically accreted to western Burma during the early stages of Mesozoic eastward subduction, and thrust westward over the Triassic flysch and metasedimentary rocks when these collided with western Burma sometime before the Albian. Subsequent further elevation of the ophiolitic rocks took place during Tertiary thrusting of the eastern belt over the western belt of the outer arc.

Chromite is confined to the slabs of serpentinite of the eastern belt, forming small sub-economic lenticular bodies in partly serpentinized dunite within the dominantly harzburgite mass. Weak copper mineralization is associated with some of the chromite.

(b) Stratiform pyritic copper sulphides, Cyprus

The massive cupriferous sulphide deposits of Cyprus described above (Ch. 4, I.B.2) were clearly generated at or near a submarine spreading centre, but the tectonic setting in which they were elevated has been the subject of numerous conflicting hypotheses. In contrast to many of these interpretations, which favour southward subduction relative to the present orientation of Cyprus (e.g. Robertson, 1977), it is possible that the Troodos Massif of ophiolitic locally imbricate rocks (Bortalotti *et al.*, 1976), representing either a spreading ridge or another oceanic topographic high, was accreted to the margin of the overriding plate beneath the southern part of the Mesaoria Plain during northward subduction of ocean floor in the Maastrictian (Turner, 1973). The 3-km-thick late Cretaceous to Miocene succession of the Mesaoria Plain to the north can be explained as the outer arc trough (Fig. 88), the related volcanic arc lying in and north of the Kyrenia Range where calc-alkaline volcanic and intrusive rocks at least partly of late Cretaceous age are present (Robertson, 1977). Olistostromes and north-dipping serpentinite sheets of the Moni Melange to the south of Troodos are interpreted as trench sediments subsequently accreted to the outer arc, and imbricate tectonic slices of Mesozoic continental margin rocks in the Mamonia Complex in southwestern Cyprus probably underthrust the outer arc

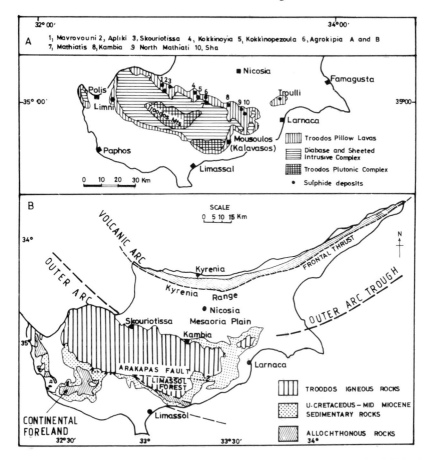

Fig. 88. Sulphide deposits and tectonic units of Cyprus. (A) Location of sulphides in Troodos pillow lavas (after Searle, 1972); (B) geological sketch map of Cyprus (after Robertson, 1977), postulated Maastrictian volcanic arc, outer arc trough, and outer arc, with fragments of continental foreland on subducting plate, according to present authors. By permission of the authors and publishers, *Trans. Instn Min. Metall.* **81**, B189–197 and *J. geol. Soc. Lond.* **134**, 269–292.

in a collision which terminated northward subduction in the region. Subsequent anti-clockwise rotation of Cyprus to its present position presumably followed rifting of the Mamonia Complex from the subducting continental margin of which it formed part.

(c) Metal-rich black shales, Southern Uplands of Scotland

These occurrences, referred to in Ch. 4, I.B.2, are mentioned here as an example of ocean floor or ocean rise pelagic metal-rich sediments elevated to form sub-economic deposits in perhaps the most widely recognized ancient outer arc, the Southern Uplands of Scotland (see Fig. 87). Nevertheless, the absence of similar deposits of economic grade in any of the eugeosynclinal belts now interpreted as accretionary prisms suggests that pelagic black shales in outer arcs are unlikely ever to be of more than scientific interest.

2 Tin, uranium and copper mineralization in outer arc granites

(a) Tin-bearing granites of the outer zone, Southwest Japan

Although the granites described here lack major mineral deposits, they are significant because they form the only example of Cenozoic tin mineralization in an outer arc setting.

Cenozoic plutons occur within the Shimanto Belt of Southwest Japan (see Fig. 104) which comprises structurally complex flysch-type sediments of Middle Triassic to Middle Tertiary age up to 15 000 m thick (Oba, 1977). The sedimentary rocks show low-grade regional metamorphism and locally mafic igneous rocks are present. The succession is commonly interpreted as ocean floor sediment offscraped during northwestward subduction which resulted in the magmatic arc to the northwest.

Plutonic rocks within the belt range from granodiorite through adamellite to granite and generally show a high K_2O/Na_2O ratio, high FeO and low magnetite content relative to the plutonic rocks from the "green tuff" volcanic arc of eastern Japan. The plutons, of Lower to Middle Miocene age, show discordant contacts with sedimentary host rocks which have suffered minor contact metamorphism, and are characterized as post-tectonic; some of the plutons are associated with ring dyke complexes and others with porphyry and rhyolites, suggesting a shallow level of emplacement. Xenoliths of pelitic and mafic rocks are abundant.

Deposits of tin and minor wolfram are mostly associated with tourmaline-bearing granites. In one mine skarn-type tin deposits are present, and in others the mineralization occurs as wolframite–quartz veins with tourmaline (Ishihara, 1973; Oba and Miyahisa, 1977).

Oba (1977) considered that the granites resulted from partial melting of geosynclinal sediments and assimilation of more felsic crustal rocks, while Marshak and Karig (1977) considered that they were probably emplaced as magmatic arc rocks above a Benioff zone related to the migrating Izu–Bonin Trench. If the outer arc sediments were deposited on continental margin rocks of the overriding plate, and subsequently imbricated, continental crust source rocks could be present at depth. However, according to the tectonic accretion hypothesis, the only rocks available for partial melting are meta-sedimentary and ocean crustal rocks above the upper mantle, suggesting that the granitic magma and concentration of tin resulted from crystal fractionation of initially more basic magma.

(b) Uranium and porphyry copper mineralization, Southern Uplands of Scotland

The Southern Uplands of northern Britain consist largely of flysch-type sediments of Upper Ordovician to Silurian age (see Fig. 87), interpreted as an

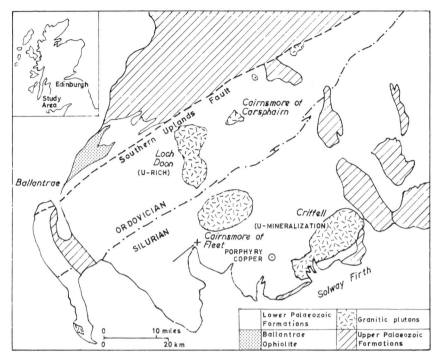

Fig. 89. Lower Devonian granites of the Southern Uplands outer arc, Scotland (from Halliday *et al.*, 1980). Location of uranium and porphyry copper mineralization also shown. By permission of the authors and publisher, *J. geol. Soc. Lond.* **137**, 329–348.

imbricate stack of deep sea fan deposits tectonically emplaced during northward subduction of ocean floor (Mitchell, 1974a; Mitchell and McKerrow, 1975; Leggett *et al.*, 1980). In the western part of the Southern Uplands, granitic plutons of Lower Devonian age intrude the flysch (Fig. 89), and were probably emplaced immediately before continental collision which terminated subduction.

Although the plutons lack economic deposits of uranium, the Loch Doon Granite, consisting largely of diorite, monzodiorite, and monzonite, shows uranium enrichment with up to 12 ppm U (Simpson *et al.*, 1979), and within the Criffel granodioritic pluton uranium mineralization occurs in veins and in a shatter zone (Bowie *et al.*, 1973). An initial Sr^{87}/Sr^{86} ratio on the Loch Doon granite of 0.705–0.706 supports an origin by partial fusion of oceanic crustal rocks at depth beneath the outer arc, yielding a monzodiorite magma which fractionated to leave granitic residual liquids (Brown *et al.*, 1979). The Loch Doon granite thus differs in source as well as setting from the collision-related uranium granites described later.

Porphyry-style copper with minor molybdenum mineralization has recently been described from granodiorite sheets, stocks and breccia pipes at Black Stockarton Moor in the Southern Uplands of Scotland. The igneous rocks

intrude Silurian greywackes at the southwestern end of the Southern Uplands
outer arc, closer to the Lower Palaeozoic trench than the Loch Doon Granite
described above. Low-grade copper mineralization with disseminated pyrite and
minor molybdenum is associated with propylitic and sericitic hydrothermal
alteration zones, and considered to have formed in a sub-volcanic environment
(Leake and Brown, 1980).

Restriction of granitic rocks to the western part of the Southern Uplands
outer arc suggests that their genesis was related to a localized thermal event,
perhaps resulting from subduction of a spreading system more or less perpendicular to the trench. However, it is not certain that granite emplacement
preceded rather than immediately followed early Devonian collision which
terminated subduction, and magma generation related to underthrusting of
continental rocks, although not supported by the Sr isotope evidence, cannot
be ruled out. A satisfactory explanation of these granites must clearly await
better understanding of the origin of similar plutons in analogous Cenozoic or
late Mesozoic settings.

The occurrence of tin in the Southwest Japan outer arc and uranium and
copper in the Lower Palaeozoic outer arc of the Southern Uplands implies that
outer arc granites are not necessarily characterized by any particular association
of metals. This may be due to differences in the nature of the source rocks
either at depth in or thrust beneath the outer arc accretionary prism, and the
presence in the Southern Uplands of porphyry copper mineralization in
particular suggests the emplacement of magmatic arc rocks anomalously close to
the trench.

3 Epigenetic deposits of antimony and mercury

Mercury ores of Pliocene to recent age are known from altered serpentinites
associated with magnesite within the Mesozoic flysch of the Coast Ranges of
California, in particular at the New Almaden mine (see Fig. 43) and in the
Ordovician of Newfoundland. Alteration of serpentinite to the silica–carbonate
host rock for the mercury ore in California is probably a result of circulation of
low-temperature meteoric and metamorphic or connate water (White *et al.*,
1973), although the origin of the metal is uncertain.

In the USSR, mercury deposits of the magnesian carbonate–cinnabar type are
considered to have formed along faults within greenstone rocks of geosynclines.
Examples have been described from Lower Palaeozoic successions of the Gornyi
Altai region, from the Tyan Shan, and from Cretaceous rocks in Sakhalin
(Kuznetzov, 1977). The association of the cinnabar with serpentinites,
"greenstones" or basic volcanic rocks, and slates or flysch-type sediments with
minor carbonates suggests an outer arc environment for most of these deposits,
although this is highly speculative in the absence of interpretations of the areas
in plate tectonic terms.

Antimony mineralization in the form of stibnite associated with quartz and calcite has been reported from fault zones within mid-Tertiary flysch of Pakistan, and has been related to the Chaman transform fault (Sillitoe, 1978). The flysch lies adjacent to an outer arc assemblage of the Axial Belt, and hence is probably underlain by rocks which include serpentinites. In view of the well-known Hg–Sb–W association elsewhere (II. B. 7), it is possible that generation of the stibnite in Pakistan may have had a similar origin to that of mercury in California.

4 Auriferous quartz veins

It has been suggested (Mitchell and Garson, 1976) that gold-bearing quartz veins present in some belts of deformed flysch or slates were emplaced as segregations at depth in outer arcs, and possibly during erosion contributed detrital gold to form placer deposits in outer arc troughs. Most flysch belts with auriferous quartz veins are of Lower Palaeozoic age or older, and significant gold consider-ed to be associated with quartz veins in recognizable Mesozoic or Cenozoic outer arcs is known only from the probable outer arc rocks of late Cenozoic age in Taiwan.

5 Outer arc rocks as sources of ore-forming fluids

The possible significance of outer arc rocks as source areas for hot mineralizing solutions has been indicated by the work of Sharp (1978) on the Ouachita Basin, possibly the largest flysch belt in North America. Sharp has argued that rapid deposition of Carboniferous flysch resulted in a reservoir of excess-pressure fluids, which in the latest Carboniferous and Permian migrated up-dip and also partly down-succession, leading by "lateral secretion" to carbonate-hosted lead–zinc mineralization in the continental foreland Cambrian strata of southeastern Missouri as suggested by Leach (1973). Graham *et al.* (1975) have shown that the Ouachita flysch (see Fig. 148) was tectonically accreted in an outer arc during southeastward subduction; Sharp's hypothesis therefore implies that migration of the fluids followed collision with the Palaeozoic and older rocks of the foreland to the northwest, as discussed later (Ch. 6, IV. B. 5). In a rather similar tectonic situation in the Persian Gulf, Dickinson (1974a) argued that hydro carbons migrated up-dip from deformed source rocks of the former continental rise of the Arabian Platform into the shelf succession to the south-west during collision with Central Iran.

6 Preservation potential

Outer arcs probably develop landward of most subduction zones not only during sedimentation in the trench or on the floor of the subducting ocean but also immediately before continental collision. This is because turbidites of the continental rise on the subducting plate as well as deep sea fan sediments deposited in the closing remnant basin will be tectonically accreted to the over-riding plate as the basin closes. However, the absence of outer arc terrains in many orogens which terminated with a collision suggests that outer arcs are rarely preserved where major plate convergence continued after the initial continent–continent or continent–arc collision. Outer arc rocks are virtually absent in the Indus Suture of the Himalayas, and in the late Cenozoic arc–continent collision belt of northern New Guinea. It seems probable that during collision the outer arc rocks either are elevated above thrusts and eroded, or more probably are overridden by rocks of the magmatic arc.

Where ancient outer arcs are present, it is probable that they occupied embayments in the margins of the colliding continents or continent and arc, and hence were preserved beside the subducting margin of remnant basins which did not completely close. They should thus occur in areas of weak collisional orogeny, but disappear along strike in zones of intense collision–related tectonism.

Mineral deposits associated with outer arc granites are more likely to be eroded than those associated with ophiolites on the inner side of the outer arc, because the former occur around the upper parts of the granites. In contrast, the latter are present within ophiolites which form steeply dipping slabs or thrust sheets; erosion of these merely exposes similar horizons within the ophiolite but at a lower topographic level.

II MAGMATIC ARCS

A Tectonic Setting

Active volcanic arcs, forming the most readily observable feature of all arc systems, include a high proportion of the world's volcanoes and lie within the zones of greatest seismicity. They form arcuate or less commonly linear to sinuous features up to thousands of kilometres in length, with volcanoes concentrated in a chain or volcanic front and more scattered volcanism in a belt which sometimes exceeds 100 km in width behind the front. Most lie in a *circum*-Pacific belt of linked arcs which includes the Aleutian, western Pacific, Indonesian, Scotia, Caribbean and Andean arc systems but there are volcanic arcs also in the Mediterranean and along the eastern margin of the Indian Ocean forming part of the Alpine-Himalayan orogenic belt. Volcanic arcs are characterized by a low positive gravity anomaly and a very high geothermal gradient. Active arcs are underlain by a zone of earthquake foci, or Benioff zone, which lies between 100 and 250 km beneath the active volcanoes and approaches the surface beneath the adjacent submarine trench or outer arc normally located on the convex side of, and 100-250 km from, the volcanic chain.

Inactive volcanic arcs are commonly deeply eroded, exposing a belt of plutons of dioritic to granodioritic composition interpreted as the root zone of the volcanoes; host rocks to the plutons may show low pressure–high temperature metamorphism of the Abukuma metamorphic facies series, with andalusite. In some arc systems a number of batholithic belts of different ages either lie in sub-parallel belts, as postulated for southern Alaska (Hudson, 1979), or may be superimposed on one another, as in the Peruvian Andes.

Volcanic arcs can conveniently be divided into continental margin arcs, situated on but near the margin of a continent, and oceanic island arcs, situated within but usually near the edge of an ocean. The reasons for these contrasting settings for volcanic arcs are discussed in 4, below, and also together with the origin of marginal basins, to which they are directly related.

1 Continental margin arcs

Volcanic chains lying on a continental margin have sometimes been termed Andean-type arcs, although the Andes are probably atypical because they have an unusually long history of subduction-related volcanism extending back to at least the mid-Mesozoic. The term cordilleran arc is also inappropriate because parts of some continental margin arcs, e.g. the islands of Sumatra and Java in the Sunda Arc, are separated from the continental interior by marine basins. Other examples of continental margin arcs are the northward continuation of the Sunda Arc through Burma, the eastern Aleutian Arc in Alaska, and the Middle America arc. It is significant that a majority of active continental margin arcs lie on the western margin of continents.

In many arcs the volcanoes mantle a major mountain range underlain by crust of greater than normal thickness due to either tectonic shortening or crustal underplating. Although the crust beneath the arc is theoretically under compression due to plate convergence, many volcanic chains undergo block faulting resulting in graben structures either parallel or oblique to the arc.

Volcanic rocks are calc-alkaline and mostly silicic or intermediate in composition, but rarely basaltic. They comprise strato-volcanoes built predominantly of either acid pyroclastics, lavas and ignimbrite sheets, or andesitic lavas, and of volcanoes of more mixed composition. Distribution of volcanoes of different composition in the Cascades of western North America has been related to segmentation of the arc by Hughes *et al.* (1980).

In general in continental margin arcs, thick successions of volcanogenic sediments accumulate on the flanks of the volcanoes, particularly in fault troughs where fluviatile sediments, air-fall pyroclastics and ignimbrites may be interbedded. Lateral variations in facies from the volcanic arc towards the submarine trench in a generalized continental margin arc have been described by Dickinson (1974b).

Between and on the flanks of the volcanoes extensive areas of pre-volcanic rocks are exposed, consisting partly of older metamorphic and uplifted sedimentary rocks, but commonly including plutons emplaced during pre-Quaternary volcanism. These are mostly of tonalite and granodiorite with minor volumes of diorite and granite. Detritus from these older rocks becomes mixed with that from volcanoes forming polymict sediments.

While most continental margin arcs form a mountain range, those situated near the continent–ocean boundary, on crust probably of less than normal continental thickness, e.g. the active Ryukyu Arc, comprise both sub-aerial and submarine eruptive rocks. The resulting successions include facies and basic volcanic rocks similar to those of oceanic arcs, although metamorphic and silicic volcanic detritus may be more common. A few continental margin arcs pass longitudinally, with related changes in magma composition, into oceanic magmatic arcs; examples are the Aleutian and Burma arcs.

2 Ocean island arcs

Oceanic volcanic arcs are separated from a continent by oceanic crust, commonly forming a marginal basin. They were first described in detail from Indonesia and the Western Pacific, but subsequently the arcuate distribution of volcanoes in the Caribbean and Scotia Arcs and in the Mediterranean was appreciated. In contrast to continental margin arcs, which occur predominantly on west-facing continental margins, the majority of active intraoceanic arcs face eastwards.

In some oceanic arcs the chain of active volcanoes is the only topographic feature above sea level; in others, e.g. the southern Lesser Antilles, an outer arc is present between the volcanic arc and the trench, and in a few arcs, for example the New Hebrides (Fig. 90) and Tonga, chains of predominantly volcanic islands older than and apparently unrelated to the present Benioff zone and volcanic activity are present between the volcanic arc and trench, forming the volcanically inactive frontal arc of Karig (1971). Remnant arcs are inactive volcanic arcs on the concave side of active arcs, from which they have become separated by creation of intra-arc oceanic crust.

The crust beneath the active volcanic arcs is mostly less than 30 km thick and may consist of normal oceanic crust, tectonically thickened oceanic crust, or older volcanic and outer arc rocks above oceanic crust. However, more complex arcs such as Japan and the North Island of New Zealand include fragments of continental rocks of Palaeozoic or Precambrian age apparently rifted off from the continental margin on their concave side.

Volcanic rocks are mostly andesitic in composition, sometimes basaltic or dacitic, but rarely include more potassic rocks. The now well-established relationship for both oceanic and continental margin volcanic arcs between K_2O content of eruptive rocks and depth to the Benioff zone was first described for late Cenozoic basaltic rocks in Japan (Kuno, 1966). The suggested evolutionary sequence in oceanic arcs from island arc tholeiites erupted at an early stage of arc development through low-potash andesites to dacites and more potassic rocks (Jakes and White, 1971) has been substantiated for only a few arc systems.

Because actively volcanic islands undergo rapid erosion as well as slow subsidence, subaerially-erupted volcanic rocks are rarely preserved. In a simple volcanic arc, e.g. the Lower Miocene sequence of Malekula in the New Hebrides

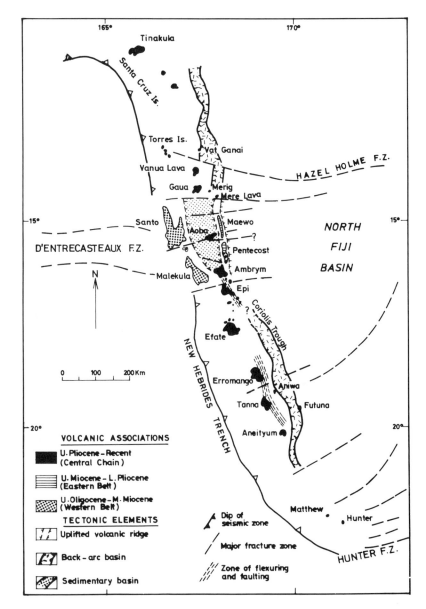

Fig. 90. New Hebrides arc system (from Carney and Macfarlane, 1979). One of the few west-facing island arcs. Note Central Chain of active volcanoes related to present subduction zone, Western Belt of Lower Miocene magmatic arc rocks, and back-arc basin. By permission of the authors and the New Hebrides Govt Geol. Surv., Port Vila.

(Mitchell and Warden, 1971), predominant facies are either basaltic pillow lavas or andesitic lava flow breccias formed by autobrecciation of submarine andesitic flows. These commonly are overlain by or interbedded with thick very coarse submarine lahar and mudflow deposits formed by erosion of subaerial volcanoes. Laterally equivalent rocks further from the volcanic chain are turbidites and mudstones; towards the volcanoes the marine volcanogenic rocks are locally interbedded with reef talus, fringing or barrier reef limestones, and lagoonal, fluviatile and fanglomerate deposits surrounding subaerial eruptive centres.

In complex oceanic volcanic arcs, e.g. Japan, the volcanogenic successions resemble those of continental margin arcs although the composition of the clastic material reflects the highly variable nature of the volcanic rocks, with tholeiitic basalts and calc-alkaline andesites, dacites and rhyolites both occurring in parallel belts and in some cases succeeding one another stratigraphically as a result of subduction-controlled changes in type of magmatism.

3 Ancient arcs

Ancient arcs, as opposed to modern active or inactive arcs, now either lie within or are tectonically welded to a continent as a result of continent–arc collision. It is usually difficult to determine whether these arcs developed on a continental margin, or were oceanic, although an oceanic setting may be indicated by the presence of abundant submarine basic volcanic rocks. The two can often be distinguished only by using criteria external to the arc itself, and requiring an understanding of the polarity of subduction and collision discussed later. In most ancient arcs plutonic rocks representing the "root zone" of the arc are more abundant than volcanic rocks.

Some ancient volcanic arc successions have been included in the "eugeosynclines" of North American authors, e.g. the Upper Palaeozoic probably oceanic arc of the Klamath Mountains (Churkin, 1974), tectonically emplaced within the Magog belt eugeosyncline of Kay (1951) during late Palaeozoic continent–arc collision. The Lower Palaeozoic geosyncline of Wales includes probably intraoceanic arc assemblages in which palaeo-volcanic islands have long been recognized (Jones, 1954).

Not all "ancient" volcanic arcs are of Palaeozoic age or older. The Eocene volcanic and associated "late geosynclinal" sedimentary rocks of the Rhodope Massif of the Hellenides, within the "geosynclinal hinterland" of Aubouin (1965), were probably deposited in a continental margin magmatic arc. These and other continental volcaniclastic rocks in the Mediterranean region have sometimes been termed "molasse" facies, in contrast to volcanogenic flysch-type sediments of oceanic arcs, and distinct from the molasse of foreland basins described later. One of the youngest ancient arc assemblages, that of Miocene age in Taiwan, has been accreted to the Asian continental margin during Pliocene collision, and in the Mediterranean the Gibraltar arc has probably been tectonically accreted to Spain and North Africa.

Other ancient examples of continental margin arcs include the Upper Silurian-Lower Devonian Sidlaw Anticline in Scotland, the Borrowdale Volcanic Group of the English Lake District, part of the Devonian succession of the Rio Tinto region of Spain, Ordovician successions of Newfoundland, and the Triassic Nicola region in Central British Columbia. Upper Proterozoic arcs of intraoceanic type have been identified in, for example, the Arabian Shield by their abundant autoclastic lavas and epiclastic andesitic rocks (Greenwood *et al.*, 1976; A.M. Al Shanti, pers. comm. 1978).

4 Volcanic arcs, subduction and drift of lithosphere

A genetic relationship between descent of subducted oceanic crust along the Benioff zone and generation of volcanic arc magmas was first suggested by Coats (1962) for the Aleutian arc. This relationship became widely accepted in the late 1960s and since 1968 a voluminous literature on the generation of subduction-related magmas has accumulated. Among the factors affecting magma generation along or above the Benioff zone are the proportion of sediment, if any, subducted with oceanic crust, the degree and nature of metamorphism and phase changes within the basaltic layer as it descends into the mantle, with consequent expulsion of water and other volatiles, the rate of subduction which controls the frictional heating between the downgoing and overriding plate, and the presence or absence of a zone of asthenosphere between the subducting crust and overriding lithosphere.

Hypotheses of magma genesis in volcanic arcs postulate three main sources: partial melting of subducting ocean crust with or without involvement of subducted pelagic sediments; partial melting of the wedge of sub-lithospheric upper mantle of the overriding plate above the Benioff zone as a result of addition of rising volatiles expelled from dehydration of the subducting crust; and a two-stage process involving generation along or above the Benioff zone, underplating of the crust, and subsequent re-melting. In all these possible processes magma may be modified by differentiation, zone refining or contamination by continental crust. Recent isotope and trace element data favour the second hypothesis, that andesitic magmas are derived largely from the mantle wedge above the subducting slab (Tarney and Windley, 1979, Hawkesworth *et al.*, 1979).

Strontium isotope data on Cenozoic volcanic rocks and plutons from the Andes and other magmatic arcs yield initial Sr^{87}/Sr^{86} ratios mostly in the range of 0.702 to 7.707, contrasting with the high values characteristic of collision-related and back-arc plutons described later. These magmatic arc plutons are broadly equivalent to the "I-type granites" of White and Chappell (1977). Francis *et al.* (1977) have shown that in North Chile where the continental crust is about 70 km thick, Sr^{87}/Sr^{86} ratios from active and recently active volcanoes range from 0.706 to 0.707, whereas in Ecuador where the crust is 40–50 km thick, and the andesites less silicic, Sr^{87}/Sr^{86} ratios are rather lower, averaging 0.7044. This led to the suggestion that in both areas the andesites are derived largely from the mantle, but that in North Chile the magma suffers isotopic contamination and undergoes greater fractionation than in Ecuador.

Ideas on the relationship of arc volcanism to subduction have been developed in particular for the Andes, where there is a gap in the volcanic arc throughout northern and central Peru and the underlying seismic Benioff zone extends to a depth of only about 100 km, rather than at least a few hundred kilometres .as is more usually the case. One explanation of this volcanic gap and associated peculiarities in the underlying seismic zone is a postulated very shallow Benioff zone with a dip of $10°-15°$ for the late Pliocene and Quaternary, resulting in compression in the overriding plate rather than extension as is the case further south (Megard and Philip, 1976); the inferred late Cenozoic flattening of the Benioff zone could have been related to a decrease in age and hence density of the subducting plate (Wortel and Vlaar, 1978). An alternative explanation of the volcanic gap is that the dip of the subducting plate is about $30°$, similar to that to the south (James, 1978), but that because the thickness of the overriding lithosphere is more than 300 km beneath central Peru, greater than that to the south, there is no layer of hot low-viscosity asthenosphere above the subducting plate, and hence no source for calc-alkaline magmas.

Support for the postulated relationship between a Benioff zone of very shallow dip and absence of arc volcanism in Peru is provided by the distribution of volcanic rocks in the central part of the Western Cordillera of North America. Very minor arc magmatism during the latest Cretaceous to Oligocene accompanied minor basement compression with block uplifts of the Laramide Orogeny, and has been related to subduction along a Benioff zone of abnormally shallow dip (Coney, 1976; Dickinson, 1978) inferred from the potash content of the scattered volcanic rocks.

The presence of active continental margin arcs mostly facing west while a majority of oceanic island arcs face east is partly related to the origin of marginal basins discussed below (p. 246). Many authors (Bostrom, 1971; Moore, 1973; Dickinson, 1978) relate this distribution of arc systems to westward drift of lithosphere with respect to underlying faster-rotating asthenosphere to which

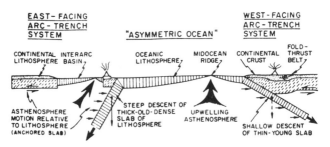

Fig. 91. Diagrammatic cross-section showing relationship of dip of Benioff zone, back-arc basin and back-arc "fold-thrust belt" to relative lithosphere–asthenosphere motion (from Dickinson, 1978). By permission of the author and publisher, *J. Phys. Earth* **26**, Suppl. S1–S19.

the arc system is anchored by the subducting slab (Fig. 91). Westward drift of the overriding continental lithosphere with respect to the subduction zone along its western margin results in a decrease in dip of the Benioff zone and consequent compressional tectonics in the continental margin arc with formation of a back-arc thrust belt (p. 218); at its eastern margin the continent tends to drift away from the arc, the Benioff zone remains relatively steep, and the continental margin arc will be under tension with formation of an extensional basin (p. 229) or marginal basin (p. 244). On north- and south-facing continental margins the arc systems may be underlain by either shallow or gently dipping Benioff zones. Consequently most continental margin arcs face west and are under compression while most of the oceanic arcs which were initiated either at a continental margin or by inter-arc rifting, face east and are at least at times under a tensional regime.

B Mineral Deposits

Magmatic arc rocks probably contain more economic metallic ore bodies per unit area than rocks formed in any other tectonic setting. Among the various types of mineralization, the greatest tonnage of extractable metal is undoubtedly that in porphyry copper deposits, but submarine exhalative sedimentary base metal deposits and gold are also important (Fig. 92), and minor but significant tin mineralization is known (Table XI).

While a voluminous literature exists on individual deposits from, and types of deposit either restricted to or characteristic of, magmatic arcs, only a few attempts have been made to describe these in the context of the arc's magmatic and tectonic history (e.g. Mitchell and Bell, 1973). In view of the numerous hypothetical origins for magmatic arc igneous rocks, which form the host to much of the mineralization, we here describe the various major types of arc deposit largely independently of discussion or arc evolution. Nevertheless, it is

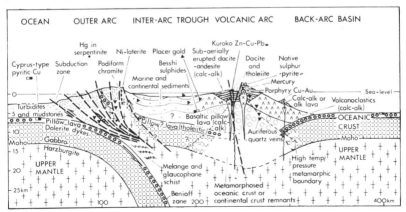

Fig. 92. Mineralization in an outer arc, outer arc trough and magmatic arc (from Mitchell and Garson, 1976). By permission of the publisher, *Minerals Sci. Engng* 8, 129–169.

TABLE XI

Mineral deposits characteristic of magmatic arcs and outer arc troughs.

Tectonic setting	Association	Genesis	Type of deposit/metals	Examples
Magmatic arcs	Tonalitic "I-type" plutons Shoshonites?	Magmatic–meteoric hydrothermal	Porphyry Cu–Au	Philippines (Tertiary)
			Porphyry Cu–Mo	Andes, W USA (Mesozoic-Tertiary)
			Porphyry Au	Vundu, Fiji (L. Cenozoic)
	Submarine rhyolitic volcaniclastics	Submarine exhalative sulphides	Kuroko-type Zn Pb Cu	Kosaka, Japan (Miocene); Venua Levu, Fiji (Pliocene); Buchans, Newfoundland (Ordovician); Captains Flat, Australia (Silurian)
	Peraluminous "S-type" granites	Magmatic–meteoric hydrothermal	Sn W	Alaska (Miocene); Inner Zone Japan (Cretaceous); E belt SE Asia (Permian)
	Andesitic caldera	Tellurides and auriferous sulphides	Au	Viti Levu, Fiji (Pliocene)

TABLE XI (Cont'd)

Tectonic setting	Association	Genesis	Type of deposit/metals	Examples
	Quartz veins in andesite	Hydrothermal	Auriferous quartz veins	Hauraki Peninsular, New Zealand (L Tertiary)
	Tonalite–diorite	Magmatic–meteoric hydrothermal	Auriferous quartz veins	Solomon Islands (L Tertiary)
Magmatic arcs	Silicic volcanic rocks	Extrusive	Magnetite–haematite apatite	Chile, Mexico (Tertiary) Scandinavia (Proterozoic)
	Basic volcanic rocks	Submarine exhalative hydrothermal	Sb W Hg	E Alps (L Palaeozoic)
	Andesitic–dacitic volcanics	Magmatic hydrothermal?	Hg (cinnabar, re-algar, orpiment), opal	Philippines; Mexico; Kamchatka (Cenozoic)
	Fluviatile sediments	Sedimentary	Placer Au	W Trough, Burma (Quaternary); Great Valley, California (Quaternary)
Outer arc troughs	Deltaic sediments	Sedimentary	Sub-bituminous coal	W Trough, Burma (Eocene)

probable that in the next few years increasing understanding of the relationship of arc magmatism to stress regimes and arc history will result in new concepts of controls on mineralization and we refer briefly to a few recent ideas on this topic in II.B.2. (f).

1 Porphyry copper deposits

The earliest observations on the relationship of mineralization to plate boundaries concerned deposits of porphyry copper, formed in magmatic arcs on overriding plate boundaries (Sillitoe, 1970; Guild, 1971), and containing more economically extractable metal than all other types of magmatic arc deposit combined. Porphyry copper deposits, usually including minor but economically significant gold or molybdenum, account for about two-thirds of the world's (outside the Soviet bloc) production of copper. Economically exploitable deposits are large or very large ore bodies invariable worked as open-cast mines, grading from 0.25% to more than 1% Cu disseminated in calc-alkaline mostly porphyritic plutons or sub-volcanic intrusions varying in composition from diorite to quartz monzonite or adamellite.

(a) Continental margin arc deposits

More than half of the known porphyry copper deposits occur in two west-facing continental margin arc systems which produce nearly half the western world's copper: the Andes, beneath which subduction continues today, and the largely inactive Upper Mesozoic and Cenozoic magmatic arcs of the Western Cordillera in North America (Fig. 93). Lower Cenozoic and older deposits in continental margin arcs since involved in continental collision are present for example in Iran

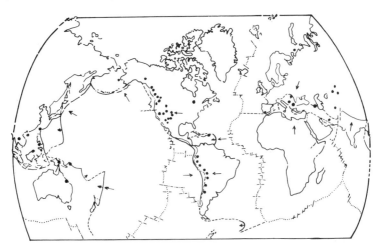

Fig. 93. Global distribution of some porphyry copper deposits; margins of major plates and relative plate motions also shown (from Guild, 1974b). By permission of the author and publisher, CTOD-IAGOD, Varna, Bulgaria.

and Tibet (Upper Mesozoic–Lower Cenozoic), in the Lesser Caucasus (Mesozoic and Upper Cenozoic), and in the Appalachians and western Tyan Shan (Palaeozoic).

Knowledge of porphyry deposits prior to the early 1970s was based largely on deposits of the North American Cordillera, and to a lesser extent on those in the Andes. Subsequently, perhaps in part stimulated by plate tectonic hypotheses, a voluminous literature has become available on these and other porphyry provinces. In the Cenozoic and Upper Mesozoic continental margin magmatic arc deposits, pre-intrusion host rocks range from volcanic rocks slightly older than the plutons through elevated sedimentary or metamorphic rocks of Palaeozoic or Mesozoic age to Precambrian rocks. The mineralized intrusions are mostly granodiorites or quartz monzonites, commonly forming small stocks or sub-volcanic intrusions which lay at depths of up to 4 km but mostly less than 2 km during mineralization.

Contacts with the pre-intrusion host rocks are generally sharp with little structural disturbance and abundant dykes and sills often project outward from the stocks. Sillitoe (1976a) noted five geometrically distinct intrusive relation-ships associated with porphyry-style mineralization in Mexico but also applicable to continental margin magmatic belts in general. The mineralized stock is normally emplaced in or above an older, but probably genetically related, larger pluton sometimes of batholith size. Porphyry-style mineralization is also associated with collapse breccia and hydrothermal breccia structures (Sillitoe and Sawkins, 1971) which form irregular bodies, sills, "pebble" dykes and well-defined tourmaline breccia pipes; ore is generated late in the mineralization sequence, in the comminuted breccia matrix characterized by distinct fluidization features.

The distribution of porphyry coppers in North America seems to have been strongly influenced by major zones of fracturing, some of which were active during pulses of mineralization, e.g. at Lornex, Valley Copper and Bethlehem in the Highland Valley, British Columbia (Hollister *et al.*, 1975). This close association with fractures is well illustrated by Hollister's (1975) diagram of the location of the northern Cordilleran porphyry deposits in relation to major faults. Lowell (1974) emphasized how porphyry deposits associated with the deeply penetrating San Pedro fracture zone in the southwestern Cordillera are related to copper-rich plutons, the locations of which were controlled by the zone. This suggests that these plutons penetrate continental crust along deep fractures some of which may have been reactivated older faults, while others were possibly initiated by vertical or more probably strike–slip movements parallel to the trench during subduction. Other areas with porphyry deposits located on major fractures include Chile where the major deposits of Quebrada Blanca, El Abra, Chuquicamata and Copaque occur on or close to a major dextral shear (Hollister and Bernstein, 1975), northern Peru where the large Michiquillay deposit lies at the intersection of the Encanada fault with another major dextral fault (Hollister and Sirvas, 1974), and Chaucha in Ecuador (Goossens and Hollister, 1973).

(b) Island arc deposits

Numerous Cenozoic and some Upper Mesozoic porphyry copper deposits are known from modern oceanic island arcs, and some Cenozoic and older deposits can be recognized as having formed in ancient island arcs subsequently welded to continents as a result of collision. Examples of deposits in modern arcs are those of Cenozoic age in the Philippines, where there are more than 30 known deposits, and in Indonesia, Fiji, the Greater Antilles, and the Aleutian arc. Porphyry coppers in ancient island arcs include those in Sardinia and the sub-economic porphyries of Miocene age in Taiwan, but in general it is difficult to distinguish ancient porphyry coppers formed in island arcs from those formed on continental margins. The young (Cenozoic) age of most porphyry deposits in modern island arcs, relative to many of those on continental margins, and consequently their relatively shallow depths of erosion, no doubt reflects the fact that arc systems with older deposits have collided with continents.

Pre-intrusion host rocks are mostly andesitic lavas and pyroclastics, e.g. Panguna in the Solomon Islands, but in some arcs they consist of flysch and serpentinite as in Sabah, of older plutons, or rarely of metamorphic rocks. Among mineralized intrusions granodiorites, quartz diorites and diorites predominate, quartz monzonites are rare. Most island arc porphyry copper deposits contain gold and some molybdenum, rather than molybdenum with little or no gold as in most continental margin arcs (Hollister, 1975), although there are exceptions and the Sipalry deposit in the Philippines has anomalously high molybdenum and low gold (Gustafson, 1978).

(c) Alteration and mineralization

Hydrothermal alteration, resulting from potassium and hydrogen metasomatism, is invariably associated with the mineralization and was first described in detail from North American deposits by Lowell and Guilbert (1970). In continental margin arcs, a central core of potassic alteration usually comprises an inner zone of orthoclase and chlorite surrounded by orthoclase and biotite; microcline may replace orthoclase (Hollister, 1975). The potassic zone is surrounded by a phyllic zone with quartz, pyrite and sericite, and between the phyllic and outermost propylitic zone is a zone of argillic alteration (Fig. 94). In the Southwestern Pacific island arcs, which exclude the continental margin arcs of Sumatra and mainland New Guinea, the hydrothermal alteration is different; there is less K-felspar, and biotite, common as an early alteration product, is not always preserved.

The primary or hypogene copper ore is associated with abundant pyrite and characteristically occurs as a disseminated accessory mineral and as fracture-filling veinlets, with disseminations largely in the potassic zone and veinlets in the phyllic zone. Ore minerals are mostly chalcopyrite and bornite with chalcopyrite dominant.

Isotopic and other geochemical studies indicate (Sheppard, 1977; Henley and McNabb, 1978) that the outer alteration zones surrounding porphyry copper

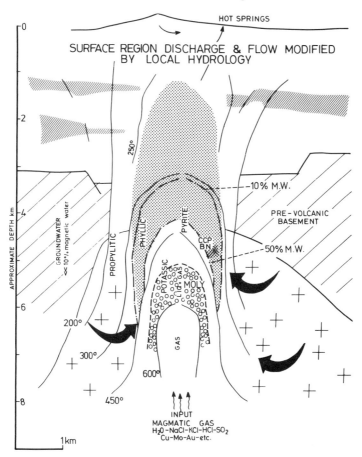

Fig. 94. Schematic summary of the plume model for a developing porphyry copper deposit showing temperature profile and alteration zones (from Henley and McNabb, 1978). By permission of the authors and publisher, *Econ. Geol.* **73**, 1-20.

deposits form at about 350°C from circulating hydrothermal systems of either meteoric ground waters or formation fluids with moderate to low salinities; the circulation is convective and driven by heat from the cooling intrusive (Fig. 94). However, transportation of most of the ore and the central potassic alteration are produced at about 750°-500°C by primary magmatic highly saline Na-K-Ca-Cl brines carrying Cu, Mo, S, etc. possibly as complex halide-hydroxyl salts. Metal precipitation results from decreasing temperature and salinity as well as induced fO_2, pH, fH_2S variations within the dispersion plume. Differences between hydrothermal alteration in continental margin and island arcs probably depend partly on the relative amounts of meteoric, magmatic and sea water involved. Phillips (1973) suggested that retrograde boiling of residual liquids in the largely consolidated igneous rock results in extensive brecciation exposing a large surface area to react with hydrothermal solutions. This brecciation includes the micro-veins and veinlets forming the characteristic porphyry crackle breccias and finally breccia pipes.

Henley and McNabb (1978) roughly estimated that formation of a 50 million tonne copper ore-body averaging 0.5% Cu required about 10^5 years, based on calculated emission rates of gas containing 427 ppm SO_2 and 0.03 ppm Cu from Showa-shinzan volcano in Japan; the gas has a Cu/SO_4 ratio close to that of hypothetical ore solutions and probably results from outgassing of a subsurface reservoir as envisaged for porphyry copper formation.

The association of porphyry deposits with calc-alkaline intrusives, both in continental margin arcs (Feiss, 1978) and in some island arcs of the Western Pacific (Mason and Feiss, 1979), depends on the partition of copper and similar metals between silicate magmas and crystallizing minerals. The Cu^{2+} ion in silicate melts is probably preferentially partitioned into sites with octahedral co-ordination rather than tetrahedral sites. As the proportion of tetrahedral sites in silicate melts increases with alkali and silica and decreases with alumina content (Burns and Fyfe, 1964), it follows that in aluminous calc-alkaline magmas with a high proportion of octahedral sites copper tends to partition into the silicate liquid, concentrating in residual liquids and becoming involved in a magmatic hydrothermal phase. In contrast, in silicate liquids with high alkali contents copper tends to partition into crystallizing minerals such as biotite, becoming "fixed" and unavailable for subsequent mineralization. Feiss (1978) showed that mineralized and non-mineralized intrusives are discriminated very efficiently using a plot of SiO_2 versus $Al_2O_3/K_2O + Na_2O + CaO$; with continued cooling there is redistribution of metals (chalcopyrite - baryte + pyrite) and the meteoric hydrothermal system tends to collapse into the central potassic (K–felspar–biotite) alteration zone, superimposing on it the argillic and propylitic (sericite-pyrite) zones.

Supergene alteration is extremely important in porphyry deposits. Percolating groundwater commonly leaches the upper part of the deposit to form a barren oxidized zone sometimes more than 100 m thick underlain with sharp contact by a zone of supergene enrichment related to the surface of the water table. Below the oxidized zone precipitation of copper results in a supergene blanket up to many tens of metres thick in which the grade of the original protore may have increased by a factor of more than four. Common ore minerals of the supergene zone are chalcocite, covellite and cuprite.

(d) Porphyry copper deposits in settings other than magmatic arcs

Although it is evident that the vast majority of porphyry copper deposits are emplaced in magmatic arcs above Benioff zones, other tectonic settings have been proposed or can be considered for a few deposits. As discussed above (Ch. 2, I.B.3. (a)), Livingstone (1973) argued that the systematic northwesterly increase in age of porphyries in the southwestern USA and Mexico was a result of movement of the continent relative to an underlying hot spot, and Lowell (1974) suggested that the mineralization was controlled by major faults and was not directly related to subduction. However, as pointed out by Sillitoe

(1975), gently dipping Benioff zones elsewhere can result in calc-alkaline magmatism far from a continental margin, and it is perhaps significant that the deposits described by Livingstone all formed during the Laramide orogeny, a period during which a shallow-dipping Benioff zone has been postulated on other grounds (e.g. Dickinson, 1978). The systematic increase in age of the deposits could perhaps be explained by southeasterly migration of a Benioff zone of suitable inclination as a result of oblique plate convergence. It appears therefore that there is no necessity to postulate an environment other than a magmatic arc for the Southwest copper porphyries.

The Boulder Batholith in Montana is one of the few examples of porphyry copper-bearing plutons unlikely to be related directly to subduction of ocean floor. As discussed in IV.B.1.(c)(ii), the Batholith lies in a back-arc thrust belt, 700 km from the late Cretaceous continental margin, and east of the late Cretaceous magmatic arc (Dickinson, 1976); its origin can most easily be related to partial melting in or beneath deep levels of the fold-thrust belt.

A further example of porphyry copper mineralization not in a magmatic arc is provided by the Black Stockarton Moor prospect in the ancient outer arc of the Southern Uplands of Scotland, described above (I.B.2.(b)).

(e) Source of metals

In porphyry deposits the metal is considered by various authors to have come from older ore bodies at depth (Lowell, 1974), volcanic rocks (Laznicka, 1976), or shales (Jensen, 1971) present in the crust beneath the deposit, from the wedge of upper mantle above the Benioff zone, and from subducted rocks. Within the subducted rocks, the suggested sources include ocean floor basalt and pelagic sediments with either abnormally high or average metal contents (Sillitoe, 1975), manganese nodules (Bromley, 1978), and ocean floor hydro-thermal sulphide deposits. It is also possible, although not widely discussed in the literature, that material removed from the base of the overriding plate by subcrustal erosion could contribute to the metal in porphyry deposits.

The subducted ocean floor source of metal appeals in particular to geochemical mass balance enthusiasts (e.g. Fyfe, 1978); this hypothesis has a superficial elegance in requiring that all metals are initially emplaced in the crust at spreading ocean ridges and are then merely cycled through the subduction zone. However, according to Hollister (1975) the distinctive lead isotopic characteristics of volcanic crust (Doe and Stacey, 1974) eliminate it as a source of both porphyry magma and metal. On the other hand molybdenum in the North American Cordillera is insignificant in porphyry copper deposits outside the limits of the thick Precambrian craton (Hollister, 1975), suggesting a crustal source for the metal. Similarly gold in the island arc porphyries was probably acquired during ascent through thick sections of volcanic rocks.

Sr^{87}/Sr^{86} isotope ratios from intrusive rocks associated with porphyry copper deposits in island arcs (Kesler *et al.*, 1975) are very low, in the range 0.703–0.704, and comparable to those of unmineralized island arc intrusive and volcanic rocks. This indicates that the magmas are derived from the subducted slab of oceanic crust, from the overlying mantle wedge, or from mantle-derived basaltic andesites, the oldest exposed and characteristically unmineralized rocks of island arcs. In continental settings (e.g. the Andes or western North America) the intrusions tend to be more potassic and the initial Sr^{87}/Sr^{86} ratios somewhat higher in the range 0.704–0.709 (e.g. Armstrong *et al.*, 1977) reflecting perhaps minor contamination with continental crust. The general geochemistry of the intrusive rocks and style of porphyry copper mineralization in continental settings are so similar to those in island arcs that the magmas in both are probably largely derived from beneath the crust, and it seems unlikely that subducted continental crust eroded from the overriding plate (p. 164) contributes significantly to the magmas or metals.

2 Kuroko-type stratiform Zn-Pb-Cu (Au-Ag) sulphides

Stratiform sulphides of copper, lead and zinc are classified in various ways, according to their tectonic setting, nature of host rocks or metal ratios (Solomon, 1976), and more recently in terms of the inferred nature of the mineralizing brine. We consider here pyritic Zn–Pb–Cu, sometimes termed polymetallic, deposits predominantly stratiform within submarine pyroclastic and epiclastic usually calc-alkaline silicic lavas of a volcanic arc; stratiform sulphides deposited in this setting form a major class of ore body, first recognized as syngenetic with respect to their host rocks and typical of island arc assemblages by Stanton (1955, 1960), and now commonly termed Kuroko-type deposits from the numerous ore bodies of Miocene age in Japan.

Although the total production of metal from ore bodies of this type is far less than that of copper from porphyry deposits, the stratiform ores are of higher grade, commonly at least 5% combined metal, and are very commonly worked as underground mines. The ores have sometimes been referred to as massive sulphides because of the very high content of pyrite and other sulphides in some deposits, but this term is now less popular. At many deposits the stratiform ore is underlain in places by a copper-rich stockwork mineralization which can also be of ore grade.

Cenozoic deposits, many of which remain more or less in the tectonic setting in which they formed, occur within the marine volcanic sequences of oceanic island arcs, e.g. in Japan and Fiji, although rather similar deposits of Mesozoic age are known from submarine volcanic rocks of what were probably continental margin arcs, e.g. the Campo Morado Zn–Pb–Cu–Ag–Au deposits of Mexico (Lorinczi and Miranda, 1978). Comparable deposits, considered to have formed in island arcs but now within continents as a result of subsequent continent–arc collision, are known from rocks of Mesozoic (Western Mine, Vancouver Island; Great Caucasus), Upper Palaeozoic (Iberian Pyrite Belt, Urals), Lower Palaeozoic (Buchans Mine, Newfoundland; Bathurst, New Brunswick, and Bathurst, New South Wales; Gjersvik and Skorovas, Norway; Avoca, Eire;

Bawdwin Mine, Burma), and Proterozoic age (Flin Flon, Canada, Umm Samiuki, Egypt).

In most Phanerozoic ore bodies the metals in order of abundance are Fe–Zn–Pb–Cu with minor but variable amounts of gold and silver. However, metal ratios vary widely, and in a few deposits, e.g. in the pyrititic ore bodies of the Iberian Pyrite Belt, copper predominates over lead and zinc. In Proterozoic deposits the lead content is commonly lower than in deposits of Phanerozoic age (Sangster, 1976b; Lambert, 1977). Gypsum and baryte often occur with the sulphides, and can form economic deposits. The high content of iron in all deposits reflects the abundance of pyrite, and the ores have been referred to as pyritic stratiform sulphides (e.g. Stanton, 1955).

(a) The Kuroko deposits of Japan

The Upper Miocene ores of Japan, within the Green Tuff Belt on Honshu and Hokkaido, are widely regarded as a "modern" analogue of older stratiform massive sulphides in silicic volcanic rocks (e.g. Sawkins, 1972). The Kuroko deposits occupy a very narrow stratigraphic interval of not more than 200 000 years (Ueno, 1975) within a succession of rhyolitic lava flows, domes, breccias and tuffs deposited below wave base and forming the upper part of a 10 Ma andesitic to felsic igneous cycle (Sato, 1977) which occupies a broad graben.

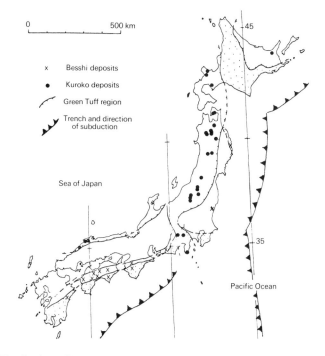

Fig. 95. Distribution of Kuroko- and Besshi-type deposits in Japan (after Tatsumi *et al.*, 1972). By permission of the authors and publisher, MMIJ-AIME, Soc. Petroleum Engrs.

More than 50 deposits are known from a belt more than 500 km long and over 60 km wide within the Miocene volcanic arc and also within the subsequent zone of Quaternary tholeiitic volcanoes (Fig. 95).

The best-described mines are those at Kosaka (Horikoshi, 1969; Urabe and Sato, 1978), where white brecciated rhyolite domes are overlain by stratiform sulphides comprising a yellow pyritic copper ore at the base overlain by black polymetallic ores with Pb, Zn and Ba; local syn-sedimentary reworking is indicated by sulphide clasts in pebbly beds. In places siliceous stockwork and disseminated ore bodies, and locally gypsum deposits, underlie the stratiform sulphides.

Each mine at Kosaka consists of a number of closely clustered ore deposits ranging in size from about 200 000 tonnes to a few million tonnes. Average ore grades from mine to mine are variable because several mines started operating in rich stratiform Kuroko ore with 1.3-7.0% Cu, 5.0-27.3% Pb, 12.3-40% Zn and 0.7-8.2 ppm Au, and then mined down to lower levels of siliceous stockwork mineralization with average grades around 1.9% Cu, 0.1% Pb and 0.2% Zn (Lambert and Sato, 1974). Most of the deposits are now nearly exhausted.

Many authors, broadly following Horikoshi (1969), relate the formation of the orebody to submarine hot spring activity following steam explosions on the flanks of the rhyolite domes (Fig. 96). The role of faults bounding the down-faulted green tuff belt in controlling mineralization has also been emphasized (Scheibner and Markham, 1976) although deposits remote from the basin

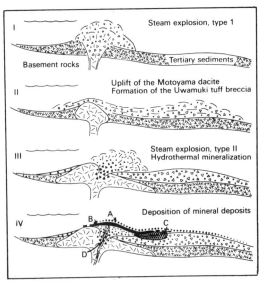

Fig. 96. Schematic cross-sections showing development of volcano and associated Kuroko deposits, Kosaka, Japan (after Horikoshi, 1969). (A) Intermediate layer of haematite, quartz and pyrite; (B) sedimentary sulphide ore; (C) gypsum ore; (D) stockwork and fissure-filling vein deposits, and ore replacing volcanic rocks. By permission of the author and publisher, *Mineral. Deposita* 4, 321–345, Springer-Verlag, Heidelberg.

margins are known. Later Cu–Pb–Zn and silver–gold veins in the Green Tuff belt are considered to have been deposited from fluids of predominantly meteoric origin implying that the Japanese magmatic arc emerged from the sea by the late Miocene.

(b) Stratiform pyritite and Zn–Pb–Cu deposits of the Iberian Pyrite Belt

The Iberian Pyrite Belt (Fig. 97), worked for over 3000 years, differs from the Kuroko belt of Japan in its predominance of pyritite deposits over Kuroko-type Zn–Pb–Cu sulphide deposits of both stratiform and stockwork type (Fig. 9A). The Belt forms the central part of a Devonian–Carboniferous basin about 35 km wide extending approximately westwards for over 230 km through southern Spain into Portugal (Strauss and Madel, 1974; Strauss *et al.*, 1977). The stratiform pyritite reserves alone are estimated to have exceeded 1000 million tonnes of which 250 million tonnes have been mined since 1850. Gold and silver mining and smelting by the Romans for 350 years is believed to have been responsible for 30 million tonnes of slag. Average massive sulphide ore grades are in the range 44–48% S, 39–44% Fe, 2–6% Cu + Zn, 0.3–0.5% As, 0.2–1.5 g/tonne Au and 5–30 g/tonne Ag. Economic copper disseminations average 0.7–1.2% Cu. The most important mines are Rio Tinto, Tharsis and La Zarza in Spain and Aljustrel and Lousal in Portugal; the large new mine at Aznacollar near Seville is scheduled to produce 12 800 tonnes of copper, 21 000 tonnes of lead and 45 200 tonnes of zinc in concentrates annually.

The Volcanic Sedimentary Complex hosting the mineralization includes

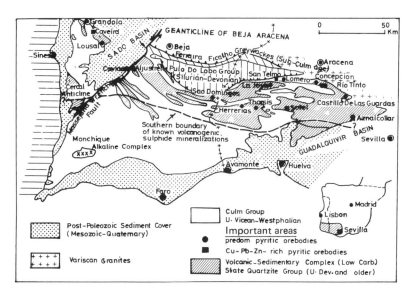

Fig. 97. Geological map of southern Spain and Portugal showing location of pyritic orebodies (from Strauss *et al.*, 1977). By permission of the authors and publisher, "Time and Strata-Bound Ore Deposits", 55–93, Springer-Verlag, Heidelberg.

Fig. 98. (A) Angled aerial photograph of mineralized hydrothermal springs with markedly enhanced values of Fe, Mn and Zn at Matupi Harbour near Rabaul in Papua New Guinea. The calc-alkaline volcanic rocks in this subduction-related environment are hydrothermally altered. Photo by Professor Peter Laznicka. (B) No 1 pit in Kuroko-type deposit in subduction-related setting at Vanua Levu, Fiji. The light-coloured foreground area consists of yellow ore (Oko) and siliceous pyrite ore (Keiko). The darkish-coloured areas on the right-hand side of the pit are pyritiferous and baryte-rich volcaniclastic sediments. White areas on the middle slopes of the pit are altered sediments, possibly dacitic tuffs, consisting now mainly of kaolinite and montmorillonite. Thick gossans are developed above the ore. The black ore (Kuroko) has been removed. Photo by Dr C. M. Rice.

Lower Carboniferous shales, frequently carbonaceous, siliceous shales and radiolarian chert. Acid volcanics are more abundant than basic volcanics and comprise quartz keratophyres with minor rhyolites and dacites. The stratiform pyritite which is closely associated with eruptive centres is considered to have accumulated during the final stages of submarine explosive acid volcanism, and either lies directly on acid pyroclastic rocks or is locally interbedded with black shales. Manganese ores either similar in age to or slightly younger than the pyritite deposits are associated with overlying tuffite shales with interbedded jasper lenses. Their spatial relationship to the pyritite is similar to that of the ochres to the cupriferous pyrite ores at Troodos in Cyprus.

The presence of massive ore lenses has been explained by the flow of sulphide mud and/or detrital sulphide material into topographical depressions situated around submarine volcanoes (Strauss *et al.*, 1977), similar to those postulated in the Buchans area, Newfoundland (Thurlow *et al.*, 1975). Stockwork mineralization of Kuroko-type forms feeders to several stratiform pyrite deposits; two types are recognized: an early stockwork consisting largely of pyrite and a later chalcopyrite-rich stockwork associated with intense chloritization of the country rocks (Pryor *et al.*, 1972).

The Iberian deposits lie within typical volcanic arc facies, and tensional conditions suggested by Munha (1979) to explain certain features indicate a similarity with the tectonic environment during deposition of the Japanese Kuroko ores.

(c) Caledonide stratiform pyritic deposits in Norway

Stratiform pyritic deposits in an island arc setting in Norway include Gjersvik (Cu > Zn) and Skorovas (Zn > Cu). These differ from Kuroko-type deposits in the absence of Kuroko-type zoning and in the predominance of early basic volcanic rocks although some andesitic and silicic lavas, agglomerates and dykes are present, especially at Skorovas. Pearce and Gale (1977) have classified the basic lavas as island arc basalts on the basis of their moderate Ti, Zr and Y concentrations and flat rare earth patterns.

The Skorovas ore body originally comprised about 10 million tonnes of massive sulphide ore including 1.5 million tonnes of essentially pyritite (Halls *et al.*, 1977) similar to that in the Iberian Pyrite Belt. The remaining pyritic ores carry about 1.15% Cu and 2.29% Zn with trace amounts of Pb, As and Ag. The orebody lies within a keratophyric pyroclastic and flow unit of the calc-alkaline part of the volcanic sequence underlain by basal basaltic pillow lava flows. Enrichment in sphalerite, chalcopyrite and locally galena with magnetite-pyrite ores at the stratigraphic top and margins of the ore lenses is interpreted by Halls *et al.* (1977) as a primary feature. Associated magnetite, chert or jasper are interpreted as colloidal iron and silica hydrosols accumulated following explosive dispersal into an oxidizing submarine environment.

Fig. 99. (A) Pyritite ore at Filon Norte open-pit, Tharsis, Spain showing graded bedding within a Kuroko-type deposit in a volcanic arc setting of Lower Carboniferous age (see Fig. 97). The Iberian Pyrite Belt extends for about 230 km through southern Spain into Portugal forming part of a Devonian-Carboniferous basin about 35 km wide. It differs from the Kuroko belt of Japan in the predominance of pyritite deposits over Kuroko-type Zn–Pb–Cu deposits. Stockwork mineralization forms feeders to several stratiform pyrite deposits. Tensional conditions (Munha, 1979) indicate a similarity with the tectonic environment of the Japanese Kuroko ores. Photo by Dr M. S. Garson. (B) Cassiterite–lepidolite pegmatite at Reung Kiet, Phangnga, south Thailand, intruded into a zone of transcurrent fracture considered to be a continental extension of an oceanic transform fault. Photo by Dr M. S. Garson.

(*d*) *Other examples of Kuroko-type deposits*

In Fiji a massive Zn-Pb-Cu sulphide deposit in Lower Pliocene dacite and rhyolite lavas and volcaniclastic rocks on Vanua Levu (Fig. 98 B) is of interest because of the preservation of various lithological ore types including breccia and stockwork mineralization; these suggested to Colley (1976) a possible genetic relationship to porphyry copper deposits. Similarity to the graben setting of the Japanese Kuroko ores is implied by evidence that the mineralization in Fiji took place during or shortly before development of a back-arc marginal or inter-arc basin (J. Carney, pers. comm. 1980).

The Buchans Mine in Newfoundland lies within a thick and extensive zone of Ordovician to Lower Silurian volcanic and volcaniclastic rocks interpreted as a Lower Palaeozoic island arc (Thurlow *et al.*, 1975). The mineralization, mostly massive and consisting of sphalerite, galena and chalcopyrite, with minor pyrite and tetrahetrite, occurs as lens-like bodies up to a kilometre in length at two horizons, each within a sequence of volcaniclastic sediments, siltstones, and dacitic flows and breccias. The mine has produced 16 million tonnes of ore containing Zn, Pb, Cu and minor Ag and Au, which consisted partly of sulphides fragmented by surface volcanic explosions and carried as rapidly moving subaqueous density flows into palaeotopographic depressions. Thurlow (1977) noted that at Buchans economic sulphide accumulations may occur as discrete transported orebodies well beyond the limits of stockwork mineralization.

At Avoca in eastern Eire, deposits containing 5-6 million tonnes of Pb-Zn-Cu pyritic ore with minor silver and gold are associated with extensive zones of low-grade disseminated ore (Platt, 1977). The host rocks are acid pyroclastics of Ordovician age within an ancient island arc complex believed to have formed above a southeasterly-dipping subduction zone. Both massive pyritic ore and disseminated sulphide ore of stringer and stockwork type have been recognized. The massive ore shows a probably primary compositional gradation with a copper-rich pyritic zone overlain by a zone of Pb-Zn-Cu pyritic ore; these zones pass laterally into clastic carbonate-rich basinal rocks. However Badham (1978) considered that reworking by slumping into a basin had caused some inversion of metal stratigraphy, and that later exhalation from a new centre filled remnant depressions with tuff, sulphides and chert.

An early-worked example of a massive Zn-Cu-Pb deposit of Upper Proterozoic age occurs in an ancient island arc setting at Umm Samuiki in the South-Eastern Desert of Egypt (Garson and Shalaby, 1976). Similar deposits occur elsewhere in this Proterozoic tract in Egypt, Saudi Arabia and the Asmara area of Ethiopia where baryte is reported in some of the ore bodies. At Umm Samiuki, the metavolcanics consist of a 8000 m thick, folded succession of andesites, rhyolites, thin basalts and pyroclastics and thick underlying pillowed basalt flows. Ore bodies are relatively small totalling 132 500 tonnes of massive sulphides grading 15.2% Zn, 1.15% Cu and 1.1% Pb, and 90 g/tonne Ag. Stockwork mineralization is also present in altered rhyolitic rocks.

In eastern Australia, stratiform sulphides of Silurian age at Captains Flat and

Woodlawn show many similarities to the Kuroko ores of Japan, although considered by Scheibner and Markham (1976) to have formed in back-arc basins (VI. B. 4). At Captains Flat mineralization occurs in shale overlying rhyolitic pyroclastic volcanic rocks and underlying dacite. Sulphides of the main ore body are pyritic, massive and banded, with the ore metals Zn–Pb–Cu–Ag–Au in order of abundance, and an average grade of 16% Zn + Pb. Beneath the ore a broad zone of disseminated pyrite contains minor chalcopyrite. Although the back- or inter-arc basin setting of the Captains Flat Trough is evidently well established, because of similarities of the sulphides to Kuroko-type ores we have included these in volcanic arc deposits.

(e) Depositional environment and source of metals

The syngenetic nature of the mineralization in Kuroko-type deposits is indicated by its parallelism to bedding in adjacent rocks by the relationship of ore thickness to topography, by the occasional presence of sulphide pebbles in conformably overlying sediments, and by analogy with active mineralized hydrothermal springs (e.g. Fig. 98A). It is now widely accepted that the ores were deposited from saline submarine hot springs, and hence the deposits are usually referred to as submarine exhalative or submarine exhalative sedimentary in origin (Sato, 1977); because of their association with volcanic rocks they have been termed volcanogenic submarine exhalative, or fumarolic deposits.

The role of the exhalative brine density in controlling the type of mineralization in sedimentary exhalative deposits has been emphasized by Sato (1977) who considered that in formation of Kuroko-type deposits the brine density is initially less than that of sea water, and that it.increases to exceed sea water density before decreasing again. Finlow-Bates and Large (1978) discussed the significance of boiling of the brines, and suggested that the copper-rich stockwork deposits are precipitated by boiling in the sub-stratiform feeder chimney, while the more soluble stratiform deposits of lead and zinc are deposited on the sea floor after boiling. This requires sufficient water depth to prevent boiling of the brine before exhalation; for the Kuroko deposits of the Kosaka mine, with an estimated temperature of 300°C and salinity less than 10%, a depth of nearly 800 m is necessary (Sato, 1977). Deposition of "massive" stratiform sulphide from submarine hydrothermal plumes was discussed by Solomon and Walshe (1979) who considered that Kuroko-type deposits accumulated in relatively deep water, while a finely-banded, thin and extensive variant of these deposits termed Rosebery-type accumulated in shallower water as a result of lateral spreading of the rising plumes at the seawater surface. Deposits which they considered resemble those of Rosebery include Captains Flat and Woodlawn in Australia, and possibly Rammelsberg (Ch. 2, II.B.2. (b) in Germany. Stratiform sulphides associated with calc-alkaline rocks have also been classified as proximal or distal according to their position relative to the associated volcanic centre (Plimer, 1978), although Large (1979) has pointed out that it is the hydrothermal exhalative centre, rather than an eruptive centre, to which the stratiform mineralization is related; the two may of course coincide in some cases.

The possible relationship of Kuroko-type ores to resurgent calderas, large circular volcanic collapse structures in which uplift of the caldera floor follows collapse, has been discussed by a number of workers, and recently applied by Harley (1979) to lenticular Zn–Pb–Cu–Ag deposits associated with felsic tuff breccias and probable rhyolite domes in the Bathurst-Newcastle area of New Brunswick, Canada. These Kuroko-type deposits, referred to as Type II deposits by Harley, occur in a sequence of volcanic and sedimentary rocks which includes basaltic lavas. They are associated with a rather different type of laterally persistent stratabound deposit (Type I), of greater economic importance, which

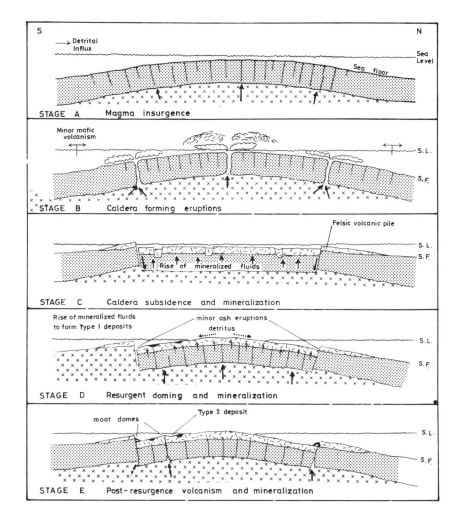

Fig. 100. Deposition of submarine exhalative sulphides associated with development of a resurgent caldera, Bathurst–Newcastle Mining District, Canada (after Harley, 1979). By permission of the author and publisher, *Econ. Geol.* 74, 786–796.

occurs in sedimentary horizons associated with porphyry and with iron formation. Harley (1979) postulated that subsidence followed by resurgent doming ruptured the volcanic pile and allowed hydrothermal fluids to rise and spread over the sea floor (Fig. 100), forming the broadly circular distribution pattern of sulphide deposits. The presence of a bimodal felsic and mafic lava sequence led Harley to support a continental setting for the mineralization, but analogy with Japan suggests that a similar sequence can form in east-facing magmatic arcs during or follow rifting from a continental margin.

The source of the metal in the Kuroko deposits of Japan has been discussed in some detail. Horikoshi (1976) considered the metals to be genetically related to the subduction-controlled generation of tholeiitic magma, the start of which coincided with the late Miocene mineralization. Either separation of aqueous ore fluids during the final stages of fractionation of magma, or leaching of older igneous rocks by brines, were processes favoured by Lambert and Sato (1974).

Sato (1976) suggested that the Kuroko metals were concentrated in the presence of water in a differentiating magma body derived either from subducting ocean floor or from the overlying mantle, and that the volume of magma and degree of differentiation controlled whether mineralization occurred. Similarly Urabe and Sato (1978) argued that in the Kosaka Mine the hydrothermal metal chloride-bearing solutions were derived from the aqueous phase of an acid magma produced by large-scale magmatic differentiation which formed the rhyolitic lava domes beneath the ore. However, Sato (1977) considered that the mineralizing brine was a mixture of magmatic hydrothermal with large volumes of sea water. Sheppard (1977) concluded from a study of stable isotopes at the Kuroko-type deposit in Fiji, that the hydrothermal solutions were of sea-water origin, and that mixing with meteoric waters is not an essential feature of Kuroko deposits.

(f) Relationship of Kuroko and porphyry copper deposits

The relationship, if any, of Kuroko-type and porphyry copper deposits has long been discussed, and in particular the presence in Japan (Ishihara, 1974) of only Kuroko deposits while in several other arcs only porphyry deposits occur has attracted widespread attention. The relative abundance of submarine environments in oceanic island arcs as opposed to continental margin arcs cannot entirely explain this distribution. Some authors (e.g. Colley, 1976) have suggested that porphyry and Kuroko ores are related to different topographic and structural levels of the same type of hydrothermal system, but there are no unequivocal examples of the two types of deposit occurring together. One possible example is the Mount Morgan pyritic copper-gold deposit of Devonian age in central eastern Queensland where stratiform sulphide mineralization is invaded by a latite porphyry stock with associated sulphide-bearing quartz veins injected into the ore of the breccia pipe.

The regional distribution of late Cenozoic deposits, with abundant porphyry copper in the west-facing Andean arc and in some oceanic island arcs while Kuroko ores are most abundant in the east-facing arc of Japan, suggests a possible relationship of mineralization to the tectonic regime. If lithosphere is drifting westwards, as discussed above, the Andean arc was under compression at least intermittently throughout the late Cenozoic, while the Japanese arc probably underwent rifting as indicated by the eruptions of late Miocene rhyolite and subsequent basalt, and by the early Tertiary opening of the Japan Sea. This would suggest that porphyry copper deposits develop in compressive settings, while Kuroko deposits are characteristic of tensional environments as argued by Mitchell (1976a) to explain the almost simultaneous late Miocene deposition of Kuroko ores over a wide area in Japan. It should be noted that in general oceanic arcs facing west also are under compression, and might be favourable for porphyry copper mineralization, while those facing east and hence under tension might be more likely hosts to Kuroko-type ores. In the Philippines emplacement of the numerous mid-Tertiary porphyry copper–gold ore bodies was probably related to eastward subduction of ocean floor, in contrast to the present westward-subducting system. Sillitoe's (1980b) suggestion that porphyry deposits are associated with andesitic and dacitic rocks erupted in strato-volcanoes, while Kuroko ores occur within rhyolitic rocks erupted in resurgent calderas situated above zones of lithospheric tension (Fig. 101), supports a broad tectonic control on the nature of the mineraliza-

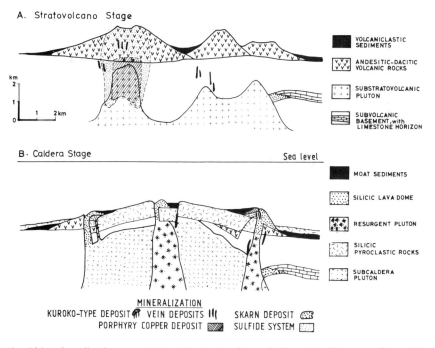

Fig. 101. Mineralization associated with strato-volcano fields and caldera complexes. (A) Porphyry copper mineralization beneath strato-volcano; (B) Kuroko-type mineralization deposited during resurgence of submarine caldera (after Sillitoe, 1980b). By permission of the author and publisher, *Geology* 8, 11-14.

tion and hence a tendency for the type of deposit to be related to the arc's orientation and consequently to some extent to its position relative to a continental margin.

Recently Uyeda and Nishiwaki (1980) discussed the relationship of porphyry copper distribution in some arc systems to the horizontal stress during mineralization, and suggested that generation of porphyry copper deposits requires a compressional regional stress environment (Fig. 102) similar to that in Chile today; conversely, they argued that Kuroko deposits were emplaced in tensional environments similar to that in the active Marianas arc system. The suggestion of Milsom (1978), that in the western Pacific complex arcs which have undergone reversals in subduction polarity are preferentially mineralized with porphyry copper deposits, is perhaps related to the fact that while most arcs face east, porphyry mineralization took place when they faced west prior to a flip in the Benioff zone.

A number of other features of the distribution of porphyry and Kuroko

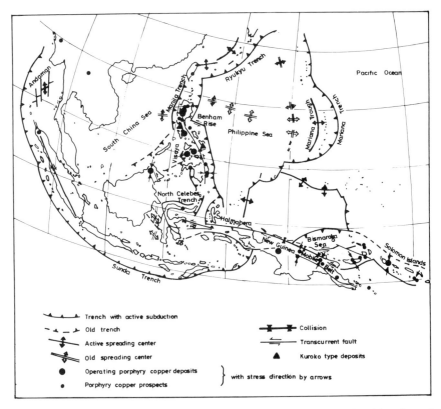

Fig. 102. Distribution of porphyry copper deposits in the western Pacific with estimated directions of maximum horizontal compressive axes at the time of mineralization (from Uyeda and Nishiwaki, 1980). By permission of the authors and publisher, *Spec. Pap. geol. Ass. Can.* **20.**

deposits are not satisfactorily explained by any hypothesis. For example, in the Peruvian Andes porphyry mineralization is abundant in the Arequipa segment of the Coastal Batholith but scarce in the adjacent Lima segment to the north (Goossens, 1976), where the composition of the plutons is broadly similar (Pitcher, 1978) but the present dip of the Benioff zone steeper. The wide distribution of Kuroko deposits in Japan but their absence in a broadly similar rift zone in North Island, New Zealand, also lacks any obvious explanation. Similarly, it is not understood why the southwest-facing arc in Sumatra lacks economic porphyry deposits. Possibly relationships analogous to that between volcanism and age of subducted lithosphere found in western South America (Wortel and Vlaar, 1978) have a bearing on some of these problems.

3 Chile-type (manto) deposits

Attempts have been made to recognize submarine exhalative or volcanogenic massive sulphide deposits which are distinct from Kuroko-type in terms of nature of host rocks, e.g. Besshi-type described above (Ch. 4, III. B. 2) and Chile-type (Mitchell and Garson, 1976), size (Fyfe, 1978), and proportion of ore metals (Hutchinson, 1973). An alternative view is that a complete range of deposits may exist (Gilmour, 1971), with features intermediate not only between those of Kuroko and Cyprus-type, but also between volcanogenic ores and non-volcanogenic deposits similar to those of for example the Kupferschiefer deposits of Europe (Wolf, 1976).

Manto-type deposits (Carter, 1960; Ruiz *et al.,* 1971), occur as stratabound up to 100 m thick and several kilometres long in calc-alkaline volcanic rocks of Jurassic to Lower Cretaceous age in Chile and southern Perú. The sulphides consist dominantly of chalcocite, bornite and chalcopyrite as vesicle and fracture fillings mainly in the upper parts of andesitic flows and ash-flow tuffs and in associated volcaniclastics and organic-rich limestones (Sillitoe, 1977). At Bueno Esperanza in Chile, one of the largest examples of manto-type deposits, 2.5 million tonnes of ore with 3% Cu and by-product silver have been produced; there are 28 mineralized horizons ranging from 2 to 25 m in thickness within a 270 m-thick andesitic sequence. The deposits are considered by Sillitoe (1977) to have been generated within a continental margin volcanic arc either subaerially or in local lagoonal environments, by circulation of meteoric water through hot volcanic rocks immediately following their accumulation.

Sillitoe (1977) suggested that volcanogenic Mexican-type deposits may have a similar origin to that of manto-type deposits. They consist of minor cassiterite, wood tin and haematite in Tertiary felsic volcanic rocks and possible rhyolite plugs in western Mexico (4 below), western USA, Bolivia and northwestern Argentina.

4 Tin and tungsten

Some of the earlier papers on subduction-related mineral deposits discussed the

presence of late Mesozoic and Cenozoic deposits of tin and tungsten on the
landward side of porphyry copper-bearing magmatic rocks in *circum*-Pacific
continental margin arc systems. Sillitoe (1972b, 1976b) ascribed this distribution
of mineral deposits in the Andes to generation of metals at different depths
along a Benioff zone of constant dip, while Mitchell and Garson (1972) and
Mitchell (1973) considered that there was a less direct relationship involving
migration and change of dip of the underlying Benioff zone. More recently it
has become apparent that many of the tin deposits described by these authors
were emplaced in back-arc magmatic belts described below (p. 218) and were
related only indirectly to the Benioff zone dip and still less to its depth.
Nevertheless, a few deposits of tin of minor economic significance appear to
have been emplaced within the magmatic arc, in some cases on the landward side
of, but different in age from, porphyry copper deposits.

The "magmatic arc" tin deposits occur in a number of late Cenozoic continental margin arcs, a tectonic setting which remains very broadly similar to that
at the time of mineralization. Examples are the tin occurrences in the Miocene
granite of the Aleutian Arc in Alaska (Reed and Lanphere, 1973), and cassiterite
in the Oligocene rhyolitic lavas of the Sierra Madre Occidental, Mexico (Swanson
et al., 1978) referred to above.

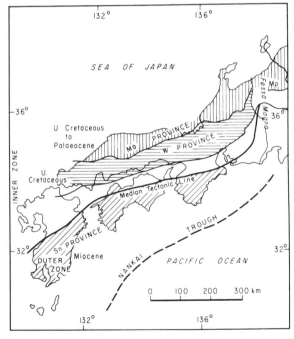

Fig. 103. Subduction-related magmatic arc (Mo and W-Sn) and outer arc (Sn-W) deposits,
Southwest Japan (after Ishihara, 1973, 1978, modified by Mitchell, 1979a). By permission
of the author and publishers, *J. geol. Soc. Lond.* **135**, 389–406 and *Bull. geol. Soc. Malaysia*
11, 81–102.

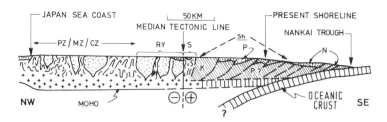

Fig. 104. Speculative crustal profile across Southwest Japan. Ry, Ryoke Belt; S. Sanbagawa Belt; Sh, Shimanto Belt; K, Cretaceous; P, Palaeogene; N. Neogene (from Dickinson, 1977). Note Cretaceous–Palaeocene magmatic arc tin granites east of Japan Sea Coast; Miocene outer arc granite with tin mineralization not shown. Note Median Tectonic Line is possible ancient thrust above which Ryoke Belt was elevated and former outer arc trough destroyed. By permission of the author and publisher, *Eos* 58, 948–952.

An example of tin mineralization within an island arc but unrelated to the present subduction system is provided by the deposits of the Inner Zone or Japan Sea side of Southwest Japan (Ishihara, 1973). A belt of late Cretaceous to Palaeocene granitic rocks extending through much of Honshu Island (Figs 103 and 104) consists largely of ilmenite-series plutons, with very small amounts of ilmenite and magnetite (Ishihara, 1977), and includes numerous occurrences of W–Sn–Cu, mostly in vein-type deposits with scheelite and some cassiterite. During the mineralization Southwest Japan was attached to the Asian continent, forming a continental margin arc, a tectonic setting very broadly analogous to that of Alaska and Mexico.

Ishihara (1977) has explained the common association of cassiterite and wolfram deposits with the ilmenite-series rocks, first recognized in Southwest Japan, by suggesting that in the magnetite-series, with which most porphyry copper deposits are associated, tin in the tetravalent state substitutes in sphene, magnetite and ilmenite, resulting in low concentrations in the residual liquid, while in the ilmenite-series tin remains available until the final stages of crystallization. Ishihara suggested that the ilmenite series magmas formed under conditions of lower oxygen fugacity than the magnetite series, as a result of their generation in continental crust and interaction with graphite before consolidation. The global distribution of some ilmenite and magnetite series plutons is shown in Fig. 105.

Economic deposits of tin in an ancient magmatic arc are represented by the ore bodies of the East Coast Belt of Malaya (Mitchell, 1977) and its probable continuation into Billiton Island in Indonesia. The mineralization in this belt differs in age, style, nature of host rocks and genesis from that in the Central Belt to the west (Ch. 6, IV. B. 1.(c)). The tin deposits, known from the Pahang Consolidated mine in Malaya and Kelapa Kampit in Billiton, occur within Permian or possibly Carboniferous terrigeneous sedimentary rocks, associated in Malaya with andesites, adjacent to intrusions of Permian or Lower Triassic granites which form part of a pre-Upper Triassic predominantly granodioritic magmatic arc. The ores are unusual in comprising the cassiterite–magnetite–

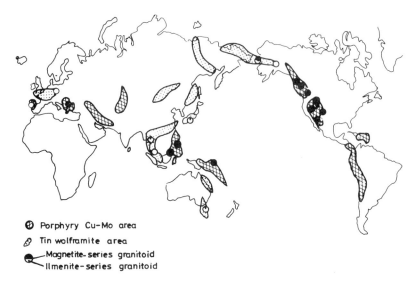

○ Porphyry Cu-Mo area

✍ Tin wolframite area

●—Magnetite-series granitoid

↖Ilmenite-series granitoid

Fig. 105. Distribution of magnetite (mostly porphyry copper–molybdenum-bearing) and ilmenite series (mostly tin-bearing) plutons (from Ishihara, 1977). Note distribution of ilmenite series plutons is not necessarily related to present east-facing continental margins because pluton belts are of various ages. By permission of the author and publisher, *Min. Geol.* **27**, 293–305.

pyrite-pyrrhotite association, and in being stratiform within the hornfelsed sedimentary host rocks. Hutchison and Taylor (1978) have suggested that while the deposits have traditionally been considered as replacement bodies related to the granite, they could be classed as volcanogenic, and hence syngenetic in origin.

5 Gold

Gold mineralization is commonly associated with magmatic arc rocks forming either economic primary deposits or resulting in economic placers. There are several different types of primary deposit, best illustrated by consideration of a number of Cenozoic examples, the distribution of which probably reflects the tendency for gold to occur in oceanic island rather than continental margin arcs.

Gold tellurides have been described in most detail from the Emperor Mine on Viti Levu in Fiji (Denholm, 1967), which is still worked intermittently. The mineralization occurs in brecciated andesitic rocks of Pliocene age, closely associated with and in part controlled by a caldera boundary fault. Tellurides and auriferous sulphide mineralization followed intrusion of trachyandesite and monzonite plugs into sedimentary rocks and andesites which fill the caldera. The caldera complex overlies rocks of calc-alkaline or island arc tholeiite composition, and hence gold mineralization post-dated the first major volcanic episode on the island. Deposits rather similar to those of the Emperor Mine have been described from Antamok and Acupan in the Philippines.

Auriferous quartz veins within andesites were described long ago by Lindgren (1933) from a thick succession of propylitized andesitic and dacitic lava flows of Lower Tertiary age in the Hauraki Peninsula of New Zealand. Migration and concentration of the gold probably accompanied metamorphism of the andesites. The gold has been mined as a primary ore and has also given rise to major placer deposits. These deposits are of interest because economic concentrations of metals, other than those related to adjacent intrusions, are rare within andesites of modern island arcs.

A sub-economic gold porphyry deposit has been described from shoshonitic rocks at Vundu in Fiji (Lawrence, 1978). Although gold-rich porphyry copper deposits apparently occur in both oceanic island and continental margin arcs (Sillitoe, 1979) it is of interest that this, the only known example of porphyry gold mineralization, lies within an oceanic island arc lacking evidence of older continental crust.

A fourth mode of occurrence of gold in magmatic arcs is that associated with quartz veins around the margins of dioritic or grandioritic plutons, as in the Solomon Islands and Philippines. Gold mineralization adjacent to a dioritic intrusion in the Baguio Gold District on Luzon in the Philippines (Fig. 106) has recently been described by Fernandez and Damasco (1980). The gold occurs in faults and fractures within Upper Mesozoic to Oligocene volcanic and sedimentary rocks of volcanic arc facies, and occupies a N-trending zone up to 8 km wide and several tens of kilometres long immediately west of an intrusive Miocene diorite. The mineralization is considered to be mostly Pleistocene in age, and hence younger than the nearby diorites and their associated porphyry copper deposits. Nevertheless, the distribution of the gold-bearing veins suggests a genetic relationship between the diorite and the mineralization.

6 Magnetite–haematite–apatite deposits

A number of Phanerozoic and Proterozoic iron deposits associated with silicic volcanic rocks and consisting largely of magnetite with minor haematite, fluorapatite and actinolite are known, and controversy continues as to whether many of these are extrusive, intrusive or exhalative-sedimentary in origin (.e.g. Parak, 1975).

In Northern Chile the Pliocene–Pleistocene El Laco deposits (Park, 1961), the tectonic setting and mode of occurrence of which are well preserved, provide a possible analogue for some of the older ore bodies, although the position of the deposits suggests a back-arc magmatic belt rather than magmatic arc. The El Laco deposits occur in flows about 20 m thick overlying ignimbrites and andesite lavas, and are more or less coeval with nearby rhyolites; they are associated with craters around the margins of a caldera. The ore contains 50% iron and reserves are estimated at one billion tonnes. Geochemical evidence suggests that the flows were derived from the underlying Palaeozoic ferruginous sedimentary rocks (Frutos and Oyarzun, 1975), presumably mobilized by intrusive rocks at depth. To the south, in a belt parallel and closer to the coast, deposits of

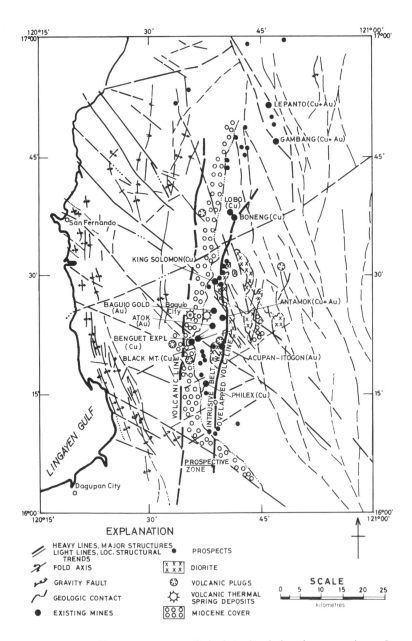

EXPLANATION

HEAVY LINES, MAJOR STRUCTURES
LIGHT LINES, LOC. STRUCTURAL
TRENDS

FOLD AXIS

GRAVITY FAULT

GEOLOGIC CONTACT

EXISTING MINES

PROSPECTS

DIORITE

VOLCANIC PLUGS

VOLCANIC THERMAL
SPRING DEPOSITS

MIOCENE COVER

SCALE

0   5   10   15   20   25

kilometres

Fig. 106. Gold mineralization adjacent to diorite intrusives in inactive magmatic arc, Luzon, Philippines (after Fernandez and Damesco, 1980). By permission of the authors and publisher, *Econ. Geol.* **74**, 1852-1868.

Mesozoic and Cenozoic age with broadly similar mineralogy are associated with predominantly andesitic rocks; most are related to intrusions and are hydrothermal in origin (Bockstrom, 1977).

In the Sierra Madre Occidental of Mexico, interpreted as a Cenozoic magmatic arc, a major source of iron ore is provided by haematite, martite and magnetite eruptive rocks within a thick succession of rhyolitic rocks in a caldera complex (Swanson *et al.*, 1978).

In contrast to the iron ores of Chile and Mexico interpreted as volcanic rocks, those of mid-Proterozoic age at Kiruna in Scandinavia, although considered by some authors (Frietsch *et al.*, 1979) to be intrusive magmatic in origin, are more commonly interpreted as sedimentary or exhalative-sedimentary deposits (Parak, 1975). The ores occur at two main horizons within and at the top of a thick sequence of kerotophyres and quartz kerotophyres, suggesting a magmatic arc environment, and Sato (1977) has suggested that they were deposited on the sea floor from hot dense metal chloride-bearing brines.

7 Antimony–tungsten–mercury deposits

Many Upper Mesozoic and Cenozoic tungsten and antimony deposits are clearly epigenetic and associated with granitic plutons, e.g. those in southeastern China. However, an important class of ore bodies, which could include possibly some of those previously thought to be epigenetic, comprises the W–Sb–Hg deposits first recognized in the Eastern Alps as stratabound and associated with volcanic rocks (Maucher, 1965).

In the Eastern Alps the ore bodies are mostly of three types: W (scheelite) with minor Mo, Cu and Bi; Sb (stibnite) with minor As, W and Cu; and Hg (cinnabar). The scheelite-type of deposit includes the recently discovered large body at Falbertal. Similar mineralization is found in Sardinia, and in Turkey where it is of Cenozoic age and associated with andesitic and dacitic eruptive rocks of a probable magmatic arc (Maucher, 1976).

Although the three types of ore occur at different localities they are all associated with broadly similar Lower Palaeozoic submarine volcanic and commonly carbonaceous sedimentary rocks, some of which are metamorphosed. Holl and Maucher (1976) and Holl (1977) consider that the mineralization was initially syngenetic and genetically related to eruption of the adjacent lavas, but that much of the ore was remobilized during subsequent deformation and metamorphism to form veins discordant to the bedding. The volcanic rocks are mostly basic but rocks of intermediate composition and more silicic porphyries are also present. Holl (1977) has interpreted the distribution of the W, Sb and Hg occurrences in the Alps in terms of their position relative to an inferred underlying Lower Palaeozoic north-dipping Benioff zone, and it appears at least possible that the mineralization took place in a broadly magmatic arc setting.

8 Mercury

Deposits of mercury associated with calc-alkaline volcanic rocks occur in the Philippines, Japan and New Zealand, and in the continental margin magmatic arcs of Mexico and Chile; they are also known from Mesozoic and Cenozoic volcanic belts in the Kuriles and Kamchatka arcs of the USSR, where they are termed volcanogenic hydrothermal deposits and consist largely of the opal-cinnabar association. Host rocks are mostly andesitic to dacitic in composition, and commonly have undergone intense hydrothermal alteration. Often associated with the mercury, which mostly occurs as cinnabar, realgar, orpiment and rarely quicksilver, are ores of gold, silver and in some cases sulphur, and semi-precious opal is also often present.

That deposits of this type form at or near the surface is indicated by the presence in the Kurile Islands of active steam and gas fumaroles around which cinnabar is being precipitated, and of present-day precipitation of mercury and opal from thermal springs in the Uzon caldera of Kamchatka (Kuznetsov, 1977).

White *et al.* (1971) suggested that mercury, separated from less volatile metals in vapour-dominated reservoirs, could be deposited above boiling brine zones in which porphyry copper deposits develop. This proposed setting is not contradicted by the evidence for near-surface or surface precipitation of mercury, its common association with andesitic to dacitic host rocks, and the strong hydrothermal alteration characteristic of the host rocks. However, in the Uzon deposits of Kamchatka the precipitation of mercury from thermal springs within areas of slightly older rhyolitic and basaltic volcanism in a major caldera (Kuznetsov, 1977) suggests a possible tensional tectonic regime more likely to be favourable for Kuroko-type rather than porphyry copper mineralization. The position of the Vyshkovo mercury deposits of Pliocene age, between the Carpathian range and the Pannonian back-arc massif (p. 238), also suggests a syn-mineralization rift or tensional setting which accompanied subsidence of the Pannonian Basin.

9 Native sulphur

Deposits of native sulphur are mined from both continental margin arcs (e.g. the Andes) and oceanic arcs (e.g. Japan and the New Hebrides). While generally characteristic of volcanoes at a late or solfataric stage of development the distribution of the deposits has little direct relationship to the Benioff zone. A sketch cross-section of a recent sulphur deposit in the northeast Japan magmatic arc is shown in Fig. 107.

10 Preservation potential

Magmatic arcs undergo erosion as a result of syn-subduction uplift and also during continent–arc collision. Porphyry copper deposits are particularly liable

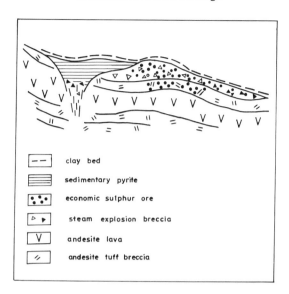

clay bed

sedimentary pyrite

economic sulphur ore

steam explosion breccia

andesite lava

andesite tuff breccia

Fig. 107. Schematic section of a single unit of typical native sulphur-pyrite deposits in Japanese magmatic arc (from Horikoshi, 1976). By permission of the author and publisher, *Spec. Pap. Geol. Ass. Can.* **14**, 121–142.

to erosion soon after formation because, although mineralization may take place up to 4 km beneath the surface, the mineralized plutons are commonly emplaced at a high topographic level in the volcanic host rocks. The large number of deposits exposed in modern Cenozoic arcs compared to those in many ancient arcs of the same surface area indicates that relatively few porphyry deposits long survive continental collision. However, that some do survive is indicated by the late Cretaceous porphyry deposits in the former cordilleran arc north of the Indus Suture in Tibet, and the numerous deposits of Palaeozoic age in the Appalachians. Tin deposits, although of minor economic importance, are perhaps more likely to be preserved because the mineralization usually takes place at a deeper level in the volcanic arc.

In contrast to porphyry deposits, stratiform sulphides have a much higher preservation potential, largely because they are deposited in marine basins within or flanking volcanic arcs and subsequently buried in thick volcanic and sedimentary successions. These basins or basin margins are less likely to be eroded than the mountain ranges within which porphyries are formed.

III OUTER ARC TROUGHS

A Tectonic Setting

Outer arc troughs or basins, the term used by Hamilton (1973) for the Mentawai Trough of the Sunda Arc, are equivalent to the "interdeeps" of van Bemmelen

(1949), to the "fore-arc basins" within the "arc-trench gap" (Dickinson, 1974b, 1976), or to the upper slope basin part of the accretionary prism (Karig and Sharman, 1975). They develop between outer arcs and "inner" or volcanic-plutonic arcs, and their successions are thus equivalent to those of some exogeosynclines, derived from and deposited beside adjacent mountain belts. Where an outer arc is not present, either as an island chain or submarine ridge, the position of the outer arc trough is occupied by a seaward "arc-trench" slope within which graben and fault troughs are filled and locally mantled in magmatic arc-derived sediments. As in outer arcs, heat flow beneath outer arc troughs is low and contrasts with the high heat flow characteristic of the adjacent volcanic arc.

Most modern outer arc trough successions are submarine, e.g. the Arika and Iquique Basins landward of the Peru–Chile Trench. Only two examples are known in which an on-land outer arc trough succession underlain by an active seismic zone is well developed, that of southwestern Alaska (Moore and Connelly, 1979) and the Western Trough of the Burma Arc; a well-developed largely submarine outer arc succession is present beneath the Mentawai Trough of the Sunda Arc (Hamilton, 1973). A generalized description of outer arcs is thus based partly on ancient successions of which examples include the Jurassic-Cretaceous Great Valley sequence of California (see Fig. 113) and the Bowser Basin and Tyaughton-Methow Trough to the north (Dickinson, 1976), part of the Mesozoic Hokonui Assemblage of New Zealand (Dickinson, 1971; Landis and Bishop, 1972), the Meso-Hellenic furrow or intra-deep of Aubouin (1965) in Greece, with up to 3000 m of Oligocene conglomerates (see Fig. 5B), and the Upper Ordovician-Silurian succession of the Midland Valley south of the Sidlaw Anticline in Scotland (see Fig. 87C).

Outer arc trough successions are characteristically thick, with up to 12 km of sediment, and mostly form either a syncline or landward-dipping homocline occupying an elongate commonly arcuate trough. In some cases the succession near the outer arc is repeated by thrusts and tectonically complex. The sediments are commonly younger than the oldest volcanic arc rocks, as described later, and may therefore be underlain by arc-derived sediments deposited on the arc-trench slope before development of the trough. The basement beneath the axial zone of the trough is rarely exposed and varies according to the previous geological history and position of the volcanic arc relative to the continental margin; it may be continental crust tectonically thinned during previous rifting, imbricate outer arc flysch overthrust by oceanic crust, or less probably in a few cases normal oceanic crust.

The contact between the outer arc trough succession and the accretionary prism of the outer arc is basically stratigraphic (see Fig. 83), with an angular unconformity caused by onlap of the trough succession onto the tectonically emplaced outer arc rocks (Karig and Sharman, 1975). However, in most arcs the boundary is complicated by thrusting during and following sedimentation, with the trough succession thrust over the outer arc, and upthrust wedges of metamorphic and ophiolitic rocks which obscure the original stratigraphic relationships; thrusting also commonly effects the succession within the outer arc trough, as in the Western Trough of Burma. Trough sediments commonly intertongue with and onlap across volcanic rocks of the magmatic arc, although a major thrust may be present near the boundary, dipping either towards the magmatic arc as in Alaska, or less commonly away from it as suggested for some other arc systems (Karig and Sharman, 1975).

There are three main possible sources of the outer arc trough sediments: elevated flysch, melange and ophiolites of the outer arc, the magmatic arc rocks and a longitudinal source which may be dominant as a result of diachronous continental collision along the length of the arc discussed later. The magmatic arc is probably the most important source, e.g. in the Jurassic to Cretaceous sequence of southwestern Alaska where arkosic and felspathic sediments pre-dominate (Moore and Connelly, 1979), in the Cenozoic succession of the Western Trough of Burma (Mitchell *et al.*, 1978), and locally in the trough west of Sumatra and Java (Moore *et al.*, 1980). Sedimentary faces vary widely both along and within different troughs from turbidite to deltaic to fluviatile, but there is a tendency for non-marine facies to increase upwards. Pyroclastic rocks are commonly a minor constituent and lavas are absent.

In general outer arc trough successions can be recognized by the presence of thick largely epiclastic volcanogenic sediments, increasing proportion of non-marine sediments upwards, and monoclinal to broadly synclinal folding. However, the basin geometry and marine to non-marine sequence are also common to foreland and back-arc compressive basin successions, and perhaps most diagnostic of an outer arc trough sequence is the presence of parallel bordering magmatic arc rocks and outer arc rocks with ophiolites.

B Mineral Deposits

Although bordering the commonly richly mineralized magmatic arc, outer arc troughs contain remarkably few significant mineral deposits (see Table XI). Despite the evidence for migration of volcanic arcs with changes in inclination of the underlying Benioff zone, outer arc troughs such as that in Alaska, the Great Valley sequence, and western Burma lack interbedded calc-alkaline volcanic rocks, and it appears that in each case the trough has remained free of arc magmatism. The mineral potential of the predominantly compositionally immature terrigeneous clastic deposits of the trough is thus very limited.

With the exception of small gold placers and coal deposits, described below, no economic deposits are known from modern outer arc troughs. It might be expected that non-marine trough successions could be favourable for sandstone-type uranium deposits, particularly where the source region includes granitic rocks of a deeply eroded volcanic arc. However, the average uranium content of most magmatic arc plutons is probably too low to provide a source for strata-bound uranium even with prolonged leaching by groundwaters.

1 Gold placers

That the outer arc trough setting is favourable for placer gold deposits is
suggested by the proximity of the adjacent rising magmatic arc in which a low-
grade epigenetic gold mineralization source may be present. However, as in the
case of foreland basin successions described below, sedimentation in outer arc
troughs is perhaps usually too rapid to allow sufficient reworking for the detrital
gold to be concentrated:

(a) Alluvial gold in the Western Trough, Burma

Gold placers in Quaternary river deposits in the northern part of the Western
Trough of Burma (see Fig. 85) have long been worked on a small scale. The
deposits lie east of the trough's synclinal axis, and are underlain by an Oligocene
to Pliocene predominantly non-marine clastic succession, which rests unconformably on an Upper Eocene coal-bearing sequence; beneath the coals is
a thick turbidite succession overlying Maastrictian conglomerates. The base of
the succession consists of Albian limestones lying unconformably on outer arc
rocks in the west and overlain unconformably by the Maastrictian sediments.

   The gold is evidently derived from elevated polymict conglomerates of
probable Oligocene age within the trough succession to the west; these in turn
were derived partly from Upper Mesozoic plutons and lavas of the volcanic arc

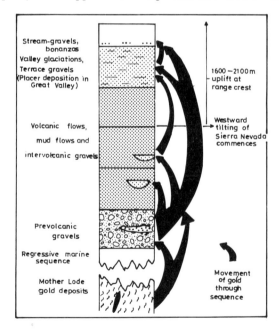

Fig. 108. Stratigraphy and tectonic events in evolution of auriferous gravels of the Great
Valley, California (from Henley and Adams, 1979). By permission of the authors and
publisher, *Trans. Instn Min. Metall.* **88**, B41–50.

to the east of the trough. However, the gold may have gone through at least one erosion-deposition cycle before it accumulated in non-economic concentrations in the Oligocene conglomerates.

(b) Alluvial gold in the Great Valley, California

The only other alluvial deposit of economic significance in an outer arc trough succession is that of the Great Valley in California (see Fig. 116), where placer deposits occur in alluvium and, to the east, in terrace gravels. Unlike the deposits in Burma, the Quaternary gold placers of the Great Valley were derived from older placers to the east (Fig. 108) protected from erosion by lava flows (Henley and Adams, 1979) and were not recycled through the Upper Mesozoic predominantly turbidite succession of the Great Valley outer arc trough sequence itself.

2 Coal

Economic deposits of coal are known from the Upper Eocene rocks of the Western Trough of Burma, where sub-bituminous coal seams up to 2m thick are worked on a small scale. The coal-bearing sequence occurs in the western limb of the synclinal trough succession, and overlies a thick succession of turbidites. The coals probably formed in a delta which terminated a cycle of subsidence and deposition begun in the Maastrictian.

As the geothermal gradient in outer arc troughs, particularly on the seaward side, is generally very low, high-rank coals might be expected only where burial under a very thick succession has occurred.

3 Preservation Potential

The absence of pre-Cenozoic mineral deposits recognized as having formed in outer arc troughs no doubt partly reflects the very poor preservation potential of these successions as a result of destruction during continental collision which normally terminates subduction. In Cenozoic continent–continent collision belts, e.g. the Himalayas, and in continent–arc collisions such as eastern Taiwan or the northern Appennines, the outer arc trough succession has largely vanished, probably due both to erosion, and to development of thrusts adjacent to the magmatic arc and tectonic overriding of the trough by the igneous or foreland rocks (see Fig. 134). Clearly placer gold deposits,which both in California and Burma are derived from older fossil placers near the top of the trough successions, are unlikely to be preserved.

Evidence of syn-subduction removal of outer arc trough sequences is provided by results of recent work in the region of the Middle America Trench, where a Mesozoic magmatic arc lies immediately landward of Miocene and younger imbricate outer arc rocks north of the trench. Unless major strike-slip faulting has occurred (I.A), the associated Mesozoic outer arc and outer arc trough have presumably been either overthrust by the Mesozoic magmatic arc or tectonically eroded (see Fig. 86), either immediately before or during the late

Cenozoic subduction which formed the still-active volcanic arc in Mexico further to the north (Scholl *et al.*, 1980).

IV BACK-ARC MAGMATIC BELTS AND THRUST BELTS

A Tectonic Setting

Landward of some continental margin magmatic arcs there is a broadly arcuate belt of elevated rocks with intrusions of predominantly granitic plutons bordered by a zone of thrusting and folding (see Fig. 81A). The magmatic belts have often been considered as part of the magmatic arc, from which they may be separated by an intermontane trough, although Dickinson (1976) referred to them as specific tectonic units related to his adjacent foreland fold-thrust belts, which are here referred to as back-arc thrust belts.

No volcanically active examples of back-arc magmatic belts are recognized, but geologically young inactive belts include the Tertiary volcanic arc of Bolivia, the Western Belt of Cretaceous to Eocene granitic plutons in Burma and Thailand, and Cretaceous and early Tertiary granites west of the Cordilleran thrust belt in western North America; some of the Upper Mesozoic plutons in the western part of the granitic belt in southeastern China may also be products of back-arc rather than magmatic arc magmatism. In general the plutons in these belts are silicic and per-aluminous and show high initial Sr^{87}/Sr^{86} ratios, indicating derivation largely from continental crust. Silicic volcanic rocks may be preserved above the plutons, as in Bolivia, but andesitic and basaltic eruptives are scarce or absent. The back-arc magmatic belt may lie as much as several hundred kilometres landward from the magmatic arc of the same age.

The host rocks to the granitic plutons are commonly elevated rocks appreciably older than the plutons, e.g. in Bolivia, in some cases showing evidence of regional metamorphism associated with the magmatism, as in the Shan Scarps of Burma.

Back-arc thrust belts lie on the landward side of the magmatic belt, and are characterized by oceanward-dipping thrusts forming, for example, the sub-Andean zone of the South American Cordillera (Fig. 109) and the foreland fold-thrust belt of the Western Cordillera in North America (Dickinson, 1976), and occupying much of the Shan Scarps zone in Burma. The thrusts correspond to some of the Type A "continental" subduction zones of Bally and Snelson (1979). Several imbricate thrusts are usually present, sometimes distributed *en echelon,* the thrust sheets and nappes involving plutons and metamorphic rocks of the back-arc magmatic belt on one side and clastic mostly continental sediments of the back-arc compressive basin, described later, on the continental side. The thrusts and related folds develop during, but are opposite in polarity to, subduction of ocean floor, and occupy a zone ranging from perhaps 20 km across as in Burma to wide zones of imbricate thrust involving substantial areas of back-arc basin rocks as in western Canada. There are few estimates of the amount of horizontal shortening in these zones but loss of up to 200 km of crust has been reported from the Canadian Rockies, and the extent of crustal shortening may approach that found in some collisional foreland thrust belts. Where a back-arc thrust belt forms a major topographic and structural feature, the orogen displays at least a superficial symmetry with thrusts of the outer arc and back-arc apparently converging at depth in the region of the volcanic arc.

The back-arc thrust belts are of significance to mineralization because it seems probable that formation of mineralized granitic plutons in back-arc magmatic belts may result from crustal thickening accompanying deformation, with generation of anatectic granites (Dickinson, 1976) by a process analogeous to that in the foreland thrust zones of collision belts described later (Ch. 6, IV. A). It is sometimes argued that crustal thickening along thrusts both synthetic to the subduction direction in the outer arc, and antithetic to subduction in the back-arc, results in a broad zone of metamorphism and granite emplacement, and that the "metamorphic core complexes" of the North American cordillera are related to this process. However, for the present we prefer to treat the back-arc magmatic belt as a discrete zone adjacent to the thrust belt, and regard the intermontane region between the magmatic arc and back-arc magmatic belt, as in the case of the Altiplano of Bolivia, as a zone of relatively minor tectonic activity.

The curvature of the thrust belts in Burma and Bolivia, both of which are convex to the continent, has led to the suggestion that the associated tin-bearing magmatic belts may develop preferentially where subduction is beneath a continental margin concave to the ocean (Mitchell, 1979a). More generally, it is apparent that the belts are largely restricted to areas behind magmatic arcs situated on west-facing continental margins and hence generally under horizontal compression, suggesting a genetic relationship to drift of lithosphere over the asthenosphere, as considered more fully in the discussion of back-arc basins below (VII.A).

Rare occurrences of alkaline granites and undersaturated rocks in or adjacent to back-arc magmatic belts are also known, and are described together with their associated mineral deposits in the following section.

B Mineral Deposits

Economic deposits of minerals generated in this setting appear to be restricted to the back-arc magmatic belt, the thrust zone being an unlikely setting for formation of any type of mineral deposit. However, pre-existing mineral deposits are commonly elevated along thrusts and exposed with their host rocks, as for example in the case of the lead–zinc deposits in Palaezoic limestones within the Cretaceous back-arc thrust belt of western North America.

Mineral deposits in the back-arc magmatic belt, like the tectonic setting itself, have been mostly described in terms of volcanic arc magmatism, and it is only recently that identification of deposits in this setting as distinct in origin from those of the magmatic arc has been emphasized (Mitchell, 1979a; Mitchell and

TABLE XII

Mineral deposits characteristic of back-arc settings.

Tectonic setting	Association	Genesis	Type of deposit/metals	Examples
Back-arc thrust belts	Peraluminous "S-type" granitic plutons and volcanics	Magmatic–meteoric hydrothermal	Sn, W	Bolivia, (Miocene); Western tin belt, SE Asia (Cretaceous–Eocene); Tungstonia granite, USA (Cretaceous)
	Adamellitic plutons	Magmatic–meteoric hydrothermal	Mo, W, Sn	Idaho Batholith, USA (L Tertiary)
	Quartz porphyry	Magmatic? hydrothermal	Cu, Au, Ag	Butte, Montana (U Cretaceous–Palaeocene)
	Fluviatile carbonaceous sandstones	Epigenetic meteoric hydrothermal	U	Wyoming (Tertiary); Khorat Plateau, Thailand (U Mesozoic); Karroo, S Africa (Permian)
	Fluviatile sediments	Sedimentary	Placer Au	Magdalena River, Columbia (Quarternary); British Columbia, (Quaternary)

TABLE XII (Cont'd)

Tectonic setting	Association	Genesis	Type of deposit/metals	Examples
Back-arc compressive cratonic basins	Mostly clastic sediments	Chemical sedimentary	Potash	Khorat Plateau, Thailand (Cretaceous)
	Deltaic sediments	Sedimentary	Bituminous coal	British Columbia, Alberta (U Jurassic–L Cretaceous)
Back-arc extensional cratonic basins	Volcanic rocks	Meteoric (magmatic) hydrothermal	Epithermal Au Ag veins	Basin and Range Province USA (Mid-Miocene)
	Fluviatile and shallow marine sediments	Placer	Sn	Andaman Sea
Back-arc marginal basins and inter-arc troughs	Harzburgite and pillow basalt	Magmatic	Podiform Cr	Within some ophiolites
	Pillow basalt	Hydrothermal exhalative sedimentary	Cu, Fe, Zn sulphides	Within some ophiolites

Beckinsale, 1981). Because few back-arc magmatic belts have been recognized, examples of mineralization from within them are scarce. Nevertheless it appears that the setting is a most important environment for the formation of Sn, W, Mo and related minerals, and that rarely porphyry copper deposits may also be generated (Table XII). The composition of the granite rocks (Beckinsale and Mitchell, 1981) suggests that uranium concentrations might be expected.

The intermontane troughs between a magmatic arc and back-arc magmatic belt are not generally characterized by specific types of mineralization, although stratabound "red bed" copper deposits are known from Tertiary sediments of the Altiplano of Bolivia (Petersen, 1970).

1 Tin and tungsten mineralization in silicic magmatic rocks

(a) Upper Tertiary tin deposits of Bolivia

Bolivia produces, mostly from primary deposits, about 15% of the non-communist world's annual tin production of around 200 000 tonnes and the southern part of the tin belt provides what some geologists consider to be the most impressive evidence for the relationship of tin-bearing magmas to subducting ocean floor. The Bolivian deposits occur in an arc convex to the continent, mostly less than 50 km wide and about 800 km in length, corresponding broadly to the Eastern Cordillera. Mineralized intrusive rocks in the north of the arc are Lower Mesozoic in age, but plutonic and volcanic rocks with associated tin mineralization are of Lower Miocene age in the centre and Upper Miocene in the southern part of the arc (Evernden et al., 1977; Grant et al., 1979). The Tertiary tin and tungsten deposits are unusual both in their common association with volcanic rocks, mostly hydrothermally altered rhyolitic to dacitic quartz porphyries, and the association of the youngest deposits with silver and base-metals; bismuth and antimony are abundant in some of these deposits (Grant et al., 1977; Turneaure, 1971). The host rocks to the plutons are Palaeozoic terrigenous sediments and Tertiary volcanics.

The tin deposits lie approximately 700 km east of the present submarine trench, and are separated from the plutonic and volcanic arc rocks of the Western Cordillera by the Altiplano basin, underplain by Upper Cretaceous and Tertiary sedimentary "molasse" and volcanic rocks. East of, and parallel to, the mineralized belt lie the Sub-Andean Ranges, a fold-thrust zone with eastward-directed overthrusts in which the youngest rocks offset are of Oligocene age (Fig. 109). The abundance of tin in the Bolivian belt contrasts with the significant but much less important scattered tin occurrences lying in a similar position with respect to the magmatic arc to the west in the northern and southern Andes.

The mineralized igneous rocks were clearly emplaced during eastward subduction of ocean floor beneath the South American continent, which has continued since at least the late Mesozoic, and the presence of tin in this setting has been related by Sillitoe (1972b, 1976b) to depth of the Benioff zone (Fig. 110). However, it seems probable that, as suggested by Dickinson (1976), magma in back-

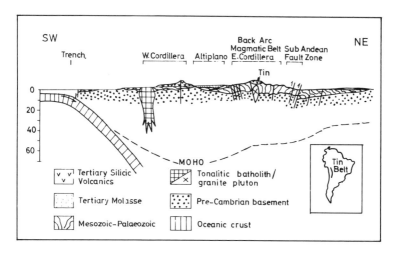

Fig. 109. Schematic cross-section through the Andes near northern end of Bolivian tin belt (after Cobbing and Pitcher, 1972); projected position of Tertiary tin deposits of southern Bolivia also shown (after Mitchell, 1979a, and Mitchell and Beckinsale, 1981). By permission of the authors and publishers, *Bull. geol. Soc. Malaysia* **11**, 81–102, and *Nature Phys. Sci.* **240**, 51–53. Copyright © 1972, Macmillan Journals.

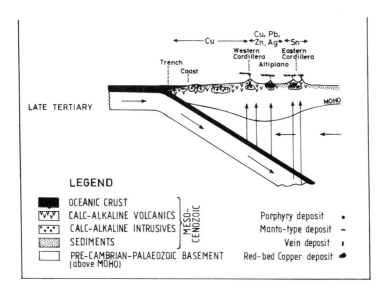

Fig. 110. Schematic cross-section through Andes in late Tertiary. Note postulated direct relationship of metal provinces to depth of Benioff zone (from Sillitoe, 1976b). By permission of the author and publisher, *Spec. Pap. geol. Ass. Can.* 59–100.

arc magmatic belts is largely anatectic and related to tectonic stacking of upper
crustal continental rocks in the adjacent thrust belt. The tin, and flourine
required to scavenge and concentrate it (Barsukov and Kuril'chikova, 1966), is
thus derived from partial melting of continental crust rather than from the subducting ocean floor. Possible Tertiary recycling of older tin concentrations,
supporting the "inheritance" concept of Routheir *et al.* (1973), is suggested by

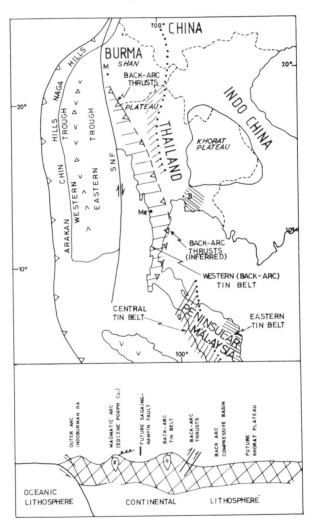

Fig. 111. Sketch map (A), and schematic cross-section (B), Western Tin belt of Southeast
Asia in Eocene; back-arc compressive basin with position of Khorat Plateau, and Central
and Eastern Tin Belts, also shown. Western Burma restored to position prior to late
Cenozoic opening of Andaman Sea. Late Mesozoic–Cenozoic magmatic arc rocks: triangle,
Quaternary; V Cretaceous–Tertiary, Λ inferred. B, Bangkok, M, Mandalay, Me, Mergui,
SNF, Sagaing–Namyin Fault, heavy dots, Upper Triassic suture.

the presence of the Triassic granites in the north of the tin-bearing belt, and of adjacent Lower Palaeozoic fossil placer deposits of cassiterite. The longitudinal variation of metals in the belt, from tungsten associated with the Triassic plutons in the north, through tin to base metals with tin in the south, is attributed by Grant *et al.* (1979) to the regional erosion level.

(b) Western Tin Belt of Southeast Asia

A belt of tin and tungsten-bearing granites of Lower Cretaceous to probable Lower Eocene age extends from southeast of Mandalay in Burma southwards to Phuket in southern Thailand, and possibly continues southwards towards Sumatra. The belt, which is slightly convex to the east and probably includes at its eastern boundary the lower Cretaceous (Beckinsale *et al.,* 1979) Mae Lama granite in Thailand and the Mawchi tungsten mine in Burma, is about 1400 km in length and mostly less than 50 km wide (Fig. 111).

The mineralization in Burma mostly consists of cassiterite or wolfram in quartz veins and greisen zones around the margins of biotite adamellites and two-mica granites; the proportion of tungsten increases towards the north. The on-land southern end of the belt in peninsular Thailand has been described by Garson *et al.* (1975). Cassiterite is restricted to biotite and two-mica granites, pegmatites, and granitic stockworks in hornfels, as well as quartz veins associated with granite margins. There is some evidence for the presence of significant tonnages of low-grade ore with cassiterite in thin quartz stringers and disseminated in adjacent granite. Cassiterite-bearing mica-tournmaline pegmatites in some places are highly kaolinized. Tungsten also occurs in greisen, pegmatites, and locally as a primary mineral in granites.

All plutons in the belt intrude either Carboniferous marine clastic sediments or metamorphosed Palaeozoic rocks, and associated volcanic rocks are absent. In the Southern Shan States of Burma a belt of eastward-directed thrusts, some of post-Cretaceous age, lies immediately east of the granites and extends southwards towards the Thai-Burma border. Evidence for more than 400 km of dextral movement on the Sagaing Fault (Mitchell, 1977) since the Middle Miocene (Curray *et al.,* 1980) indicates that in the Cretaceous and early Tertiary the tin belt was bordered to the west by the western Burma granodioritic magmatic arc of similar age. The tectonic setting during mineralization, although partially obscured by subsequent tectonic events, was thus similar in many ways to that of the back-arc magmatic belt of Bolivia.

It has been suggested (Mitchell and Garson, 1972) that the tin was concentrated by fluorine rising from a subduction zone migrating westward during opening of the Andaman Sea marginal basin. However, the more recent evidence for the greater age of the mineralized granites relative to that of the Andaman Sea, and the discovery of a thrust belt within and east of the tin belt in Burma, suggests that generation of the mineralized granites was genetically related to the thrusts within continental crust (Mitchell, 1979a) and hence related only indirectly to the subduction of ocean floor.

(c) Tungsten and tin mineralization in Nevada and Idaho

Several types of mineral deposit are associated with intrusive rocks lying in or west of the Cordilleran back-arc thrust belt in the western USA. Very recently, Miller and Bradfish (1980) described the distribution of late Mesozoic to mid-Tertiary peraluminous muscovite granites or adamellites within this belt and noted their location at the western margin of Precambrian crust (Fig. 112). It may also be noted that the eastern margin of the Cordilleran thrust belt is a potential major source of petroleum, in particular gas.

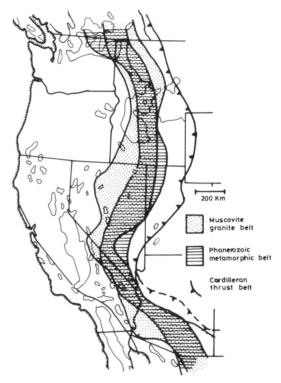

Fig. 112. Inner Cordilleran muscovite-bearing plutonic belt in relation to Cordilleran thrust belt and belt of Mesozoic and Cenozoic regional metamorphism (after Miller and Bradfish, 1980). By permission of the authors and publisher, *Geology* 8, 412–416.

(i) Tungstonia Granite, Nevada. The Tungstonia Granite of Nevada (Best *et al.*, 1974), with which several abandoned tungsten workings are associated, is the largest of a group of plutons lying in a belt of elevated metamorphic rocks within the mid-Cretaceous Sevier Orogenic Belt. The latest Cretaceous age of the granite indicates that it was emplaced during the maximum development of the back-arc cratonic basin east of the Orogenic Belt (Dickinson, 1976) and hence was possibly related to thrusting in the back-arc thrust belt developed on the former Sevier orogen between the plutons and the back-arc basin.

The Tungstonia Granite is unusual in consisting largely of a two-mica granite with large muscovite phenocrysts, surrounded by a leucocratic and commonly aplitic border facies. Initial Sr^{87}/Sr^{86} ratios (maximum 0.725) suggest an origin by anatectic melting of Precambrian rocks at depth, an origin supported by evidence of Mesozoic high-grade metamorphism of Precambrian rocks elsewhere in the metamorphic core of the Cordillera. Best *et al.* (1974) suggested that the muscovite is primary, although it is presumably not impossible that emplacement of later granites in the same pluton could have resulted in hydrothermal circulation and associated muscovitization.

(*ii*) *Molybdenum, tungsten and tin mineralization, Idaho Batholith.* Mineralization in granitic rocks of Palaeocene and Eocene age within the predominantly Upper Cretaceous Idaho Batholith has recently been described by Bennett (1980). The Tertiary intrusions are mostly adamellites and granites with U, Th and K^{40} contents significantly higher than in the main Cretaceous batholith. Mineralization associated with the granites includes epigenetic deposits of Au, Mo, W, Ag, Pb and Zn, and many plutons contain disseminated Mo, W and Sn mineralization. Bennett (1980) suggested that vertical jointing and possibly thrust faults favoured the development of massive hydrothermal convective cells which resulted in mineralization. On a broader scale the generation of the Tertiary granites and their metal content were probably related to movement on the early Tertiary back-arc thrusts lying east of the batholith, although Dickinson (1976) shows the Idaho Batholith as part of a magmatic arc in the early Tertiary.

(d) Tungsten–antimony of southeastern China

The extensive belt of Mesozoic granitic rocks in southeastern China, which includes the world's most important tungsten province, has been interpreted as a continental margin magmatic arc within which granitic belts of Mesozoic and possibly Palaeozoic age were emplaced above northwestward-subducting ocean floor (e.g. Bor-Ming Jahn *et al.*, 1976). However, the 700 km width of the belt and presence to the west of eastward-dipping thrusts suggest that plutons in the western part of the belt, with which most of the tungsten and antimony are associated, may have been related to back-arc thrusting and folding. If so, the belt is unusual in that it developed inland of an easterly, rather than westerly-facing continental margin. Possibly concentration of the metals in the predominantly anatectic granites (Mei-Zhong Yan *et al.*, 1980) resulted from mobilization of mineralization associated with less silicic granites.

2 Vein-type deposits of Butte, Montana

The Butte Cu–Au–Ag deposit of Montana, one of the largest vein-type deposits in the world, lies on the western margin of the Upper Cretaceous Boulder Batholith to which it is genetically related (e.g. Dixon, 1979). The batholith, predominantly a Cretaceous hornblende–biotite quartz monzonite, is cut by latest Cretaceous quartz-porphyry dykes with which some of the mineralization is associated, and overlain by co-magmatic rhyolites and dacites. The ore body shows some similarities particularly in hydrothermal alteration to porphyry copper deposits, typical of calc-alkaline magmatic belts. Nevertheless it lies within a thrust zone of the late Cretaceous back-arc thrust belt of the Cordillera, and the associated dykes and rhyolites, if not the quartz monzonite, could be interpreted as back-arc magmatic belt rocks, although the two-mica granites of Palaeocene age to the southwest of the Boulder batholith have a composition more typical of plutons emplaced in this type of setting.

3 Mineralization associated with carbonatites and alkaline plutons

On the landward side of a few continental margin arc systems there are occurrences of alkaline granite and undersaturated rocks including nepheline syenite and carbonatite, which were evidently emplaced during ocean floor subduction. Examples of alkaline rocks and carbonatites in probable back-arc magmatic belt settings include the Cerro Manomo carbonatite in Bolivia, and possibly the Great Glen carbonatite and Loch Borolan alkaline complex in Scotland.

In Bolivia the Cerro Manomo carbonatite and at least two nepheline syenite plutons lie immediately east of the Cordillera Real belt of tin-bearing granites between La Paz and Oruro. A belt of alkaline granites occurs in a similar tectonic setting in Peru. Little is known about mineralization at the Bolivian carbonatite apart from the presence of long-fibre asbestiform riebeckite of economic interest.

The Great Glen carbonatite of Lower Devonian age in northern Scotland is situated on a major transcurrent fault, possibly an extension of a transform fault, towards the continental side of a magmatic arc associated with a northerly-dipping subduction zone which lay beneath the Southern Uplands. Fenites associated with the carbonatite carry asbestiform riebeckite, locally abundant but with fibre too short to be of economic interest. Similar material occurs at the Loch Ailsh alkaline complex in northern Scotland. At the nearby Loch Borrolan alkaline complex (Notholt, 1979) there are basal pyroxenite layers rich in apatite (up to 17%) and titaniferous magnetite (5–90%).

Van Breemen *et al.* (1979) stated that the Loch Borrolan and Loch Ailsh alkaline complexes were emplaced during crustal arching in response to orogenic compression and that a belt of syn-magmatic thrusts directed towards the foreland migrated across the alkaline centres. They considered that the change from alkaline to more widespread (late Caledonian) granite magmatism evidently coincided with the end of thrust movements and beginning of post-orogenic uplift. The zone of thrusting described by van Breemen *et al.*, could be interpreted as a back-arc thrust belt with associated alkaline magmatism, subsequently engulfed in late Caledonian magmatic arc granites.

4 Preservation potential

The preservation potential of back-arc magmatic belts during continental collision is probably considerably greater than that of magmatic arcs, as the back-arc belt lies further from the collision zone and so is less likely to be elevated and eroded during the collision-related orogeny. Nevertheless, it is possible that thrusting in the back-arc thrust belt may continue or be re-activated during continent–continent collision. An example is provided by the Cenozoic movement on southward-dipping thrusts in Tibet, far to the north of and contemporaneous with the India–Asia collision. Collision-related convergence in the thrust belt, if not accompanied by pluton emplacement, could result in elevation and erosion of the upper mineralized levels of the magmatic belt. Other reasons for the scarcity of recognizable back-arc magmatic belts are their restricted distribution compared to that of magmatic arcs, and the fact that in interpretations of ancient arc systems the back-arc magmatic belt has probably been included within the magmatic arc.

V BACK-ARC COMPRESSIVE CRATONIC BASINS

A Tectonic Setting

Back-arc compressive basins (see Fig. 81A) lie landward of the back-arc magmatic and thrust belt described above, on continental crust depressed in front of the thrust. They have also been referred to as retro-arc basins (Dickinson, 1974b) and back-arc foreland basins (Dickinson, 1976) but the latter name is not used here because the term "foreland" can more appropriately be applied to the subducting continental plate in collision belts, which corresponds in many orogens to the foreland of geosynclinal terminology. The back-arc basins themselves are equivalent to some of the "exogeosynclines" of ancient orogens.

Modern examples of back-arc compressive basins include largely marine basins such as that of the Malacca Straits east of Sumatra, although the back-arc thrust belt is here poorly developed, and continental basins of which that east of the Andean cordillera in Bolivia forms an example.

Dickinson (1976) has discussed the evolution of back-arc compressive basins with particular reference to the ancient Rocky Mountain basin succession of Upper Jurassic to Eocene age in the eastern part of the Western Cordillera of North America (Fig. 113). Sedimentation in part of this basin, with the upward stratigraphic change from marine to continental "molasse" facies, has been described by Eisbacher (1974). Another ancient example of a back-arc compressive basin is provided by the continental red beds of Cretaceous age and probably the underlying turbidites and Jurassic non-marine sandstones within and east of the back-arc thrust belt in the Shan States of Burma. The influence of subduction-related tectonics far into the continent is illustrated by the

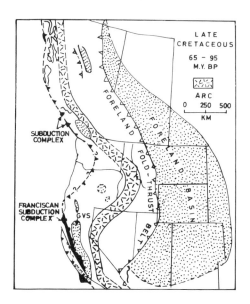

Fig. 113. Back-arc compressive basin ("foreland basin") and back-arc thrust belt ("fold-thrust") in western Cordillera of North America. GVS - Great Valley Sequence (after Dickinson, 1976). By permission of the author and publisher, *Can. J. Earth Sci.* **13**, 1268–1287.

position of the North American basins up to 1200 km east of the continental margin.

In most basins the predominant sources of sediment are the uplifted sedimentary, metamorphic and often granitic rocks of the back-arc thrust belt, although in some arc systems the volcanic arc may contribute a significant amount of detritus. In the Malacca Straits, back-arc basin sediment is derived from both the active Sunda volcanic arc in Sumatra, and the much older uplifted predominantly granitic Main Range of Malaysia on the continental side of the basin. Fluviatile sediments probably predominate in most back-arc basins (Dickinson, 1976), although shallow marine sediments clearly can occur, and in the north-central part of the Canadian Cordillera accumulation of a Lower Jurassic flysch sequence evidently took place in a cratonic basin east of the magmatic arc (Eisbacher, 1977).

On their oceaniç or magmatic arc side, back-arc compressive basin successions commonly undergo thrusting, antithetic to subduction, during and following sedimentation, with the axis of maximum sedimentation migrating towards the continent. They thus resemble in this respect as well as in the composition of the sediments and the sedimentary facies, the foreland basins of collision belts such as the Alpine molasse, described later (Ch. 6, V. A). The basement under the basins is usually relatively old continental crust, although in some cases it can include rocks tectonically accreted to the continent during collisional orogenies immediately preceding development of the arc system. As in the case of back-arc thrust belts, there is some evidence that back-arc compressive basins mostly develop on the western side of continents, in particular where the magmatic arc is convex to the continent.

B Mineral Deposits

Back-arc compressive basins with their clastic sedimentary fill and absence of volcanic and intrusive rocks contain not only significant coal deposits, but also important placers and sandstone-type uranium mineralization (Table XII).

1 Sandstone-type uranium deposits

(a) Sandstone-type uranium in Wyoming

Large reserves of sandstone-type uranium mineralization are being mined from a number of basins in a back-arc continental setting in Wyoming (Fig. 113), east of the "foreland" fold thrust belt of Dickinson (1976). One of these uranium districts of major importance is the Southern Powder River Basin, where uranium ore occurs in Palaeocene fluviatile sediments (Raines *et al.*, 1978; Dahl and Hagmaier, 1974).

Uranium mineralization is concentrated within poorly consolidated red sandstones most of which form point-bar deposits, within a succession of sandstones, clays, lignites and minor tuffs, with economically exploitable coal seems higher in the sedimentary succession. The richest uranium deposits occur where the point-bar sandstones grade into carbonaceous back-swamp or floodplain environments. These deposits are good examples of roll-type uranium mineralization. They are epigenetic and considered to be related to circulation of ground water, with uranium carried as carbonate complexes until precipitated by reduction at the oxidized–reduced sandstone boundary, on contact with organic and inorganically produced pyrite.

The source of the uranium in the Southern Powder River Basin is considered to be intraformational tuffs (Dahl and Hagmaier, 1974) but it has also been suggested (Stuckless and Nkomo, 1978) that much of the sedimentary uranium in Wyoming basins was leached from uranium-bearing Precambrian granites to the east, elevated in the late Cretaceous.

(b) Sandstone-type uranium, Khorat Plateau, Thailand

Minor uranium mineralization has been described (Shawe *et al.*, 1975) from the Khorat Plateau in eastern Thailand (see Fig. 111A), a region which has subsequently been the target of a number of exploration programmes.

The Plateau consists of a very gently folded succession of mostly non-marine clastic sediments of Triassic to Cretaceous age, the Cretaceous rocks forming much of the Plateau surface and the Triassic lying unconformably on Palaeozoic rocks. Uranium is confined to the Upper Jurassic sequence of locally red sandstones and siltstones, and rare conglomerates with carbonaceous layers, exposed near the western margin of the Plateau. Uranium minerals and copper occur in small lenticular bodies within sediments associated with silicified wood and mammal bones.

The tectonic setting of the Plateau, east of the Shan Scarps–western Thailand back-arc thrust belt with abundant tin-bearing granites described above (IV. B. 1. (b)), indicates that it occupies the position of a back-arc cratonic basin, and Shawe *et al.* (1975) have compared its tectonic setting to that of the Colorado Plateau. There is some uncertainty as to whether the Khorat Plateau was a back-arc basin during either Upper Jurassic sedimentation or the presumably subsequent mineralization, because development of the back-arc thrust belt to the west may be no older than Cretaceous when deposition of red bed clastic carbonates took place in and immediately east of the thrust zone. However, the Plateau was undoubtedly back-arc with respect to the northern part of the Sunda magmatic arc in the Jurassic, regardless of any other tectonic features which may have controlled sedimentation.

(c) Uranium mineralization in the Karroo Supergroup, Southern Africa

Occurrences of uranium in the Upper Palaeozoic to Lower Mesozoic Karroo Supergroup of South Africa were first described by von Backstrom (1974). Subsequently other occurrences have been reported from the Karroo elsewhere in Southern Africa and have led to systematic exploration in several areas. In the South African occurrences the mineralization occurs along bedding planes and joints with grades mostly below 0.05% U. It is reportedly both sedimentary and diagenetic or epigenetic in origin, usually associated with channel-fill facies and organic remains. In Argentina, Permian continental sandstones similar in age and lithology to part of the Karroo succession contain a number of uranium occurrences of which at least one is economic.

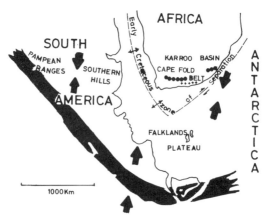

Fig. 114. Southwestern Gondwanaland in early Mesozoic, showing Cape Fold Belt. Dark shading, Andean magmatic arc; crosses, axis of gravity anomaly; solid circles, electromagnetic anomaly; arrows indicate convergence (modified from Lock, 1980). By permission of the author and publisher, *Geology* 8, 35–39.

The gently folded Karroo Supergroup is up to 5000 m thick and predominantly non-marine, although turbidites are present in the lower part. The succession includes fluvial and lacustrine sediments with local laval flows, and is cut by doleritic dykes; economic coal beds occur in places. The nature of the tectonic setting within Gondwanaland in which Karroo sedimentation took place has been uncertain. However, recently Lock (1980) interpreted the Cape Fold Belt, lying immediately south of the Karroo succession, and its extension into South America to the west and Antarctica to the east (Fig. 114), as a back-arc thrust belt, developed in the latest Palaeozoic and early Mesozoic prior to the break-up of Gondwanaland. The associated magmatic arc was situated on the Falklands Plateau to the present south, and was underlain by a Benioff zone dipping gently northward from a subduction zone bordering the southern margin of the continent (Fig. 115). According to this model, the Karroo sediments, and stratigraphic equivalents in Argentina, accumulated in a back-arc cratonic basin lying at least 2000 km north of the subduction zone and more than 1000 km north of the magmatic arc. Evidence for a southerly source for at least part of the Karroo Supergroup indicates that elevation of the thrust belt accompanied sedimentation.

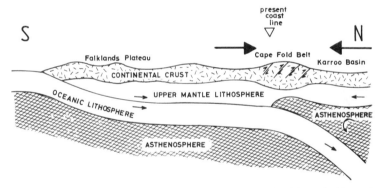

Fig. 115. Schematic cross-section of Southwestern Gondwanaland in early Mesozoic (after Lock, 1980). Note Cape Fold Belt interpreted as early Mesozoic back-arc thrust belt. By permission of the author and publisher, *Geology* **8**, 35–39.

It should be noted however that Karroo sediments with uranium mineralization occur further north in Zambia, Malawi and Tanzania within basins considered to be incipient rifts related to the initial break-up of Gondwanaland. The rifting was perhaps a back-arc event related to the underlying Benioff zone (VI. B. 3).

2 Alluvial gold and tin

Henley and Adams (1979) recently discussed the tectonic setting of some Upper Mesozoic and Cenozoic major gold placer deposits in terms of tectonics related to continental margins. They emphasized the importance of rejuvenation of

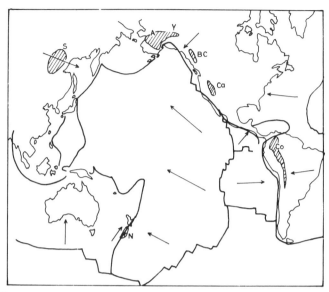

Fig. 116. Giant gold placers, including those here interpreted as back-arc compressive basin (Columbia and British Columbia) and outer arc trough (Great Valley California) deposits (after Henley and Adams, 1979). A, Alaska; Y, Yukon; BC, British Columbia; Ca, California; Co, Columbia etc; N, Otago; S, Siberia. By permission of the authors and publisher, *Trans. Instn Min. Metall.* 88, B41–50.

older orogenic belts with epigenetic gold and multiple episodes of erosional reworking to form large placer accumulations. Among the deposits referred to are those in the Magdalena basin of Columbia, east of the Andes, and in the back-arc basin of British Columbia in Canada (Fig. 116).

Detrital gold in the Magdalena River basin, east of the Cordillera Central with Upper Cenozoic strato-volcanoes, was first deposited in the late Miocene within "molasse" facies up to 6 km thick (Van Houten, 1976a). These sediments were reworked during the Pliocene and Pleistocene uplift of the Cordillera Central, and late Tertiary gravels are now at elevations of more than 2000 m. Erosion and reworking of these gravels has resulted in the rich deposits of the present river system. The Magdalena Valley basin deposits have been affected by movement on eastward-dipping thrusts within the basin, unlike other back-arc compressive basins where the thrusts are antithetic to the ocean floor subduction direction.

Predominantly alluvial and drowned alluvial cassiterite deposits of the western coast of the Malay Peninsula form the largest placer tin ore bodies in the world. Their accumulation on the eastern margin of the compressive back-arc marine basin of the Malacca Straits is largely a result of Tertiary uplift of the Main Range granite source in Malaya combined with favourable Quaternary sea level changes (Batchelor, 1979). Neverthless the Tertiary uplift of the Main Range and subsidence of the Malacca Straits are both related to subduction beneath the magmatic arc of Sumatra to the southwest.

3 Evaporites of Khorat Plateau, Thailand and Laos

The Khorat Plateau succession (see Fig. 111A), with uranium mineralization in Jurassic sandstone described above (V. B. 1. (b)), has recently been found to include evaporites with thick potash beds which form a major addition to the world's reserves (Hite and Japakasetr, 1979). The salt occurs in three layers within the Cretaceous Maha Sarakham Formation, beneath both the Khorat and Sakon Nakhon Basins of the Plateau and underlying an area which may extend over 30 000 km^3. The Lower Salt overlain by a clastic unit probably represents the thickest halite layer in the world, locally exceeding 400 m in thickness, and includes in its upper part the economic potash deposits in the form of carnallite and locally thick sylvite lenses.

Hite and Japakasetr (1979) suggested that anhydrite layers which underlie and overlie the salt layers possibly thicken westward, implying a long marine connection with an ocean to the west. They also compared the mineralogy of the Khorat deposits, with abundant tachyhydrite as a constituent of the carnallite, to those of Cretaceous age bordering the South Atlantic (Ch. 3, I. B. 1). However, it seems probable that the Khorat salt basin and related seaway were related to subsidence east of the back-arc magmatic belt referred to above, rather than to initial emplacement of oceanic crust in a rift zone as in the case of the Angolan deposits.

4 Coal

The Rocky Mountain region of British Columbia and Alberta provides the best example of coal deposits formed in a back-arc compressive basin, and is of interest here because it clearly indicates the relationship of coal rank to tectonic setting. In the Rocky Mountain Coal Belt, a major source of coking coals, deposits are confined to the Upper Jurassic to Lower Cretaceous Kootenay Formation of Eisbacher (1974), deposited to the east of the back-arc thrust belt. The Rocky Mountain coals, highly deformed in the Laramide Orogeny when the thrust belt migrated eastwards, are low and medium-volatile bituminous coals

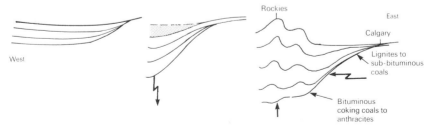

Fig. 117. Relationship of coal rank to depth of burial beneath back-arc thrust belt, Rocky Mountain coal basin, Canada (after Falcon, 1977). Stages of development of basin in sequence from left to right. By permission of the author and publisher, *Minerals Sci. Engng.* **9**, 198–217.

and anthracites; to the east in the less deformed Foothills region are lower-rank high-volatile bituminous coals, while further east in the Plains region are sub-bituminous coals and lignites (Fig. 117). The rank of the coals is presumably largely related to temperature during burial, but their value as coking coals is to some extent determined by their position of formation within the Mesozoic delta (Rushton, 1977).

A further although relatively insignificant example of coals deposited in a back-arc compressive basin is provided by the folded and thrust sub-bituminous deposits in Jurassic sandstones lying along the eastern edge of the back-arc magmatic and thrust belt east of the Sagaing Fault in Burma. The composition of the sandstones indicates a partly granitic source, presumably situated in the back-arc magmatic belt to the west; associated with the deltaic sandstones are recumbently folded turbidites indicating major tectonic deformation in the thrust belt.

5 Preservation potential

The presence in western North America of extensive back-arc compressive basins with thick Upper Mesozoic to Lower Cenozoic sedimentary successions suggests that basins of this type are likely to be preserved until a major continent-continent collision occurs. However, it is possible that with a sudden decrease in dip of the Benioff zone during subduction the magmatic arc will sweep across and be superimposed on the older basin, resulting in uplift and erosion accompanying emplacement of plutons. During collision, in which the basin will normally lie on the overriding plate, it may be elevated and destroyed, but perhaps more probably sedimentation will continue after collision in what has become a hinterland basin (Ch. 6, VI) with respect to the collision belt. With syn-collision landward extension of the back-arc thrust belt the axis of maximum sedimentation in the basin migrates in front of the advancing thrusts. Ultimately with cessation of plate convergence the basins probably undergo isostatic uplift and eventually the sedimentary succession is largely or entirely eroded.

VI BACK-ARC EXTENSIONAL CRATONIC BASINS

A Tectonic Setting

In contrast to compressive basins developed behind back-arc thrust belts, an important class of basin is that formed on the continental side of magmatic arcs in a predominantly tensional regime (see Fig. 81B), where thrust belts are either absent or inactive during sedimentation. These basins, characterized by high heat flow and thin crust, appear to develop in two main plate boundary-related settings. In one setting, typified by the Basin and Range Province of western North America, extensional tectonics followed cessation in subduction and arc volcanism, and development of a transform fault, probably due to attempted subduction of a spreading centre. In the other setting, the extensional basins appear to be incipient back-arc marginal basins, described later and characteristic of eastward-facing arc systems, which failed to develop into major areas of oceanic crust; typical of these is the Pannonian Basin on the concave or western side of the eastward-facing Carpathian arc.

Evidence of extensive late Cenozoic extension in the Basin and Range Province is provided by the development in the Miocene of normal faults and basalt-rhyolite "bi-modal" volcanism. These events are now attributed by some authors to arrival at the subduction zone in the early Miocene of a ridge system trending oblique to the trench, with consequent initiation of the San Andreas transform fault and development of an expanding triangular area beneath which ocean floor subduction had ceased (Dickinson and Snyder, 1979). This area included by Quaternary time the Colorado Plateau and Basin and Range Province (Fig. 118).

Another explanation of extension in the Basin and Range Province considers it to be part of a 700 km long rift system which includes feeder dykes of the Columbia River basalts and the graben of the western Snake River Plain; the extension is considered to have been initiated in a back-arc setting as a result of

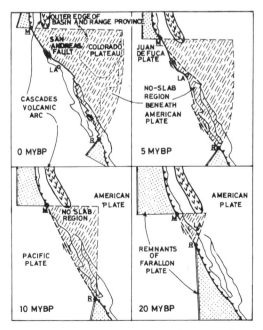

Fig. 118. Sketch map showing evolution of San Andreas coastal transform system and relationship of Basin and Range Province and Colorado Plateau to Miocene termination of subduction; M. Mendocino triple junction; R, Rivera triple junction; LA, Los Angeles (from Dickinson, 1978). By permission of the author and publisher, *J. Phys. Earth* **26**, Suppl. S1-19.

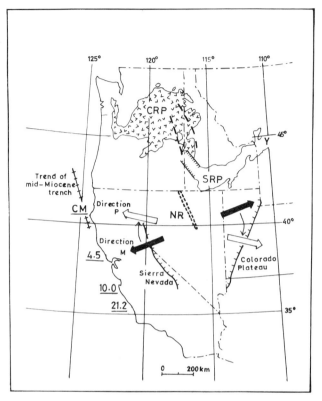

Fig. 119. Map of western USA showing mid-Miocene (M) and present (P) extension direction. CRP, Columbia River flood basalt province; SRP, Snake River Plain; Y, Yellowstone National Park. Northern Nevada rift (NR) is outlined by short dashes. Dates (in millions of years before present) off coast of California mark the locations of Mendocino triple junction; CM, Cape Mendocino, present location of the triple junction (after Zoback and Thompson, 1978). By permission of the authors and publisher, *Geology* 6, 111–116.

increase in dip of the Benioff zone, with a consequent change in stress which allowed the Yellowstone hotspot (Fig. 119) to penetrate the American plate (Zoback and Thompson, 1978). This hypothesis agrees broadly with the ensialic basin concept of Scholz *et al.* (1971) who suggested that emplacement of a back-arc mantle diapir above a steeply dipping Benioff zone resulted in thinning of the continental crust beneath the Province.

A third possibility is that basaltic volcanism, which probably coincided with the start of Basin and Range development, pre-dated the cessation of subduction, and that faulting in the Province resulted from a reduction in the absolute western motion of North America (Cross and Pilger, 1978). Finally, Livaccari (1979) has attempted to explain the Miocene evolution of the Basin and Range province in terms of a broad zone of dextral shearing related to movement on the San Andreas fault.

In the Pannonian Basin (Boccaletti *et al.*, 1973; Boccaletti *et al.*, 1976)

Miocene subsidence accompanied calc-alkaline volcanism in the magmatic arc to the east and was followed by deposition of marine sediments 3 km thick mostly in the Miocene; in the late Pliocene and Quaternary basaltic volcanism took place. The crust of the basin is less than 30 km thick, and normal faults in the pre-Cenozoic basement suggest late Cenozoic extension. High heat flow is attributed to emplacement of a hot mantle diapir beneath the basin axis. Failure of the basin to develop into a major oceanic area was possibly related to continental collision in the Carpathians, although the basaltic volcanism and high geothermal gradient in the basin evidently post-dated collision.

While basaltic volcanism could be expected in back-arc areas in a tensional regime, it is of interest that along the eastern side of the southern Andes there are extensive areas of plateau-forming alkali olivine basalts of Tertiary age lacking evidence of extensional tectonics. Charrier *et al.* (1979) consider that the basalts, of Palaeocene and Lower Miocene to Pliocene age, and lying up to 600 km east of the submarine trench and 300 km east of the magmatic arc, are unrelated to subduction of the Chile Rise. Their back-arc location in an area

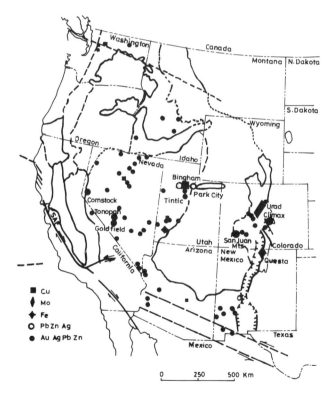

Fig. 120. Post-Laramide ore deposits of western United States; SAF, San Andreas Fault (from Guild, 1978b). By permission of the author and publisher, *J. geol. Soc. Lond.* **135**, 355–376.

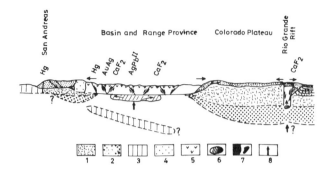

Fig. 121. Schematic cross-section through Basin and Range Province and adjacent areas of western USA in the Miocene, showing proposed relationship of mineral deposits to plate tectonic phenomena. 1, crust; 2, mantle; 3, oceanic lithosphere; 4, intrusive rock; 5, volcanic rock; 6, magma; 7, ore; 8, energy (heat), not to scale (after Guild, 1978b). By permission of the author and publisher, *J. geol. Soc. Lond.* **135**, 355–376.

lacking a back-arc thrust belt suggests that the basalts could be related to incipient crustal extension.

B Mineral Deposits

The mineral potential of back-arc extensional cratonic basins is difficult to assess as there are few clearly recognizable examples; the Basin and Range Province is an important mineral belt, and while there are several epigenetic sulphide deposits bordering the Pannonian basin, the basin itself, lacking the bi-modal basalt–rhyolite volcanism of the Basin and Range, is apparently virtually unmineralized. Fluorite and antimony in northern Thailand were probably emplaced in an analogous setting to the Basin and Range Province.

1 Epithermal gold-silver of the Basin and Range Province, western USA

The Basin and Range Province of Nevada and adjacent areas developed in the Middle Miocene across a former widespread magmatic arc terrain which itself had been emplaced across an older back-arc thrust belt (Dickinson, 1976). Mineralization probably accompanied mid-Miocene normal faulting and basalt–rhyolite bi-modal volcanism which characterized the back-arc basin development (Figs 120 and 121). The area is well known for its numerous deposits of gold and silver (O'Neil and Silberman, 1974) which include the Comstock Lode, the most important epithermal precious metal deposit in the USA. The metals occur in epithermal veins of middle to late Tertiary age, emplaced after the start of Basin and Range Faulting. Mineralized veins occur not only in the Upper Tertiary volcanic rocks, mostly andesite and dacite, but also in older sedimentary and metamorphic rocks. Oxygen and hydrogen isotope analyses on hydrothermal minerals indicate that meteoric water was the dominant mineralizing fluid, although in deep levels of the Comstock Lode there is evidence of a large magmatic component (O'Neil and Silberman, 1974). Mercury and fluorite mineralization also took place in the Province during the Miocene and Pliocene (Guild, 1978b). Circulation of the meteoric water which concentrated the precious metals (Guild, 1978b) was possibly produced by high heat flow above the thinned crust, but the fluorine was presumably derived from a sub-crustal source.

2 Porphyry molybdenum deposits of the western cordillera, North America

Some economic porphyry molybdenum deposits possibly developed in continental margin magmatic arcs and can be considered as molybdenum-rich end-members of porphyry copper-molybdenum ore bodies, others formed in intracontinental rifts (Ch. 2, II.B.1.(e)), and a few possibly formed in back-arc magmatic belts (IV.B.1(c)(ii)). However, there is evidence that in the western Cordillera of North America the major deposits of this type, often associated with tungsten mineralization, were emplaced in a back-arc rift setting.

The porphyry molybdenum or molybdenite deposits of the western Cordillera mainly occur within the limits of the Precambrian craton (Hollister, 1975), the largest molybdenite tonnages occurring in the areas of thickest crust (Woodcock and Hollister, 1978). The deposits are characterized by very fine-grained molybdenite occurring alone or with quartz and pyrite in fractures and open spaces of stockworks and breccias, associated with acid to intermediate rocks, mostly fine-grained and porphyritic. Deposits range from less than 50 000 tonnes to over 250 000 tonnes molybdenum grading from 0.06 to 0.29% Mo. The largest deposits in the south-western USA, including Climax, Henderson and Questa, have more than 250 million tonnes of ore.

The molybdenite porphyry band cuts across geological provinces and extends from the Intermontane Belt of British Columbia through the Rocky Mountain region and Idaho–Montana mineral belt to Colorado, New Mexico and Texas. Compared to many porphyry copper deposits, porphyry molybdenums show little evidence of major fault control and tend to occur in groups, e.g. the Climax–Urad–Henderson deposits, and several others within the age range of late Jurassic to Miocene lie within the NE–trending Colorado Mineral Belt extending at least 1000 km across the Rocky Mountains. There is a striking scarcity of porphyry molybdenite deposits in the Basin and Range Province, an area rich in porphyry copper deposits formed before Basin and Range faulting.

Porphyry molybdenite deposits resemble porphyry copper deposits with regard to their relationships to intrusive rocks and association with large hydrothermal zones. They differ however in their low copper content (generally less than 22 ppm), in their high tungsten/molybdenum and low rhenium/molybdenum ratios and in having sporadic tin and high fluorine (fluorite and topaz) content.

Most porphyry molybdenite deposits are more deep-seated than porphyry coppers and rarely occur within extrusive volcanic phases; stocks, mostly from 0.5 to 1 km diameter, are generally more acid and intrusive rocks include alaskite, granite and quartz monzonite to quartz diorite intruded at different

times (Woodcock and Hollister, 1978). The argillic alteration zone is less con-
spicuous and in general hydrothermal biotite is not as abundant as in some
porphyry copper deposits. Molybdenite occurs largely within the potassic
alteration zone with a pyrite halo extending beyond the propylitic zones. Traces
of chalcopyrite form a halo just outside the molybdenite zone, and tungsten,
generally introduced later than the molybdenite as wolframite and scheelite, also
tends to occur outside the molybdenite zone. Where tungsten coincides with
molybdenite as at the Climax Lower Ore body, it is recovered as a by-product.
The late stage of mineralization in many stocks takes the form of polymetallic
sulphide veins carrying galena, sphalerite, pyrite, chalcopyrite and generally
lead-bismuth sulphosalts accompanied often by fluorite or carbonate.

While the tectonic setting in which the Upper Mesozoic and Lower Cenozoic
porphyry molybdenum deposits formed remains uncertain, the Urad and Climax
mines of Oligocene age (see Fig. 120) and possibly the Miocene Questa deposit,
lie adjacent to the Rio Grande Graben, and Guild (1978b) suggested that they
were emplaced following cessation of subduction beneath the region during the
early stage of a tensional alkalic environment.

3 Coal rank and flood basalts of the Parana, Karroo and Antarctic Basins

Reconstructions of Gondwanaland indicate that, as in the case of the older
Karroo Supergroup and equivalents referred to above (V.B.1.(c)), the flood
basalts and associated dykes of Lower Jurrassic age in the Karroo Basin in
southern Africa and the margin of Antarctica, and of Lower Cretaceous age in

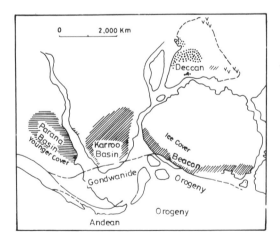

Fig. 122. Gondwanide orogeny along part of Pacific margin of Gondwanaland, with back-
arc basalts (after Cox, 1978). "Gondwanide orogeny" includes back-arc thrust belt of
Fig. 115. Horizontal hatching, early Cretaceous basaltic volcanism; diagonal hatching,
early Jurassic basaltic volcanism (Deccan Traps and other Indian volcanism not considered
here). By permission of the author and publisher, *Nature, Lond.* **274**, 47–49. Copyright
© 1978, Macmillan Journals.

the Parana Basin of South America, occupied a more or less continuous belt, before late Mesozoic fragmentation. Cox (1978) has pointed out that, as postulated by du Toit (1937), these basalt provinces were bordered by an orogenic belt to the south, and argued that the basalts developed in extensional back-arc basins developed above a north-dipping Benioff zone (Fig. 122). He also suggested that late Mesozoic fragmentation of Gondwanaland was possibly related to the back-arc magmatism, and was analogous to the formation of the marginal basins of the western Pacific. Since Lock (1980) has implied that the Upper Palaeozoic and Lower Mesozoic sediments of the Karroo Supergroup were deposited in a back-arc compressive basin (V.B.1.(c)), it follows that in the Permian or Triassic the tectonic setting of the basins changed from a compressional to a tensional environment.

In contrast to the predominantly compressive basins characteristic of areas behind west-facing arcs, and tensional basins typical of east-facing arc systems, the Gondwana basins thus provide some evidence that both types of basins may develop at different times behind arcs facing neither east nor west. This hypothesis is of interest in view of Mackowsky's (1975) suggestion that the variations in rank, from anthracite to semi-anthracite to fat coal, of Permian coal in the Karroo Basin of South Africa, could be due to intense volcanism and dyke injection during the break-up of Gondwanaland.

4 Stratiform sulphide deposits in extensional inter-arc basins, eastern Australia

The volcanic ridges and intervening basins in the Lower Palaeozic rocks of eastern Australia have been interpreted as a system of back-arc extensional cratonic and marginal basins related to westward subduction of ocean floor beneath the continent. Rather than a single major basin similar to the Basin and Range Province of USA, in eastern Australia a complex of basins is present, perhaps resembling the Pannonian and Sylvania Basins separated by the volcanic Apuseni Mts west of the Carpathian volcanic arc.

Mineralization associated with the cratonic basins and reviewed by Scheibner and Markham (1976) includes the Silurian stratiform sulphides at Captains Flat and Woodlawn. Similarities with the Kuroko ores of Japan include the presence of associated rhyolite, the metal ratios and a zone of black ore, although the yellow ore zone is absent.

5 Preservation potential

Back-arc extensional basins which become inactive before they develop into marginal basins may survive a collision and undergo continued sedimentation. This is perhaps partly due to subsidence accompanying cooling of the thinned crust. Since they probably develop above a Benioff zone of steep dip and hence relatively slow rate of subduction, plate convergence following collision is unlikely to be significant and the basins may escape the uplift which might be expected as a result of syn-collision crustal thickening beneath the overriding plate.

VII BACK-ARC MARGINAL BASINS AND INTER-ARC TROUGHS

A Tectonic Setting

Most active island arcs are separated from the continent or from another arc by a deep marine back-arc basin referred to respectively as a marginal basin (Menard, 1967) or inter-arc basin (Karig, 1971). Marginal basins are particularly well developed around the western rim of the Pacific, where they include the Sea of

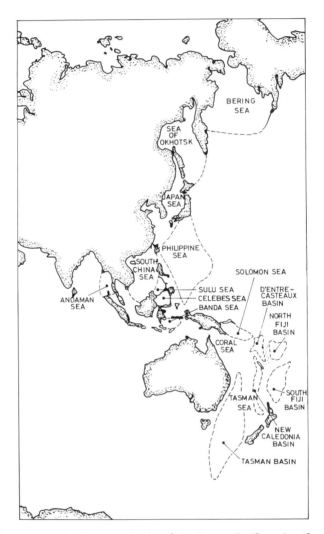

Fig. 123. Some marginal and inter-arc basins of the Western Pacific region (from Packham and Falvey, 1971). By permission of the authors and publisher, *Tectonophysics* 11, 79–109, Elsevier, Amsterdam.

Japan and South China Sea (Fig. 123), in the Caribbean area, and in the Mediterranean, e.g. the Ionian Sea. Inter-arc basins are largely restricted to the western Pacific, where they include the Mariana Trough on the western or concave side of the Mariana arc (see Fig. 102). Heat flow in late Cenozoic marginal inter-arc basins is characteristically high but with scattered low values.

In the late 1960s the Pacific marginal basins were widely interpreted as either oceanic crust with a thick sedimentary layer (Menard, 1967), suggesting that they were segments of Pacific ocean floor trapped behind a developing island arc, or as areas of former continent "basified" during subsidence and in a stage of development transitional to ocean crust, the latter origin remaining in favour with Russian geologists until recently (Sololviev *et al.*, 1977). Subsequently it was shown by deep-sea drilling that the crust of some marginal basins and inter-arc troughs was younger than the adjacent Pacific Ocean floor in front of the magmatic arc, and several authors suggested that the basins had developed by a process of Cenozoic back-arc spreading of oceanic-type crust with horizontal migration of the magmatic arc away from the continent (Karig, 1971; Matsuda and Uyeda, 1971; Packham and Falvey, 1971). It may be inferred that the back-arc marginal basin spreading is normally preceded by development of back-arc extensional basins described above (VI.A).

Various hypotheses were evolved to explain the back-arc spreading and the high heat flow characteristic of back-arc basins, most of them related to the subduction of ocean floor along the inclined Benioff zone which underlies Cenozoic basins behind active volcanic arcs at depths exceeding 150 km. Matsuda and Uyeda (1971) favoured a convective system in the upper mantle driven by frictional heat along the underlying Benioff zone (Fig. 124), while Sleep and Toksoz (1971) suggested that spreading was related to a mechanically driven circulation system induced by the drag of the downgoing ocean floor on the overlying upper mantle (Fig. 125). More recently, it has been suggested (Molnar and Atwater, 1978) that back-arc spreading is dependent on the age of the subducted lithosphere, and that where this is old, thick and relatively dense

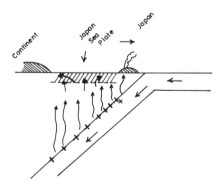

Fig. 124. Formation of Japan Sea marginal basin (from Matsuda and Uyeda, 1971). By permission of the authors and publisher, *Tectonophysics* **11**, 5–27, Elsevier, Amsterdam.

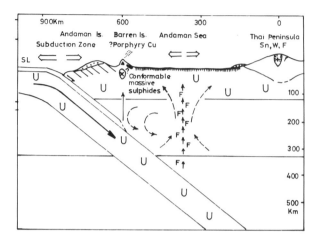

Fig. 125. Cross-section showing development of Andaman Sea marginal basin by back-arc spreading due to drag-induced upper mantle circulation of Sleep and Toksoz (1971) (after Mitchell and Garson, 1972). By permission of the publisher, *Minerals Sci. Engng* 8, 129–169.

its rate of descent is fast relative to plate convergence, leading to oceanward migration of the flexure in the downgoing slab and hence to migration of the arc system. Uyeda and Miyashiro (1974) suggested that the Japan Sea opened as a result of subduction of a spreading ridge, and Eguchi *et al.* (1979) favoured a similar origin for development of the Andaman Sea in the Miocene. It has also been suggested that back-arc rifting is associated with strike–slip faulting, parallel and adjacent to the magmatic arc, and that the Semangko Rift in Sumatra is an incipient marginal basin.

Other authors (II.A.4) noted the predominance of marginal basins along the eastern edge of continents beneath which ocean floor was being consumed, as in the western Pacific and Caribbean and behind the Scotia arc, as opposed to the continental margin magmatic arcs with local back-arc thrust belts above eastward-dipping Benioff zones of the largely inactive western north American and active Andean and Sunda Arcs, and postulated that development of the marginal basins was related to westward drift of lithosphere (see Fig. 91). As pointed out by Dickinson (1978), this drift could result in migration of spreading ocean ridges towards the eastern side of ocean basins, with consequent subduction of younger crust in the east than in the west. Subduction of older more dense crust beneath the western margin of the ocean would then reinforce the tendency to outward migration of the east-facing arcs. According to this hypothesis, the Andaman Sea bordered by a west-facing arc would be anomalous and presumably related to either the spreading ridge subduction or "leaky" transform fault processes referred to above.

Although many marginal basins are now considered to have formed by back-arc spreading, some, e.g. the Bering Sea, probably are segments of ocean floor "trapped" by development of an oceanic magmatic arc. At least one marginal basin, the Sea of Okhotsk, has been interpreted as a microcontinent which collided with the Asian mainland in the early Cenozoic following westward subduction of ocean floor (Den and Hotta, 1973).

Inter-arc basins, such as the Mariana Trough west of the volcanic arc, are considered to have formed by rifting within or at the rear of a magmatic arc, rather than at a continental margin, resulting in the active frontal volcanic arc and inactive remnant arc of Karig (1970). Several examples are known from the western Pacific and Indonesia, and while most lie behind eastward-facing volcanic arcs, a few, e.g. the Lau Basin in Tonga, evidently developed above eastward-dipping Benioff zones.

Most back-arc basins lack distinct or easily mappable magnetic anomalies typical of ocean basins created at mid-ocean rises, leading to early suggestions that back-arc crust is formed by rise of mantle diapirs at scattered locations (Matsuda and Uyeda, 1971). However, Curray *et al.* (1980) have shown that the Andaman Sea, one of the few marginal basins formed behind a west-facing island arc, developed since the mid-Miocene by spreading along an E-trending rise system with numerous transform offsets (see Fig. 82), perhaps related to the Sagaing–Namyin transform fault described later. Recently the West Philippine Basin has also been interpreted as the result of sea-floor spreading from a mid-basin ridge analogous to that at ocean rises (Andrews, 1980). Nevertheless, evidence for extensive volcanism away from the postulated spreading ridge in the Skikkoku Basin indicates that the spreading process in marginal basins is by no means fully understood.

Development of a marginal basin by subduction-related rifting and migration away from the continent as postulated for Japan, requires that rifting was initiated behind or within a continental margin magmatic arc. Remnants of this magmatic arc on the continental margin are rarely preserved, and unlike young rifted or Atlantic-type margins bordering major oceans, described above, the continental border of marginal basins rarely shows trilete rift junctions with failed rift arms, evidence of distinct pre-rift doming, or half-graben faulted successions; there is also little evidence of continental rift-related tholeiitic or alkaline volcanism. Nevertheless, it has been suggested that the Seoul–Wonsan graben in Korea developed as a Miocene failed arm of a trilete system which resulted in opening of the Sea of Japan (Burke, 1977). The Khlong Marui Fault system and graben in the Kra Isthmus of southern Thailand (Garson *et al.,* 1975) could also be interpreted as a Cenozoic failed arm with a component of strike-slip motion, related to a rift system on the other two arms of which continued movement resulted in Miocene opening of the Andaman Sea; however, there is some evidence that the Khlong Marui Fault was active in the late Mesozoic, long before the opening of the Andaman Sea (Ch. 7, II. B.6).

Marginal and inter-arc basins commonly close as a result of subduction, resulting in continent–arc or arc–arc collision, and it is argued by many authors that tectonic emplacement of ophiolites is in most orogens the result of closure of a marginal basin rather than of a major ocean. However, there is little firm evidence for this. Geochemical differences between oceanic and marginal basin crust are not well established (Ch. 4, I.B.2.(a)), and whether or not emplacement of major ophiolites on continental crust requires the presence of a transform fault immediately prior to collision as suggested by Karson and Dewey (1978), ophiolites could originate in both ocean and back-arc oceanic basins.

The nature of the sedimentary fill in back-arc basins has been used as evidence that ophiolites and associated deep marine sediments originated in marginal rather than ocean basins. Near the magmatic arc marginal basins are characterized by thick volcaniclastic aprons forming submarine fan complexes (Fig. 126); these pass towards the basin centre into pelagic brown clay with minor volcanic material and ultimately into calcareous pelagic oozes (Karig and Moore, 1975). The continental margin is highly variable, with prograding margins where large rivers deposit terrigeneous sediments, as in the South China Sea, but minor sedimentation where rivers are absent as in the Sea of Japan.

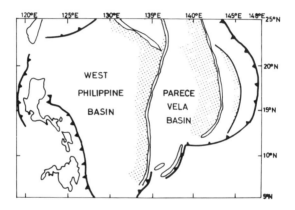

Fig. 126. Areas of thick volcaniclastic aprons adjacent to inactive or remnant arcs and the active Marianas arc in the Philippine, Parece Vela and Marianas back-arc basins (from Karig, 1975). By permission of the author and publisher, US Govt Printing Office, Washington.

Distinction of these sequences from those of remnant oceanic basins, when imbricated in a continent–arc collision, is difficult, and can rarely be used to prove a marginal basin origin.

B Mineral Deposits

Marginal basins, and to a far lesser extent inter-arc basins, have been considered in a rather general way as favourable locations for formation of mineral deposits. This is no doubt partly a result of the extensive and in some cases successful exploration in recent years for petroleum in continental margins of the Western Pacific marginal basins, and partly due to the realization in the early 1970s that ophiolitic rocks can be generated in the basins and that calc-alkaline volcanism occurs in the island arcs on their seaward margins.

1 Types of deposit

Economic mineral deposits either formed in or related to the development of back-arc basins could include those deposited during the initial stage of rifting of the arc system from the continent, those formed following rifting on the trailing continental margin, and hydrothermal deposits of the crustal spreading system within the basin.

There is little indication that any known mineral occurrences can be related to the few possible rift zones, for example the Wonsan Graben in Korea, identified as being associated with back-arc basin development. This lack of deposits in failed marginal basin rifts no doubt reflects the apparent absence of alkaline rocks and thick rift-fill successions which are so characteristic of intra-continental rift zones.

Because continental margins bordering marginal basins commonly subside as the basins open, forming, for example, the extensive drowned shelf areas in the Andaman Sea, they might be considered as favourable sites for formation of mineral deposits similar to those Atlantic-type continental margins, e.g. phosphorites and carbonate-hosted lead–zinc deposits in the shelf successions. The lack of such known deposits from the continental margins bordering the Western Pacific marginal basins partly reflects the fact that the shelf successions in this region are still largely beneath the sea and hence unexposed. Other factors are the absence of upwelling cold currents favourable for phosphorite deposition, and the limited development of major carbonate successions favourable for lead–zinc mineralization due to the inflow of continent–derived terrigeneous sediments.

One of the few continental margins in this setting with significant known sedimentary mineral deposits is that bordering the Andaman Sea in southern Burma and Thailand. Quaternary placer deposits of cassiterite are mined from drowned river valleys and from recent deposits undergoing active reworking on the sea floor. However, the setting for these deposits is unusual in that they result from erosion of Cretaceous to Eocene mineralized granites interpreted as a back-arc magmatic belt; marine deposition of granitic detritus is a result of northward displacement of the western Burma volcanic arc along the Sagaing Fault with associated opening of the Andaman Sea in the late Cenozoic (see Fig. 111).

Formation of cassiterite or other placer deposits on continental margins bordering most marginal basins is however improbable. For example, major alluvial gold deposits are unlikely to occur in settings comparable to that of the Andaman Sea tin, because the most probable source area of alluvial gold, a granodioritic magmatic arc, is normally rifted away from the continental margin during marginal basin opening.

No sulphides have been reported from the sea floor and inferred spreading centres in modern marginal basins or back-arc troughs, although hydrothermal activity and oxide deposits are known, e.g. from the Mariana Trough where heat flow is exceptionally high (Hussong *et al.*, 1978). Nevertheless many ophiolites with stratiform pyritic copper sulphide deposits and in some cases podiform chromite are interpreted by some authors as having originated in marginal basins; examples include those of Cyprus, the Oman ophiolite, the Betts Cove deposits in Newfoundland, and the Løkken-type deposits in Norway, described in Ch. 4.

There are probably at least two reasons for the scarcity of observed mineralization in modern marginal basins relative to that in supposed ancient marginal basin crust preserved as ophiolites. First, marginal basins by definition lie between continents and volcanic arcs and so receive considerable volumes of terrigeneous sediment and ash which could blanket metallic ores formed on the basaltic layer of the crust. Secondly, as pointed out above, since marginal basin crust cannot reliably be distinguished from that of ocean basins, it is at least equally probable that most mineralized ophiolites originated at an oceanic centre rather than in a marginal basin.

Economically significant deposits of manganese nodules have not been reported from marginal basins, probably because the small size of the basins inhibits deep-sea currents which are necessary to sweep the sea floor clear of detrital sediments and volcanic ash.

2 Preservation potential

Most marginal basins eventually close as a result of subduction of their oceanic floor, the related Benioff zone being inclined either oceanward beneath the magmatic arc, or landward beneath the continental margin. In the former case the continental margin rocks and related mineral deposits will be preserved in the foreland thrust belt of the resulting collision as described later, and may be overridden by ophiolitic rocks of the basin floor with associated copper sulphide or chromite deposits; these in turn may be overthrust by island arc rocks. In the latter case, a magmatic arc will develop across the continental margin, and during collision ophiolites may be thrust onto the island arc system which formerly lay at the oceanward boundary of the marginal basin. In both cases rapid erosion will follow the collisional orogeny, but because of the imbricate tectonic stacking of the various rock units portions of each may be preserved even after prolonged erosion.

In a few cases marginal basins may be preserved in their original form without complete closure by subduction. Examples have been reported from the Lower Palaeozoic of New South Wales and from the Arabian Shield. These relict marginal basins become infilled with successions of terrigeneous and volcanogenic sediment, but unless the basin crust is elevated by some later tectonic event, it is unlikely that minerals associated with ophiolitic rocks will be exposed.

CHAPTER 6

Deposits of Collision-related Settings

Long before the plate tectonic hypothesis, Wegener and later proponents of continental drift recognized that collision between continents took place as a result of their relative movement, and that India collided with Asia to form the Himalayas was appreciated by Argand (1924). It is now generally accepted that collisions result from continued subduction of oceanic crust, which, unless matched by creation of new crust in the subducting ocean, leads to the approach and eventual tectonic juxtaposition of either island arc or continental crust on the subducting plate with either an active island arc or continental margin arc on the overriding plate. Since about 1970 collision has been recognized as the most important single cause of mountain building and orogeny, and ancient collision belts or suture zones have been identified within orogenic belts of Mesozoic to late Cenozoic age and, although not without dispute, of Palaeozoic and Proterozoic age also. Moreover, it has become evident that some types of mineral deposit form in collision belts, and that the present tectonic position within the orogen of many previously formed deposits can be explained by collision tectonics.

Some of the combinations of island arc, continent and subduction zones involved in collision were illustrated by Dewey and Bird (1970). It is now apparent that the two most significant in terms of mountain building and mineralization are those involving a passive continental margin on the subducting plate. This may collide either with an active continental margin arc on the overriding plate forming a continent-continent collision (Fig. 127), as in the Himalayas (see Fig. 139), in the Eocene, or, more commonly, with an island arc on the overriding plate forming a continent-arc collision; examples of the latter are the early Tertiary collision in the Appenines following westward subduction (Kligfield, 1979) and in Central New Guinea following northward subduction Hamilton, 1979), and late Tertiary collision in Taiwan following eastward subduction (Biq Chingchang, 1978) and in Timor following northward subduction (Audley Charles *et al.*, 1979). An older example of a continent-continent collision is the late carboniferous collision of northern with southern Europe in Southwest England and the Erzgebirge, while the late Palaeozoic Antler orogeny in western North America provides a probable ancient example of a continent-arc collision. Other collisional situations arise either where an ocean closes by simul-

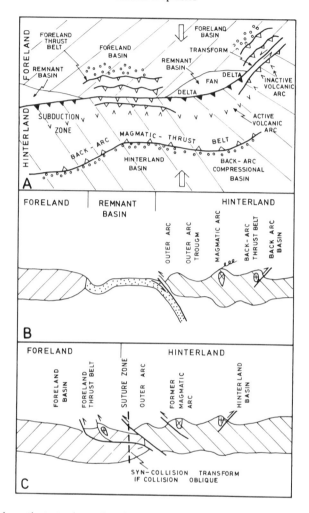

Fig. 127. Schematic tectonic settings in continent-continent collision belt. (A) Plan view, remnant basins and collisional orogen; (B) cross-section through remnant basin and arc system before collision; (C) cross-section through collisional orogen.

taneous subduction at opposite margins along divergent Benioff zones, e.g. in northern New Guinea (Hamilton, 1979) and in the Molucca Sea area in the Philippines where the Sangihe and Halmahera island arcs are colliding (Fig. 128), or where an inactive island arc on the subducting plate collides with an active continental margin. In most collision belts, which resulted from subduction along only one Benioff zone, it is convenient to define the polarity or subduction direction in terms of the geosynclinal terminology applied in particular to the Alpine chains of Europe (Aubouin, 1965), where the continent on the subducting plate is the foreland and the continent or island arc on the overriding plate is the hinterland.

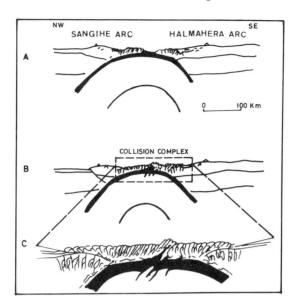

Fig. 128. Schematic Molucca Sea collision zone. (A) Initial collision, (B) present structure, (C) expanded view of collision complex (from Silver and Moore, 1978, *in* United Nations, 1980). By permission of the authors and publisher, *J. geophys. Res.* **83**, 1681–1691.

Although continental crust cannot because of its low density be subducted to great depths, during collision the continental foreland may underthrust relatively thin sheets of the original overriding plate, and, more significantly, undergoes thrusting itself; with development of successive thrusts further towards the foreland interior, hundreds of kilometres of crustal shortening can occur, as in the Appalachians and Himalayas. In general syn- and immediately post-collision deformation is largely concentrated in the underthrusting foreland, with less intense thrusting in the adjacent former arc system and in the former back-arc thrust belt in the hinterland of the overriding continent; the similarity of deformation in Taiwan or the northern Appenines to that in the Himalayas (see Figs. 135 and 139) suggests that on the foreland this deformation is independent of whether a continent or island arc lies on the overriding plate.

During the foreland thrusting, successions of the continental shelf and rise and ophiolite sequences of the ocean floor are commonly tectonically juxtaposed, with generation and emplacement of anatectic granites. Thrusting of volcanic arc rocks across the foreland shelf has taken place in some collision belts, e.g. in the early Tertiary collision in central New Guinea (Hamilton, 1979); in some other collision belts the arc was first welded to the foreland during continent–island arc collision, and subsequently thrust across the foreland during a further collision involving closure of a marginal basin behind the arc, as postulated for the Kohistan area of northern Pakistan (Tahirkheli *et al.*, 1979).

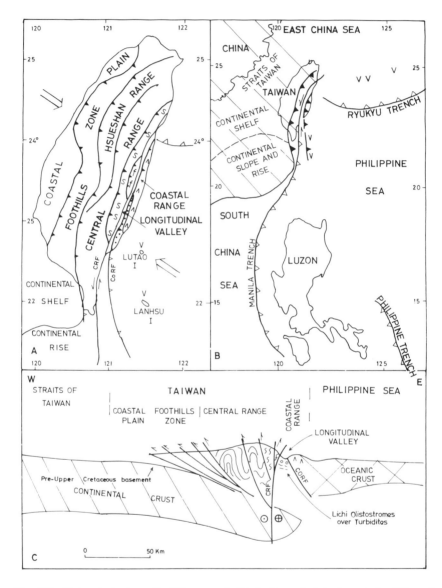

Fig. 129. Major tectonic units (A), present distribution of plate boundaries (B), and speculative cross-section (C) of the Pliocene continent–arc collision belt in Taiwan, showing syn-collision transform Central Range Fault (CRF) developed as a result of oblique plate convergence. S, schists; f, ultrabasic and glaucophane schists; Λ, inactive volcanic arc; V, active volcanoes; CoRF, Coastal Range Fault. (A) Modified from Biq Chingchang, 1978; (C) after Murphy, 1973. By permission of the authors and publishers, *Acta Oceanographica Taiwanica* **8**, 22–42 and *Bull. geol. Soc. Malaysia* **6**, 27–42.

Where pre-collision subduction of ocean floor is oblique rather than perpendicular to the trench, a major transform fault may develop in the collision zone, with oblique convergence resolved into a strike–slip component on the transform and thrusting within the foreland. A modern transform fault in this setting in the Longitudinal Valley of Taiwan (Fig. 129) forms the boundary between the foreland to the west and the "suture zone" or rocks of the outer and magmatic arc to the east (Biq Chingchang, 1971, 1978); the Markham–Ramu Fault in New Guinea probably formed in a comparable setting during late Cenozoic collision of New Guinea with the Schouten Island – New Britain arc to the north (Hamilton, 1979). In the Eastern Alps also there is evidence for a major component of strike–slip motion during the Eocene collision. An ancient analogue is probably the Highland Boundary Fault bordering the Dalradian foreland to the north in Scotland (Mitchell, 1981 b), interpreted as an end-

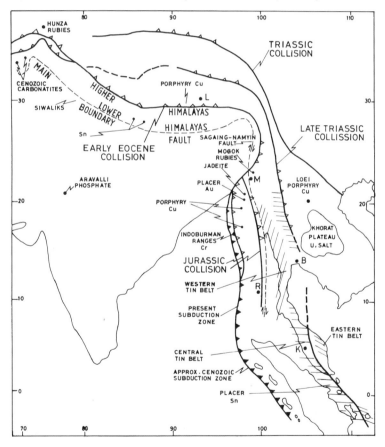

Fig. 130. Mesozoic and early Cenozoic suture zones in Southeast Asia and Tibet showing some mineral occurrences and deposits in various tectonic settings referred to in text. India and western Burma restored to Eocene position soon after initial India–Asia collision. B, Bangkok; K, Kuala Lumpur; L, Lhasa; M, Mandalay; R, Rangoon.

TABLE XIII

Plate or global tectonic terms with equivalent geosynclinal terminology; collision, or late orogenic stage of development of Aubouin (1965).

Tectonic setting terms	Foreland basin	Foreland shelf	Foreland thrust belt	Suture	Hinterland margin	Hinterland basin	Intramontane trough
Alternative "plate tectonic" terms	Continent on subducting plate				Overriding continental plate		
	Peripheral basin		Foreland thrust belt				
	Externides					Internides	
Equivalent geosynclinal terms	Exogeosyncline, molasse foredeep	Miogeosyncline		Eugeosyncline, eugeanticlinal ridge		Zeugogeosyncline	Taphrogeosyncline
						Exogeosyncline	Molasse back-deep

Cambrian sinistral transform developed during diachronous collision of the foreland with an oceanic arc to the southeast.

Mineralization in collision-related settings includes important hydrothermal deposits associated with immediately post-collision granitic intrusions, as well as tectonically emplaced and exposed deposits formed previously in a variety of tectonic settings, including continental shelf and particularly the highly mineralized magmatic arc and ocean ridge environments. Consequently, because continental collision is the major cause of mountain building, most of the less deeply eroded mountain ranges contain abundant mineral deposits. The often-quoted lack of economic deposits in the Himalayas (Fig. 130) largely reflects the poor mineralization in the late Precambrian and Phanerozoic foreland succession, most metallic deposits no doubt occurring in the magmatic arc of the Transhimalayas to the north.

Continental collision is broadly equivalent to the late orogenic stage of development defined by Aubouin (1965) in the Alpine geosynclines (Table XIII) and to the orogenic stage of Russian geologists. The orogenic stage in Russia is characterized by formation of granites with pegmatites and deposits of Sn, W, Mo, Bi and F.

Associated with collision, seven main tectonic settings can be distinguished (see Fig. 127C): pre-collision remnant and residual ocean basins, syn-collision suture zones, the "hinterland" margin of the overriding plate, syn- to post-collision foreland thrust belts, foreland basins, hinterland basins, and intramontane troughs of the foreland and hinterland.

1 REMNANT AND RESIDUAL OCEAN BASINS

A Tectonic Setting

1 Remnant basins

Remnant basins are closing or subducting basins bordered on one side by a subduction zone and on the other by the continental foreland or inactive island arc on the subducting plate (see Fig. 127). They mostly do not form significant tectonic settings with regard to mineralization, but an understanding of their genetic relationship to continental collision helps to explain other collision-related geological processes. Remnant ocean basins correspond in tectonic setting to the sedimentary basins of the flysch stage of development of Aubouin's (1965) elementary geosyncline, and the flysch-type sediments deposited within the basins and in and bordering the trench form the predominant lithology of his eugeosynclinal realm. They develop adjacent to collision zones wherever there are embayments in either the foreland or overriding plate, and also where the margin of the continental foreland is oblique to that of the overriding plate.

Passive continental margins are commonly irregular because their shape is determined partly by the angular intersection in trilete junctions of the initial intracontinental rift zones. As a subducting ocean basin progressively closes, the leading parts of the continental margin on the subducting plate approach and eventually enter the subduction zone, underthrusting the outer arc of the over-riding plate during initial collision. With continued subduction the loci of incipient collision migrate as embayments in the continental margins are pro-gressively closed, resulting in diachronous collision along the length of the original ocean basin. The closing "remnants" of the ocean basin within the embayments, and wedge-shaped basins beside a migrating collision zone, termed remnant basins by Graham *et al.* (1975), are significant tectonic settings in terms of both sedimentation and subsequent orogeny.

Following entry of the foreland margin into the subduction zone, uplift of the outer arc begins as a result of continental underthrusting beneath it and iso-static adjustment. Drainage from the rising arc supplies abundant detritus to the adjacent subducting remnant ocean basin, resulting in a major delta system and associated deep sea fan which migrate laterally together with the diachronous collision zone (Graham *et al.*, 1975). With progressive subduction the submarine fan turbidites are off-scraped in the subduction zone and accreted to the outer arc, elevation of which provides a continuous source of clastic sediment to the fan. Thus diachronous collision of irregular continental margins and consequent elevation of the outer arc leads to sedimentation in the closing remnant basins and tectonic accretion of this sediment to the adjacent outer arc, all of which processes are similarly diachronous. Theoretically, outer arc-derived sediments could be recycled several times through submarine fans and the subduction zone during closure of a major remnant basin.

Most remnant basins close during progressive continental collision, the basin succession first becoming accreted above the subduction zone to form an outer arc, and subsequently undergoing further deformation during continental collision. It has also been suggested that during collision nappes of outer arc rocks may be not only thrust towards the foreland, but also expelled laterally into adjacent remnant basins occupying embayments in the continental margins. Identification of remnant basin deposits within ancient outer arcs and collision belts has so far been possible in only a few orogens, e.g. in the Ouachitas Mountains within the Appalachians (Graham *et al.*, 1975), and in part of the Southern Uplands of Scotland, but it is probably that many imbricate flysch belts in orogens developed as a result of tectonic accretion of remnant basin deposits.

2   Residual basins

In some remnant basins subduction ceased before collision, and isolated marine successions are preserved, underlain by oceanic crust but surrounded by crust of continental thickness. Only a few of these ancient remnant basins, sometimes termed residual basins, have been recognized. One example, the North Caspian depression, is underlaid by about 14 km of sediment including an evaporite sequence, and is overthrust in the east by the western margin of the Ural Mountains; if the basin is underlain by oceanic crust, it may be the oldest non-tectonized ocean crust in the world (Burke, 1977). Characteristic features of non-tectonized ancient remnant or residual ocean basins are their position between a magmatic arc and shelf or coastal plain of the continental foreland, and a thick sedimentary succession passing upward from pelagic and submarine fan deposits into shallow marine or delta facies and finally into continental deposits.

B Mineral Deposits

Neither tectonized remnant basin successions preserved in outer arcs nor relatively undeformed residual basin sequences normally include mineral deposits, although presumably coals formed during the deltaic sedimentation might be preserved in either (see Fig. 35A). It is also possible that at an early

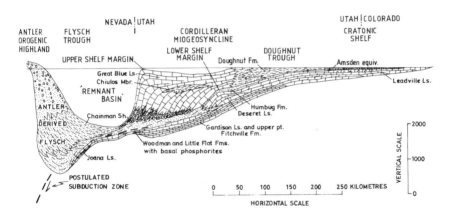

Fig. 131. Generalized regional stratigraphic cross-section through western USA in Mississippian (after Rose, 1976), showing phosphorite horizon. Position of closing remnant basin and postulated Benioff zone also shown. By permission of the author and publisher, *J. Res. U.S. Geol. Surv.* **4**, 449–466.

stage of collision remnant basins could form stagnant environments with slow accumulation of carbonaceous sediments and phosphorites, seaward of the continental shelf of the foreland on the subducting plate. A possible example is provided by the phosphorites west of the Lower Carboniferous carbonate shelf in western USA (Rose, 1976), deposited in a remnant basin prior to collision with arc complexes of the Antler orogenic belt to the west (Fig. 131). Following continental collision, remnant basin sediments accreted to form an outer arc during basin closure could provide a source of hydrothermal and perhaps saline ore-forming fluids able to migrate into depositional sites within tectonically juxtaposed shelf successions (IV.B.5).

II SUTURE ZONES AND COLLISION-RELATED OPHIOLITE SHEETS

A Tectonic Setting

1 Suture zones

In most collision belts a tectonically complex zone of olistostromes, melanges, pelagic sediments, ophiolitic rocks mostly with abundant serpentinites, and glaucophane or lawsonite schists is commonly present on the hinterland or overriding plate side of the mountain range or foreland thrust belt described later. These zones show evidence of thrusting towards the foreland, and magmatic arc rocks of the overriding plate may be tectonically juxtaposed with the other rocks. Examples in Cenozoic collision belts are the Indus–Tsangpo line in the Tibetan Himalayas, the Coastal Range of eastern Taiwan, a major dislocation north of the Zagros Thrust in Iran, and possibly the thrusts at the base of the Kolbano facies in Timor (Carter *et al.*, 1976). Older examples include the early Mesozoic "Bentong-Raub" line in Malaya, the late Palaeozoic "Meneage crush zone", adjacent flysch and Lizard ophiolite in Southwest England, and the mid-Palaeozoic Coolac Ultramafic Belt of New South Wales.

The zones have been termed sutures by Dewey and Bird (1970) and are commonly interpreted as the boundary between rocks of the original overriding and subducting plates. In some collision belts they probably indicate the trace of the subduction zone along which ocean floor was lost immediately before collision. In others, it is apparent that the "sutures" are the landward belt of an outer arc (Ch. 5, I. A), and hence formed part of the overriding plate during closure of the remnant basin; the actual "suture" in this case lies between continental rocks of the subducting continental foreland and the tectonically accreted flysch, including continental rise sediments, of the structurally higher outer arc, and its exact location can rarely be fixed precisely. It is also evident that in some collision belts rocks of the overriding plate, landward of the magmatic arc, are thrust across the "suture zone" onto the subducting foreland, e.g. in the Eastern Alps. In general where the margins of the plates involved in a collision are irregular, an outer arc preserved in zones of minor post-collision convergence will pass along strike into a narrow suture zone which includes serpentinites and glaucophane schists.

Some suture zones consist largely of previously accreted outer arc rocks further tectonized in the collision, together with magmatic arc fragments. The Indus Suture, for example, includes wedges of volcanic rocks and locally Triassic flysch which may represent tectonically juxtaposed volcanic and outer arc rocks of a Mesozoic island arc system largely subducted during the collision with India in the early Eocene.

2 Ophiolitic sheets on continental forelands

The main significance of suture zones to mineral deposits is that, as pointed out by Dewey and Bird (1971), they are commonly the source or "root zone" of large allochthonous ophiolitic sheets, with their associated chromite and cupriferous sulphide ore bodies, thrust or obducted onto continental margins. There is a voluminous literature reviewed fairly recently by Coleman (1977) on the tectonic settings for and causes of emplacement of ophiolites, in particular the large ophiolitic sheets of the Oman, New Caledonia, and Papua, often involving the disputed concept of thrusting onto the overriding plate during ocean floor subduction (Coleman, 1971; Smith and Woodcock, 1976), a process for which we consider there is very little evidence. That ophiolites are thrust onto the foreland of the subducting plate during collision is indicated by the presence of the Kiogas nappe on the Indian foreland in the Himalayas, 50 km south of the Indus Suture zone, and emplaced either in the Cretaceous during collision of India with an island arc to the north, or during the India–Asia collision. Johnson (1979) has recently shown that the Papuan ophiolite of Papua New Guinea was thrust southwards onto a continental fragment on the

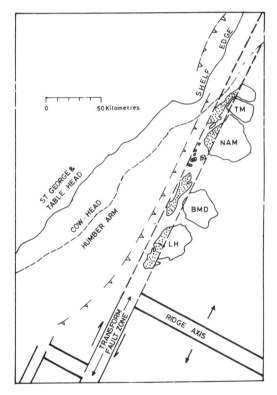

Fig. 132. Bay of Islands and Coastal Complex area, Newfoundland, during Tremadoc, showing non-transform domain (stippled). BMD, Blow-Me-Down Mt.; LH, Lewis Hills; NAM, North Arm Mt.; TM, Table Mt. (after Karson and Dewey, 1978). Postulated subduction zone also shown. By permission of the authors and publisher, *Bull. geol. Soc. Am.* **89**, 1037–1049.

subducting plate during collision with an island arc to the north in the Miocene. We consider that syn-collision sutures, together with syn-subduction outer arcs (Ch. 5, I. A), are the only tectonic settings in which large ophiolitic bodies are elevated and from which they may be expelled as thrust sheets. After tectonic emplacement, the ophiolites may in some cases undergo gravity gliding towards the foreland, depending on the location of uplift related to foreland thrusting.

Karson and Dewey (1978) suggested that the Lower Ordovician ophiolites of the Coastal and Bay of Islands Complexes of western Newfoundland formed the external or non-transform portion of a ridge–ridge oceanic transform fault, thrust onto a passive continental margin following the start of subduction beneath the transform zone (Fig. 132). This process, broadly similar to the transform obduction of Brookfield (1977), can be reconciled with syn-collision ophiolite emplacement if the transform system develops within a small subducting ocean basin. Following attempted subduction of the transform and its accretion to the overriding plate, the ocean closes and the transform is thrust

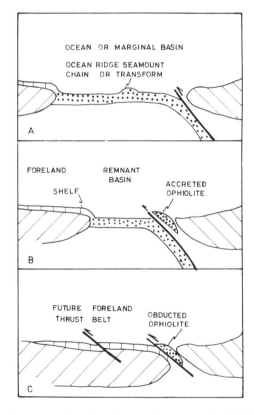

Fig. 133. Schematic cross-section showing emplacement of ophiolite in continent–arc collision. (A) subduction of ocean floor; (B) tectonic accretion of oceanic topographic high to overriding plate; (C) closure of remnant basin and collision with obduction of ophiolite: note absence of outer arc flysch compared with Fig. 84.

onto the passive continental foreland on the subducting plate. A similar process of ophiolite emplacement would occur if instead of the transform system a linear oceanic ridge or seamount chain entered and blocked the trench (Fig. 133).

The same process could explain the ophiolitic fragments interpreted as part of a mid-Cenozoic transform zone in the Coastal Range of Taiwan. The Taiwan ophiolite was emplaced, probably from the east, either immediately before or during Pliocene collision of the Asian foreland to the west with an island arc to the east (see Fig. 129). The magmatic arc rocks of the Coastal Range, structurally above the ophiolites, occupy a very limited area, and the absence of preserved magmatic arcs of appropriate age in analogous ancient settings, e.g. in the Coastal Belt of Newfoundland, could be explained by transform faulting, thrusting, or rifting subsequent to the collision.

Where transform zones or other oceanic ridge topographic features are tectonically accreted to the overriding plate long before collision, they will normally be elevated in an outer arc (Ch. 5, I. A) landward of subsequently accreted flysch. The outer arc flysch depresses the subducting ocean floor, and inhibits accretion of further oceanic ridges or seamounts. Syn-collision ophiolite obduction onto the foreland thus takes place where the closing ocean basin lacks a thick turbidite sequence to be accreted to the overriding plate, allowing tectonically accreted ophiolites to be thrust directly onto the foreland shelf succession. Many ophiolite sheets on continental forelands were emplaced within 10 or 20 Ma of their formation because ophiolites generated and tectonically accreted to the overriding plate long before collision would commonly be bordered by a major outer arc flysch belt.

The topographic expression of the suture zone is usually a narrow valley in which ophiolites, and turbidites and olistostromes deposited in the early stages of collision, are overlain by post-collision conglomerates, e.g. the Indus molasse in

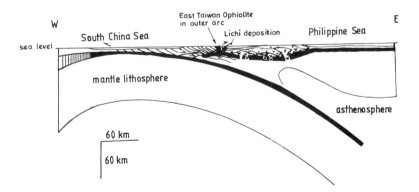

Fig. 134. Cross-section through Taiwan in Pliocene showing thrusting of outer arc rocks towards magmatic arc along basal detachment plane (from Suppe and Liou, 1979). Note East Taiwan Ophiolite emplaced in outer arc before collision with continent to west. Compare with alternative interpretation, Fig. 129C. By permission of the authors and publisher, *Mem. geol. Soc. China* **3**, 147–154.

the Himalayan suture. Subsequently the zone may remain a negative topographic feature and hence the root zone of the ophiolite nappes is commonly blanketed in younger sediments.

While in most Cenozoic sutures thrusting is dominantly synthetic to the subduction direction, in Taiwan Suppe and Liou (1979) have inferred that the foreland thrust belt and former outer arc have become detached from the underlying subducting crust and undergone thrusting towards the overriding plate (Fig. 134). This process could explain evidence for overthrusting of the overriding plate by ophiolites and absence of an outer arc trough in some other suture zones, e.g. the Himalayas (see Fig. 139).

B Mineral Deposits

With the exception of deposits preserved in tectonic slivers of magmatic arc or other rocks, mineral deposits in suture zones consist of those previously formed within and emplaced with either ophiolites or high pressure metamorphic rocks (see Table XIV).

1 Tectonic settings for mineralized ophiolites

Mineral deposits in ophiolites obducted onto the foreland during collision are apparently identical in type and of similar abundance to those in ophiolites of outer arcs, and have been described in Chapter 4. Some of the mineralized ophiolites in the foreland setting clearly originated in marginal basins but the ocean ridge or marginal basin origin cannot be identified from the deposits themselves.

As discussed above, obducted ophiolites are interpreted by some authors as oceanic transform fault zones or their extensions, and it has been suggested that associated Cyprus-type sulphides had formed in the transform zone rather than at an ocean rise. However, in the absence of criteria for distinguishing sulphide or chromite deposits formed at spreading centres from those formed at transforms, this argument remains somewhat academic.

The Cyprus-type sulphides of Oman and Newfoundland provide two examples of mineralization in foreland ophiolites emplaced during collision, and some features indicative of their deformation are indicated below. Other deposits emplaced on continental forelands include the podiform chromite of Oman, Cuba, Turkey and New Guinea.

The Semail ophiolite of Oman is probably one of the best exposed and least deformed complexes of its type (Alabaster *et al.*, 1980), and was emplaced onto continental slope and shelf carbonates of the Arabian Platform in a late Cretaceous collision following northeastwards subduction of ocean floor. The main sulphide deposits are located within a zone of seamount centres considered to have formed off the axis of the spreading ridge. The bodies are roughly saucer- or cigar-shaped and are only slightly folded and tilted as a result of collisional tectonics. They are therefore preserved virtually as formed in bathymetric depressions within the undulating ocean floor surface, beneath later basaltic pillows erupted from the seamounts (see Fig. 72).

In Newfoundland, where emplacement of ophiolites onto the continental margin has been discussed by Karson and Dewey (1978), ore-bodies of Cyprus-type are in general much more intensely deformed than those in the Oman nappe. Locally, underlying pillows are highly flattened and schistose with epidotization and silicification. Sulphide bodies however are generally elongated roughly parallel to the host rock schistosity, with remobilized sulphide veinlets, although sedimentary textures in the ores are fairly well-preserved. On Pilley's Island, sedimentary structures and pillow orientations show that the deposits are generally not inverted and have variable shallow to steep southward dips (Upadhyay and Strong, 1973). However stratigraphic interpretation is complicated by deformation which increases westward away from the suture zone. In the Springdale Peninsula chalcopyrite ore is enriched within hinges of late folds (Kennedy and De Grace, 1972) possibly related to syn-collision emplacement.

2 Jadeite and nephrite

The metamorphic minerals of economic importance associated with suture zones are jadeite and nephrite. These are mostly found in rocks interpreted as melanges, occurring either in outer arcs within collision belts or in the suture zone itself, and nephrite is much more common than jadeite. Pressures of more than 6 kbar are necessary for the formation of these minerals, and they commonly occur in association with glaucophane or lawsonite schists of the lawsonite–glaucophane–jadeite facies, formed at depths of at least 25 km (Winkler, 1967). Their elevation probably takes place along thrusts within the outer arc active both during subduction and also following collision. Only one example of economic gem-quality jadeite *in situ* is known, in the form of albite–jadeite dykes of pre-Albian and probably Mesozoic age within serpentinite in northern Burma (Soe Win, 1968).

3 Preservation potential

Suture zone rocks are commonly preserved at various structural levels and as noted by Dewey and Bird (1971) provide a key to understanding ancient orogens. Nevertheless, because they commonly form only a narrow zone between rocks of the originally overriding and subducting plates their significance in terms of economic deposits is much less than that of the associated obducted ophiolite sheets. The presence of the latter in orogens of Lower Palaeozoic age and younger provides strong evidence that plate tectonic processes operated throughout the Phanerozoic, and their scarcity in Proterozoic terrains has been used as evidence against Proterozoic plate tectonics. However, most extensive ophiolites are of Upper Mesozoic or Cenozoic age, and in Lower Palaeozoic orogenic belts only a few are sufficiently large to produce significant chromite or copper sulphide ore. It seems probable that, as could be expected with thrust sheets emplaced over isostatically depressed continental forelands, post-obduction uplift normally results in erosion of the ophiolite sheets within a period of not more than 100 Ma or so, except where subsequent large-scale tectonic events result in their burial by, and temporary preservation beneath, sedimentary rocks or thrusts. The ophiolite root zone or actual suture extends into the mantle as a narrow usually sinuous belt and can be detected geophysically even where buried by later sediments.

III HINTERLAND MARGIN OF THE OVERRIDING PLATE

A Tectonic Setting

The former continental margin arc or island arc on the overriding plate or hinterland commonly undergoes less deformation than the foreland during collision, and evidence for the presence of collision-related anatectic granites in Cenozoic examples of this setting is disputable. Nevertheless a continental overriding plate in collision belts characteristically undergoes considerable uplift, and evidence from the Iranian and Himalayan collisions suggests that uplift is proportional to the rate of plate convergence; it is also probable that the elevation depends on the volume of continental crust subducted during the collision (Ben Avraham and Nur, 1976).

In some cases the overriding plate margin undergoes thrusting synthetic to the subduction direction, and the outer arc, outer arc trough and sometimes the magmatic arc may be either thrust onto the foreland and eroded, or more commonly underthrust along faults developing behind the magmatic arc. Where the outer arc is thrust under the overriding plate, tectonic juxtaposition of magmatic arc plutons and volcanic rocks with either suture zone ophiolites or rocks of the foreland can occur, as is the case locally in the Indus suture of Tibet and in Taiwan. The development of shallow-dipping thrusts in an analogous position within the overriding plate of the Eastern Alps has been discussed at length by Oxburgh (1972).

Beyond the magmatic arc, rocks of the overriding plate rarely show much evidence of syn- to immediately post-collision metamorphism or tight folding, but there are exceptions and the overriding plate margin in Pakistan, between the Karakorum magmatic arc to the north and suture zone bordering Kohistan to the south, evidently suffered deformation as well as rapid uplift following late Cenozoic collision (Tahirkheli *et al.*, 1979). Moreover, it is argued by some geologists that the overriding plate throughout the Tibetan Plateau is undergoing intense crustal shortening and metamorphism at depth as a result of the India-Asia collision, although direct evidence for this is limited as discussed below

(VII. A). Another possible example of syn-collision metamorphism on the overriding plate is provided by the Ryoke Belt of Japan, where Cretaceous metamorphism and emplacement of two-mica granites perhaps took place on an overriding plate margin during collision with the Shimanto Belt to the southeast. However, the Ryoke Belt with its characteristic Abukuma-type low pressure regional metamorphism (Winkler, 1967) is generally considered to be more typical of a magmatic arc root zone.

B Mineral Deposits

As might be expected in a zone dominated by strong uplift and local thrusting but normally lacking magmatic activity or sedimentation, no specific types of mineral deposit are considered to be characteristic of the overriding plate margin in collision belts, other than those previously emplaced during subduction. However, in view of a possible relationship between gemstone deposits and overriding plate margins, we discuss briefly below the ruby deposits of Kashmir, Afghanistan and Burma as their distribution suggests that either their formation or exposure is related to the Cenozoic tectonic setting.

1 Gemstone deposits of Kashmir, Afghanistan and Burma

While there is no obvious reason why deposits of gem-quality corundum should be associated with collision belts as speculatively suggested by Mitchell and Garson (1976), it is perhaps more than a coincidence that three of the world's few ruby deposits lie on but near the southern margin of the overriding Asian plate (see Figs 130 and 147) and were elevated and exposed as a result of the India–Asia collision.

In the Hunza area of Kashmir, within the northwestern part of the Karakorum range, gem-quality rubies are mined from several localities (Okrush *et al.*, 1976). They occur within calcite–marble interbedded with sillimanite gneisses and schists which are cut by pegmatites and aplites; the metamorphic rocks are considered to represent Palaeozoic sediments metamorphosed during the early Tertiary under temperatures of around 700°C and pressures up to 7 kbar. The marbles lie south of the central Karakorum batholith belt of mid- to late Tertiary predominantly granodioritic plutons, and north of a probable suture zone; south of the suture lies the Kohistan belt of possible island arc rocks which collided with and was thrust southwards onto the Indian plate in the early Tertiary. Collision along the suture, following northward subduction and generation of the Karakorum granodiorites in a magmatic arc, probably took place in the late Tertiary (Tahirkheli *et al.*, 1979).

In the Sarobi area of eastern Afghanistan near the Khyber Pass, gem-quality rubies and red spinels have been mined from Archaen to Lower Proterozoic marbles intruded by Oligocene granitoids (Dronov *et al.*, 1973). The marbles are from 1 to 5 m thick forming a 5-km-long belt within a series of sillimanite-garnet and cordierite–garnet gneisses. These rocks occur in a complicated zone of thrust blocks and ophiolitic serpentinites at the boundary between the Kohistan belt and the former overriding continental plate of Central Afghanistan to the north, a setting comparable to that of the Hunza area.

In the Mogok Belt of Burma, the world's most famous deposits of ruby associated with spinel and other gemstones occur near the contact of nepheline syenites and alaskites intrusive into marble; sapphires are produced both from the marble and the nearby pegmatites. The marble is interbedded with sillimanite-grade schists and gneisses of uncertain metamorphic age but there is some radiometric evidence that the alkaline intrusives were emplaced in the Tertiary. The metamorphic rocks of the Mogok Belt are overlain to the east without an obvious tectonic or stratigraphic break by Lower Palaeozoic sediments and are thrust westwards over ophiolitic rocks of probable Mesozoic age; there is evidence that elevation of the Belt took place in the Tertiary as a result of collision of the northeastern margin of India with northern Burma, before northward movement of the Indian plate along the Sagaing–Namyin Fault.

Metamorphism which resulted in the formation of the Kashmir, Afghanistan and Burma deposits, associated in Burma with emplacement of the undersaturated alkaline rocks, may have taken place either at deep levels in the magmatic arc during ocean floor subduction, or during the subsequent continental collision. In either case exposure of the deposits was the result of rapid uplift of the former overriding plate following collision.

IV FORELAND THRUST BELTS

A Tectonic Setting

In most collision belts involving a continent on the subducting plate the major topographic feature is a belt of folded and thrust sedimentary or metamorphic rocks, lying between the suture zone and foreland basin. The thrusts, forming many of the type A "continental" subduction zones of Bally (1975), and first recognized as zones of crustal shortening by Holmquist (1900), are invariably

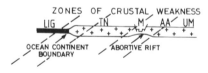

Fig. 135. Zones of crustal weakness in Northern Appenine continental margin inherited from Upper Triassic and Jurassic rifting (after Kligfield, 1979). Note aborted rift is favourable site for foreland thrust belt ("Alpi Apuane Shear Zone") in Fig. 136. LIG, Ligurian Zone, TN, Tuscan nappe, M, Massa, AA, Alpi Apuane, UM, Umbria. By permission of the author and publisher, *Am. J. Sci.* 279, 676–691.

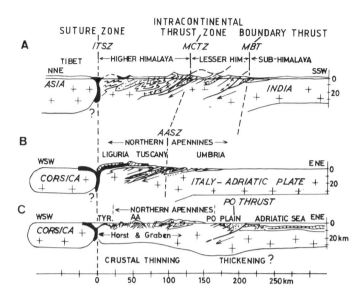

Fig. 136. Comparative sections between Himalayas and Northern Appenines. (A) Present day Himalayas: ITSZ, Indus–Tsangpo Suture zone, MCTZ, Main Central Thrust zone, MBT, Main Boundary thrust; (B) Oligocene–Miocene section from Corsica to the Adriatic: AASZ, Alpi Apuane Shear zone; (C) Messinian to Recent section: TYR, Tyrrhenian Sea, AA, Alpi Apuane region (from Kligfield, 1979). By permission of the author and publisher, *Am. J. Sci.* **279**, 676–691.

directed towards the foreland, and occupy a belt up to 200 km in width, although the zone of most intense thrusting is usually, but not invariably, within 50 km of the suture zone. Kligfield (1979) suggested that in the Appenines the position of the post-collision mid-Tertiary foreland thrust belt (Fig. 136) was determined by a zone of crustal weakness corresponding to an Upper Triassic failed rift (Fig. 135).

In almost all Cenozoic collision belts it is evident that the thrusts developed earliest near the suture and progressively later towards the foreland, with related uplift above the thrusts migrating in the same direction. In some foreland thrust belts, in particular the Alps, evaporites of the former continental shelf form preferred horizons for thrusting, and there is evidence in the Himalayas for movement at deeper structural levels along graphitic schist horizons. Crustal thickness in the collision belt is usually greatest beneath the thrust belt, exceeding 55 km beneath the Lepontine area in the Alps and 65 km beneath the Higher Himalayas.

The thrust sheets mostly consist of either foreland shelf rocks, as in the Alps and Himalayas, or foreland continental shelf and rise rocks as in the Central Highlands of New Guinea, Taiwan, and the Northern Appenines. Major thrust sheets of metamorphic rocks are also present in some foreland thrust belts; in the Pennine Zone of the Central Alps, in the Northern Appenines, and in the "Central Gneiss" zone of the Himalayas they are interpreted as elevated base-

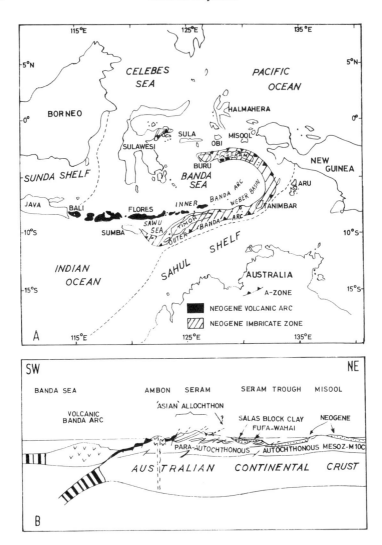

Fig. 137. Collision zone between Sahul Shelf and Banda arc system, Eastern Indonesia. (A) Location map; (B) cross section. Dotted line is edge of Australian continental shelf (from Audley Charles *et al.*, 1979). Note post-collision volcanism on Ambon. By permission of the authors and publisher, *J. geol. Soc. Lond.* **136**, 547–568.

ment of the foreland which underwent further metamorphism or "basement remobilization" during collision. In the Northern Appenines, in Taiwan, probably in the Outer Banda Arc collision belt (Fig. 137), and particularly in the Himalayas, metamorphism of the foreland sedimentary rocks is associated with thrusting following the collision, and in the Himalayas syn-thrusting plutons are present. Elevation and major lateral translation of metamorphic rocks is also recognized in the late Palaeozoic continent–continent collision in

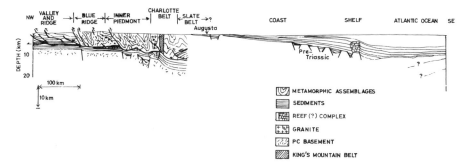

Fig. 138. Generalized cross-section of Southeastern United States (from Cook *et al.*, 1979) showing thrusting related to late Palaeozoic collision. By permission of the authors and publisher, *Geology* 7, 563–567.

the southern Appalachians (Fig. 138), where a sheet of metamorphic rocks up to 15 km thick moved at least 260 km westward along a basal thrust plane across continental shelf rocks of the foreland (Cook *et al.*, 1979). In the Austro-Alpine zone of the Eastern Alps, a thin sheet of metamorphic basement of the overriding plate, lacking remobilization features, is thrust across the ophiolitic and associated sedimentary rocks of the Bundnerschiefer onto the foreland in a major nappe, largely eliminating the suture zone (Dietrich, 1976).

Syn-thrusting and hence syn- to early post-collision anatectic granites are common in or on the suture side of most pre-Cenozoic foreland thrust belts. Syn-collision plutons in the southern Appalachians include the Stone Mountain Granite, a peraluminous leucocratic two-mica tourmaline-bearing quartz monzonite with an initial Sr^{87}/Sr^{86} ratio of 0.725 (Whitney *et al.*, 1976). However, the most direct evidence for collision-related metamorphism and pluton emplacement is from the Himalayas; in the Lepontine zone of the Alps post-collision early Miocene metamorphism is not associated with major pluton emplacement although scattered Neogene bodies such as the Bergell tonalite and Novate granite are clearly collision-related.

In the Himalayas (Le Fort, 1975; Powell, 1979), northward subduction of Tethys was followed by collision of India with Asia along the Indus–Tsangpo suture line in the early Eocene. South of the suture, the presence of extensive basaltic detritus in mid-Cretaceous sandstones (Bordet *et al.*, 1971) suggests an ophiolitic source, which in turn implies Lower Cretaceous collision with an arc system to the north; this arc, possibly formerly continuous with that of western Burma, has subsequently been largely destroyed, probably by underthrusting beneath Tibet. Following the early Eocene continental collision, the Main Central Thrust developed during the Oligocene in the continental plate of the Indian foreland 150 to 220 km south of and parallel to the suture (Fig. 139). The Higher Himalayas, elevated above the thrust during the mid-Tertiary, consist of Precambrian "basement" gneisses or "central gneisses", overlain by a thick Phanerozoic "Tethyan" sedimentary succession, part of which was metamor-

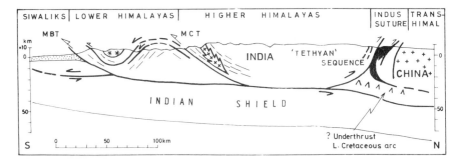

Fig. 139. Schematic cross-section of Himalayas (modified from Le Fort, 1975, and Powell, 1979). MBT, Main Boundary Thrust; MCT, Main Central Thrust; *, Cambrian granite of klippen in Lower Himalayas; +, Tertiary leucogranite of Higher Himalayas. By permission of the authors and publishers, *Am. J. Sci.* **275A,** 1–44 and Geol. Surv. Pakistan, Quetta.

phosed in the Tertiary and includes local gneiss domes. Nappes of the central gneisses and Tethyan succession were transported southwards above the Main Central Thrust, and are now preserved as klippen overlying Precambrian to Upper Palaeozoic low-grade meta-sedimentary rocks in the Lower Himalayas.

Major granite bodies and pegmatites of mid-Tertiary age, described below, intrude the "basement" gneisses and the overlying Tethyan succession of the Higher Himalayas between the Main Central Thrust and the Indus Suture. Rb/Sr dating on the Higher Himalayan Manaslu Granite yielded, subject to minor correction due to experimental error, a late Oligocene age of emplacement and an exceptionally high initial Sr^{87}/Sr^{86} ratio of 0.74 (Hamet and Allègre, 1976), indicating an anatectic origin. Association of metamorphism with movement on the Main Central Thrust is indicated by the reversed metamorphic isograds in the Barrovian-type facies series yielding mid-Miocene K/Ar ages beneath it, reported long ago by Gansser (1964), and by Tertiary mineral ages on schists beneath central gneiss klippen in the Lower Himalayas. Initial movement in the Miocene on the Main Boundary Thrust, located further south in the foreland and up to 270 km from the suture, is suggested by uplift of the Lower Himalayas indicated by clasts in the Upper Miocene Siwalik foreland basin sediments (V. A). Continued or renewed movement on the Thrust and splay faults to the south in the Quaternary is indicated by uplift of the Lower Himalayas and of Siwalik sediments of Pliocene age, and by overthrusting of the Siwaliks by the Lower Himalayan metamorphic rocks and by slices of Permo-Carboniferous "Gondwana" sediments.

The cause of the metamorphism and generation of granites in foreland thrust belts is not entirely understood. Andrieux *et al.* (1977) considered that in the Himalayas shear heating on the Main Central Thrust resulted in metamorphism and partial melting at depths of 30 - 40 km to form the granites, which rose as sheets along the foliation, but it has been argued (Toksoz and Bird, 1977) that convergence rates were too slow to generate sufficient heat. However, there is little difficulty in generating partial melts within tectonically thickened continental crust, and analogy with the Himalayas suggests that in many older orogens high-grade regional metamorphism and simultaneous emplacement of silicic anatectic granites took place during thrusting within the foreland which immediately followed collision.

The role of gravity as opposed to thrusting in development of foreland nappes has sometimes been emphasized but recent work suggests that gravitational sliding in this setting is largely confined to high-level nappes of sedimentary rocks.

The tectonic setting most similar to a foreland thrust belt is undoubtedly the back-arc thrust and magmatic belt behind some continental margin arc systems (Ch. 5, IV. A). However, one fundamental distinction is that the foreland thrust belt is separated from the magmatic arc of the hinterland by "suture zone" ophiolitic rocks, flysch and pelagic sediments, while the back-arc magmatic belt lies on the same continental plate as the magmatic arc, and intervening ophiolites and pelagic sediments of appropriate age are absent. During continent–continent collision, deformation in the back-arc thrust belt may be intensified, with consequent metamorphism and generation of granites, but conditions are so similar to those obtaining in the belt during subduction that they are not considered further here. Prolonged syn- to post-collision movements on these back-arc thrusts, antithetic to the subduction direction, results in a   superficial symmetry to the orogen, as opposed to the marked polarity characteristic of some (Fig. 137) but not all (see Fig. 134) continent-arc collisions.

B Mineral Deposits

Foreland thrust belts can be important sources of ore because they include both tectonically emplaced pre-existing deposits and mineralization generated during and hence genetically related to the collision (see Table XIV).

Because of the often intense crustal shortening and tectonic juxtaposition of rock units formed in different settings, a wide variety of previously generated mineral deposits can be elevated and tectonically juxtaposed with each other during collision, and exposed by subsequent erosion in the foreland thrust belt. These commonly include deposits of the continental shelf of the under-thrusting continental foreland, e.g. carbonate-hosted lead–zinc deposits and phosphorites, and mineralization associated with ophiolitic rocks of the ocean floor such as cupriferous sulphides and chromite. The shelf successions may be underlain by either rift-related half graben successions or older orogenic belts, which may be exposed with their associated mineralization during collision.

It is often argued that metamorphism can lead to concentration of previously minor or disseminated sulphides. While sulphides can become concentrated in fold hinges during deformation, it appears that during collision-related regional metamorphism sulphides may undergo recrystallization with an increase in grain size, but that metamorphism itself results in no increase in ore grade.

The types of mineral deposit most directly related to the formation of foreland thrust belts are those associated with collision-related granitic magmatism, although the thrust belts have not until recently been considered as significant settings for generation of metallic mineral deposits. This is no doubt partly because in the Himalayas, the only Tertiary collision belt with significant exposed collision-related magmatic rocks, there are no known currently economic metallic deposits associated with the plutons. However, granitic plutons occurring in ancient orogens and considered to have formed in similar tectonic settings to the Himalayan granites contain important deposits of the tin–tungsten–fluorite–niobium association and also uranium, here termed pluton-associated hydrothermal deposits, as well as mineralized carbonatites.

1 Pluton-associated hydrothermal deposits of tin and uranium

Major deposits of tin and associated tungsten, fluorite and niobium, and magmatic uranium deposits, are associated with granitic plutons in some ancient orogens in tectonic settings comparable to that of the collision-related granites of the foreland thrust belt in the Higher Himalayas. Examples include the tin-bearing granites of the Main Range in Malaysia, and the major uranium deposit at Rossing in Namibia. Analogy between the older deposits and the Himalayas is based largely on the tectonic setting rather than on similarities in type of mineralization because the Higher Himalayas contain only traces of tin–tungsten mineralization and significant uranium mineralization has not yet been reported. As the Himalayas provide the clearest example of a Tertiary continent–continent collision belt, and are also deeply eroded, a description of the Himalayan granites and their setting is therefore given below and followed by brief accounts of what are considered to be analogous mineralized granites elsewhere.

(a) Tin and uranium occurrences in Tertiary granites, Higher Himalayas

Granitic bodies in the Higher Himalayas of Nepal, north of the Main Central Thrust, have been described by Le Fort (1973, 1975) and others. The plutons, which include the Manaslu Granite referred to above (IV. A), consist almost entirely of massive to foliated muscovite-biotite granites and leucogranites with or without tourmaline. Aplites and pegmatites are widespread. Most granites are in the form of large concordant slabs or sheets parallel to the northward-dipping bedding or foliation in the host rocks (Fig. 140), and thin gently dipping and folded granitic sheets have been reported. Meta-sedimentary xenoliths are common, and the granites have many features typical of the S-type granites of Chappell and White (1974) and White and Chappell (1977), considered to be anatectic and derived from partial melting of meta-sedimentary rocks. Host rocks range from Precambrian to Cretaceous in age and the Phanerozoic rocks near the intrusions mostly show low-grade regional metamorphism.

The granites of the Higher Himalayas are comparable in many ways to the highly fractionated granites in some of the world's major tin fields. Similarities

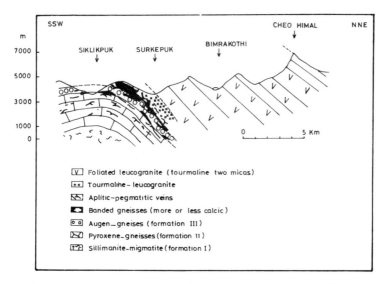

Fig. 140. Geological cross-section through the mid-Tertiary Manaslu Massif of the Higher Himalayas (after Le Fort, 1975). By permission of the author and publisher, *Am. J. Sci.* **275A**, 1–44.

include the predominance of leucocratic mostly peraluminous granite with muscovite and biotite and widespread tourmaline, common aplites and pegmatites, high initial Sr^{87}/Sr^{86} ratios, local hydrothermal alteration with greisen zones, and absence of associated volcanic rocks. Differences from tin granites are the commonly inclined slab or sheet-like form of the Himalayan granites, as opposed to the generally accepted batholith form with a deep "root" of many tin granites elsewhere, and the scarcity of significant tin within the Himalayan rocks, although cassiterite has been reported from pegmatites in the higher nappes of Bhutan (Poulose, 1975).

The possible importance of post-magmatic hydrothermal solutions in forming economic deposits (e.g. Beckinsale *et al.*, 1979) in Southeast Asia and Southwest England is referred to below. The absence of significant known economic mineralization in the Higher Himalayan granites could be explained as a result of their mid-Tertiary age and lack of time for development of subsequent hydrothermal convective systems. On the other hand, late Cenozoic mostly sub-volcanic silicic rocks in Bolivia are extensively mineralized with tin, (Ch. 5, IV. B.1. (a)), presumably due to either late-magmatic or subsequent meteoric hydrothermal circulation.

Minor uranium mineralization has been reported from pegmatitic granites of unknown age immediately above the Main Central Thrust in Nepal, and although no economic deposits of uranium are yet known from anywhere in the Himalayas, it is of interest that the Manaslu leucogranite contains unusually high background values of uranium, comparable to those of the mineralized granites in the Massif Central of France (P. Le Fort, pers. comm. 1980).

(b) Lower Palaeozoic tin-bearing granites, Lower Himalayas

In the Lower Himalayas, particularly in Nepal, granitic bodies occur within meta-sedimentary klippen which overlie rocks of lower metamorphic grade and are probably stratigraphically equivalent to part of the Higher Himalayan succession (see Fig. 139). In Central Nepal (Stocklin, 1980) the plutons are predominantly massive but locally foliated porphyritic granites and leucogranites with biotite, biotite and muscovite, or biotite, tourmaline and muscovite, mostly lacking foliation and with aplites and pegmatites near their margins. All form sheet-like bodies broadly parallel to bedding but in places cutting sharply across stratigraphic units; schist xenoliths are widespread. In places gneisses and migmatites are associated with the margins of plutons. In Pakistan, Le Fort *et al.* (1980) have described the Manserah Pluton, a two-mica cordierite-bearing porphyritic granite and monzogranite in a tectonic setting analogous to that of the Nepal granites.

In the Lower Himalayas of western Nepal primary deposits of Sn, Mo, Ta, Nb and W have been described from greisenized fine-grained leucocratic granite and pegmatites forming a north-dipping sheet within a thrust sheet or klippen, and cassiterite–sulphide bodies are present in schists adjacent to the granite. In Central Nepal Talalov (1977) described mineralization from the Palung Granite, an undeformed to foliated muscovite–tourmaline granite and leucogranite with apogranite, aplites and pegmatites. The Granite is locally greisenized and reportedly contains scheelite, cassiterite, fluorite, topaz and tantalo-columbite as accessories. In Pakistan, tungsten mineralization has been reported from the Manserah Pluton.

The "Lower Himalayan" granites of Nepal show some similarities in composition to the Manaslu and other granites of the Higher Himalayas, and have been considered to be of similar age and origin (Andrieux *et al.*, 1976; Stocklin, 1980), leading Mitchell (1978a) to suggest that they formed a Cenozoic collision-related analogue for older tin granites in some ancient orogens. However, a recent Rb/Sr whole rock isochron on the Manserah Pluton of Pakistan has yielded an Upper Cambrian age and an initial Sr^{87}/Sr^{86} ratio of 0.719 (Le Fort *et al.*, 1980) and there is now radiometric evidence for an Upper Cambrian age for the Palung Granite (R. Beckinsale, pers. comm. 1980), indicating that these plutons are clearly unrelated to the Himalayan collision orogeny. Moreover, Le Fort *et al.* (1980) argue that the granites are part of a cordierite granite belt, distinct in composition from the Higher Himalayan granites.

The anatectic nature of the granites suggests that they were emplaced in either a back-arc magmatic belt or foreland thrust belt, rather than in a magmatic arc. Evidence that part of the western Southeast Asia – southern Tibet block lay adjacent to India (Fig. 141) before rifting away during the early Permian Panjal Trap volcanism suggests that the Lower Himalayan granites were emplaced in a foreland thrust belt following Cambrian collision of southern Tibet with India (Mitchell, 1979b, 1981a). The inferred Cambrian thrusts of this belt may have been reactivated following the India–Asia collision to form some of the Cenozoic thrusts of the Himalayas.

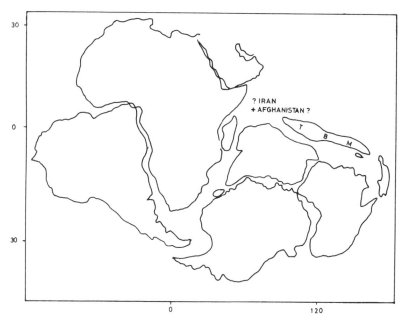

Fig. 141. Position of S Tibet – W Southeast Asia Block in Gondwanaland. Reconstruction of other continents from Powell *et al.*, (1980). B, eastern Burma; M, Malay Peninsula; T, southern Tibet. By permission of the authors and publisher, *Tectonophysics* **68**, 13–29, Elsevier, Amsterdam.

(c) The Central Tin Belt, Southeast Asia

Tin-bearing granites of the Indonesian "tin" island of Bangka, the Main Range of the Malay peninsula, and northwestern Thailand can be considered as one belt, formed in a single tectonic setting in the late Triassic, and sometimes referred to as the Central Tin Belt of Southeast Asia (Mitchell, 1977). This belt provides about half the western world's tin production, largely from placer deposits. To the west lie the younger back-arc magmatic belt tin-granites of Burma, westernmost northern Thailand, and peninsular Thailand (Ch. 5, IV. B. 1.(b)), and to the east the older Eastern Belt granites including the major primary cassiterite-sulphide deposits of Pahang Consolidated in eastern Malaya and of Billiton Island in Indonesia (Ch. 5, II. B. 4). Erosion of the Main Range of Malaya and its inferred southwestward continuation into Indonesia has given rise to what are perhaps the most productive alluvial tin deposits in the world, referred to in Ch. 5, V. B. 2.

Primary mineralization (Hosking, 1973) is associated with mostly peraluminous two-mica granites and adamellites largely of late Triassic age which intrude isoclinally folded and probably thrust Palaeozoic sedimentary and metasedimentary rocks. The only significant primary tin ore mineral is cassiterite although malayaite and stannite also occur. Much of the cassiterite, particularly in the west of the belt, is pleochroic indicating the presence of tantalum and niobum minerals.

The deposits in Malaya can be grouped in four main types: pegmatites, skarns, lode and xenothermal deposits. Tin-bearing pegmatites with niobium-tantalum are associated with granites west of the Main Range, while those lacking niobum–tantalum are widespread within the Range. Tin-bearing skarns contain malayaite, cassiterite or both within calc–hornfels. At different localities stratiform, vein-like and pipe-like mineralized skarns are present. Lode or vein deposits are common in non-calcareous metasediments adjacent to granites; some veins occur in swarms and are major sources of ore. Xenothermal deposits, which are highly telescoped and contain antimony and mercury, are best developed in the Kinta and Selangor areas, where they are mostly associated with limestones and occur as veins or pipes.

It has often been supposed (e.g. Hosking, 1977) that at least part of the mineralization in the Main Range is of Cretaceous age and resulted from assimilation of older deposits by Cretaceous granites and subsequent deposition. However, radioactive age determinations provide little evidence of Cretaceous igneous activity adjacent to many of the ore bodies, and it is now clear that a large part of the mineralization is either Upper Triassic in age or at least associated with Upper Triassic granites. The possibility that tin mineralization in western Thailand resulted from circulation of post-intrusion hydrothermal meteoric water, which re-set K/Ar ages of micas, has been discussed by Beckinsale *et al.* (1979).

Most of the tin-and tungsten-bearing granites of northwestern Thailand are of similar age (Teggin, 1975; Beckinsale *et al.*, 1979) to those of the Main Range in Malaya. Production is largely from primary deposits and numerous small tungsten mines are present. Initial Sr^{87}/Sr^{86} ratios of the late Triassic granites (Beckinsale *et al.*, 1979) are high, comparable to those of similar age in the Main Range.

Intrusion of the granites in the late Triassic into similar host rocks throughout the Central Belt indicates a similar origin, and it has been suggested that they were emplaced in the subducting foreland of western Southeast Asia during collision with eastern Thailand and eastern Malaya following Triassic eastward subduction of ocean floor (Mitchell, 1976b, 1977; Hutchison, 1978). The postulated suture zone extends from east of Bangka Island in Indonesia through the zone of ultrabasic and metamorphic rocks in medial Malaya to beneath the Gulf of Siam (Fig. 142); it continues northwards from west of Bangkok through Chieng Rai in Thailand into easternmost Burma where a large body of ultrabasic rock has recently been reported. There is some evidence that the suture extends further through western Yunnan and the Nukiang or Salween suture in eastern Tibet into southern Tibet (Mitchell, 1981a). Major granitic plutons lie west of the suture in the Northern Shan States and in western Yunnan (Li *et al.*, 1979) but unlike the granites in an analogous tectonic setting in Thailand and Peninsular Malaysia there is no evidence that these contain economic tin deposits.

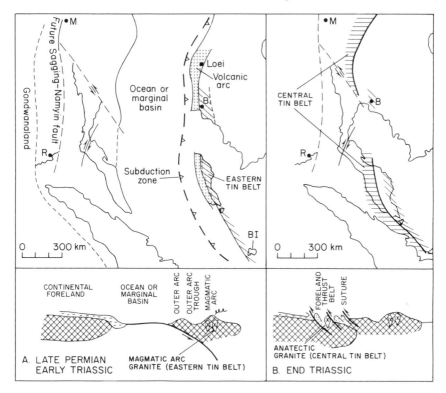

Fig. 142. Tectonic settings for mineralization in Eastern and Central Tin Belts, Southeast Asia (after Mitchell, 1977). (A) Late Permian to Triassic eastward subduction of ocean floor beneath eastern Malaya and northeastern Thailand. Porphyry copper mineralization at Loei and tin mineralization in eastern Malaya and Billiton Island within magmatic arc. (B) Late Triassic collision of western Southeast Asia block with northeastern Thailand–eastern Malaya; tin mineralization associated with granites of foreland thrust belt in Main Range Malaysia and Thailand. B, Bangkok, BI, Billiton Island, M, Mandalay, R, Rangoon. By permission of the publisher, *Bull. geol. Soc. Malaysia* **9**, 123–140.

The pre-collision Permian to Triassic magmatic arc of the overriding plate, which includes the Eastern Tin Belt (Ch. 5, II. B. 4) is preserved in eastern Malaya, and in central northern Thailand where it lacks tin mineralization but includes the porphyry copper deposit at Loei (Hutchison and Taylor, 1978). Inclusion within the magmatic arc of Billiton Island in Indonesia is supported by the style of mineralization (Ch. 5, II. B. 4), but not by lead and strontium isotope data which indicate that granites associated with the tin were derived from continental crust in the late Triassic (Jones *et al.*, 1977).

Significant differences between the Central Belt granites and those of undoubted collision origin in the Himalayas are the absence in Southeast Asia of associated major metamorphic nappes and of inclined granitic sheets. Nevertheless stratigraphic data and the radiometric ages of the plutons indicate a close relationship between collision and granite emplacement in the continental rocks of the subducting foreland.

(d) Hercynian tin granites, Southwest England

The end-Carboniferous to Lower Permian post-tectonic granites of Southwest England contain what were formerly very major deposits of tin with associated large copper deposits and fluorite and tungsten mineralization. It is estimated that more than 2 million tonnes of tin have been produced from the region. Granites of similar age, composition and mineral association are present in the Erzgebirge of East Germany and Czechoslovakia. The predominant plutons in Southwest England are andalusite and cordierite-bearing adamellites with lesser proportions of two-mica adamellite and granite, commonly porphyritic, and mostly forming stocks and steep-sided cupolas probably related to a major batholith at depth. Tourmalinized granites, greisen zones and extensive areas of later kaolinization exploited for china clay are widespread.

The tin mineralization mostly occurs in quartz–greisen lodes and veins which are usually but not invariably adjacent to granitic stocks and cupolas. In places metal zoning is present with tin surrounded by a copper and lead zone. Some of the granites are enriched in uranium, present in zircon in primary biotite; a few greisen contain significant but non-economic quantities of uranium and scattered pitchblende-bearing veins occur (Simpson *et al.*, 1979).

The plutons intrude a folded and probably thrust succession of Devonian to middle Carboniferous flysch, minor grits and basic volcanic rocks which include a melange and have suffered low-grade regional metamorphism. In the south the Gramscatho turbidites show north-facing folds and are overlain by wildflysch or olistostromes while in the north paralic sediments are juxtaposed with turbidites and overlain by non-marine sediments of the Bude Formation. Along the south coast ophiolitic rocks (Kirby, 1979) of the Lizard Complex are thrust northwards over the flysch. North of the granites of North Devon, South Wales and southern Ireland Upper Carboniferous clastics are cut by southward-dipping thrusts.

The tectonic setting for emplacement of the granites has been interpreted as an end-Carboniferous continental collision (Dewey and Burke, 1973; Mitchell, 1974b) following southward subduction of part of the Rheic ocean beneath a magmatic arc in Europe to the south (Fig. 143), and the granites have been compared to those of the Higher Himalayas (Mitchell, 1974b; Mitchell and Garson, 1976). An anatectic and hence possibly collisional origin of the granites is suggested by their similarities to the S-type granites of Chappell and White (1974). The granites have also been explained either as an Andean-type magmatic arc (Nicolas, 1972; Badham and Halls, 1975) or as the result of foreland crustal thickening developed on the overriding continental plate (Bromley, 1976) during northward subduction. Simpson *et al.* (1979) favoured an Andean-type magmatic arc for granite emplacement, and argued that an initial Sr^{87}/Sr^{86} ratio of 0.706 determined on one pluton was lower than that of most

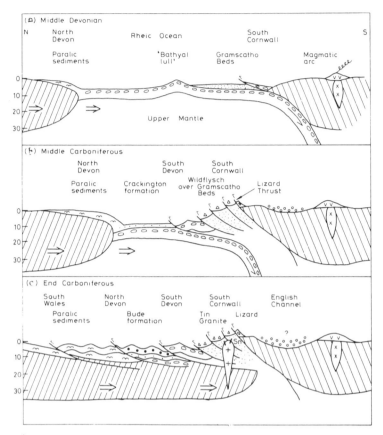

Fig. 143. Southwest England tin granites in relation to the end-Carboniferous collision (from Mitchell, 1974b). By permission of the publisher, *Trans. Instn Min. Metall.* 8, B95–97.

collision-related anatectic granites and indicated a predominantly mantle origin. They suggested that the concentration of uranium and incompatible elements in the granites is related to scavenging by fluorine resulting from the breakdown of phlogopite during dehydration of subducting oceanic crust.

A subduction-related magmatic arc origin for both the granites and mineralization, is difficult to reconcile with the location of the granites in deformed flysch-type host rocks and with their apparently syn- to post-collision age. The back-arc magmatic belt setting for the granites implied by Bromley (1976) could be reconciled with evidence for Devonian northward subduction beneath a magmatic arc in Brittany (Le Fort, 1979) but would explain neither the end Carboniferous – earliest Permian age of the Southwest England granites nor the presence of the Lizard ophiolites and the melanges. Interpretation of the granites as outer arc plutons analogous to those of Southwest Japan or the Aleutian arc (Ch.5., I.B.2.(a)) is unsatisfactory because of their large volume, abundant tin mineralization, and syn-collision age, although it could perhaps explain the low

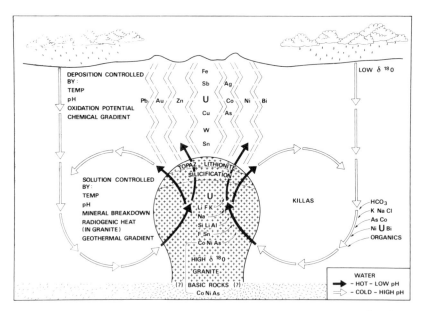

Fig. 144. Model for interaction between granite magma enriched in metals, fluorine and chlorine, and meteoric hydrothermal convective system to produce vein-type mineralization (from Simpson *et al.*, 1979). By permission of the authors and publisher, *Phil. Trans. R. Soc. Lond.* **291A**, 133–160.

initial Sr^{87}/Sr^{86} ratio. Pending more substantial evidence from additional Sr^{87}/Sr^{86} ratios, we favour a collisional origin for the granites, and suggest that following southward subduction and closure of an oceanic basin beneath an outer arc comprising the Gramscatho turbidites and associated wildflysch, the outer arc was thrust northwards onto continental crust of southern Britain. The non-marine Bude Formation then accumulated in either a remnant or more probably foreland basin north of the outer arc. Granites were generated by partial melting of either outer arc and ocean floor rocks, or of underthrust continental crust, and rose into the former outer arc, north of the Lizard ophiolitic belt. The absence of nappes of the continental foreland analogous to those of the Himalayas remains a problem and we suggest that they remain unexposed beneath the outer arc. Alternatively, it is possible that the granites were intruded into outer arc rocks during continental collision in a setting for which there is no precise Cenozoic analogue.

Mineralization and associated greisenization were possibly largely younger than emplacement of the water-undersaturated granites, and related to meteoric hydrothermal water circulating in a convecting system (Sheppard, 1977), the tin and minor uranium being concentrated by fluorine-bearing solutions as suggested by Simpson *et al.* (1979) (Fig. 144). As Southwest England became tectonically stable after the end-Carboniferous collision, there is no obvious cause of the subsequent heating events required to explain evidence of Mesozoic tin and uranium mineralization episodes. One possibility is that intermittent tensional stresses allowed circulation of meteoric waters which underwent radiogenic heating in and adjacent to the granites.

(e) Tin and tungsten in late Hercynian granites, Portugal

A detailed account of tin–tungsten deposits at Panasqueira, Portugal, has recently been given by Kelly and Rye (1979). The deposits, in which the ore minerals are mostly wolfram and cassiterite, comprise the major source of tungsten in western Europe, and occur in numerous flat-lying quartz veins within schists overlying a greisenized two-mica granite cupola. Formation of the vein openings is considered to have been related to hydraulic dilation due to excess of fluid over lithostatic pressure during mineralization. The granite is late Hercynian (around 290 Ma) in age, and is considered to have provided structural conduits for introduction of slightly younger vein fluids which also formed a hydrothermal silica cap to the cupola. Kelly and Rye (1979) consider the granite to have many features of the anatectic S-type granites of Chappell and White (1974); as the granite is also one of the younger (approximately end-Carboniferous) post-tectonic granites of the Hercynian orogen, it may reasonably be interpreted as a collision-related pluton, perhaps emplaced in a tectonic setting analogous to that of the Southwest England granites.

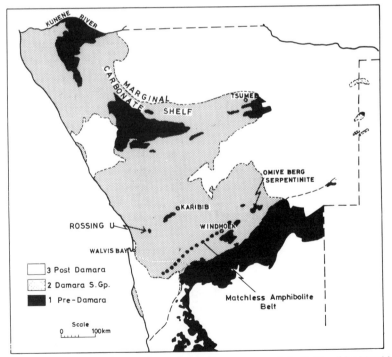

Fig. 145. Location of Rossing uranium mine in Damara orogenic belt, Namibia; Matchless Amphibolite Belt also shown (modified from Martin, 1978). By permission of the author and publisher, "Mineralization in Metamorphic Terranes", 405–415, Van Schaik, Pretoria.

(f) "Magmatic" uranium in the Rossing Mine, Namibia

The Rossing mine in Namibia is the world's largest uranium mine, the largest and richest magmatic or granite-hosted uranium ore body, and is also the oldest significant deposit of this type; 1979 production at Rossing approached 5000 tonnes uranium metal.

Descriptions by Berning *et al.* (1976) and Jacob (1978) indicate that the deposit lies within the central part of the Damara Belt, within the thick Damara Group of shallow water metasediments which underwent regional meta-morphism and isoclinal folding in the earliest Cambrian "Pan African" event (Fig. 145). Syn- to late-tectonic granites of the Red and Salem Granite suites intruded the metasediments in the zone of most intense metamorphism, the latter forming porphyritic gneissic granite, porphyritic biotite granite and leucogranite mostly as conformable bodies occupying synclines. A red-coloured alaskitic pegmatitic granite, one of a number of late- to post-tectonic intrusions, mostly forms pods and dykes of fine to very coarse-grained or pegmatitic granite partly within the Red Granite and elsewhere confined largely to a single strati-graphic unit. Uranium, in the form of uraninite and secondary minerals, is present in concentrations of several hundred ppm in the alaskitic rocks and occurs preferentially in alaskites underlying and in contact with carbonate beds of the Rossing Formation, suggesting that the marbles trapped the melt (Jacob, 1978).

The Red and Salem Granites are considered to have formed by partial melting of underlying pre-Damara gneisses and deeply buried stratigraphic equivalents of the Damara Group host rocks. The mineralized alaskitic granite is either a late-stage differentiate of one of the Salem Granites, or derived directly as an anatectic water-saturated melt; in both cases uranium was concentrated from trace amounts in either the Damara Group source rocks or underlying basement.

The Damara Belt is commonly interpreted as an ensialic intracratonic sedi-mentary basin (e.g. Martin, 1978). However, the presence south of the Rossing area of a major greywacke facies and of a possible suture zone represented by the linear Matchless amphibolite belt 300 km in length with at least one large body of serpentinite to the northeast, suggest an origin involving continental collision (Watters, 1976). Regional metamorphism and deformation of the Damara Group with emplacement of syn- to late-tectonic uranium-bearing leucogranites can be interpreted in terms of crustal thickening accompanying collision, although the position of the amphibolite belt to the south together with the southward direction of thrusting would suggest that Rossing lay on the overriding plate rather than on the subducting plate as implied by Watters (1976) and inferred from analogy with the Himalayas. Alternatively, it is possible that the granites were generated in a back-arc magmatic belt rather than a collision belt, a tectonic setting analogous to that of the Cape Fold Belt to the south developed in the late Palaeozoic (Ch.5, V.B.1(c)), but this is unlikely because serpentinites of the Matchless belt are not typical of back-arc setting.

(g) Uranium mineralization in the Western Massif Central, France

Uranium ore in the Saint Sylvestre granite of the Massif Central, where more than 20 000 tonnes of uranium metal are proved, has recently been described by Le Roy (1978). The deposit is of interest as its tectonic setting during mineralization has been compared specifically with that of the Himalayan collision granites.

The Saint-Sylvestre granite complex was emplaced into schist and gneisses during the Middle Devonian. It consists of the foliated biotite–sillimanite Brame granite and the non-foliated mostly coarse-grained two-mica Saint-Sylvestre granite, interpreted respectively as a lower and an upper facies of the same pluton. The upper facies with quartz–biotite–muscovite–orthoclase–albite is thought to have formed from late- and post-magmatic alteration including muscovitization and albitization.

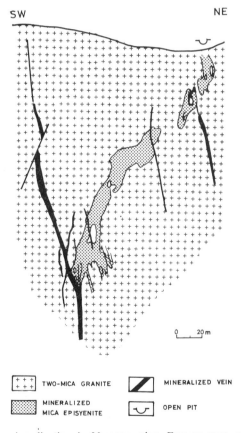

Fig. 146. Uranium mineralization in Margnac mine, France; cross-section showing mica episyenitic pipe cut by a mineralized vein (from Le Roy, 1978). By permission of the author and publisher, *Econ. Geol.* **73**, 1611–1634.

Uranium mineralization with associated muscovitization and loss of silica mostly occurs in the form of pipes at the intersection of older E- and N-trending barren faults, and in NW-trending veins related to younger faults (Fig. 146). The mineralization and muscovitization are considered to be of the same age as the intrusion of microgranitic and lamprophyric dykes which are parallel to the older fault trends and have been dated at 285 ± 10 Ma or earliest Permian, coinciding with the end of the Hercynian orogeny. The uranium occurs near the base of the Saint-Sylvestre granite which in this zone contains up to 20 ppm uranium.

Mineralization in the pipes and veins is of two types, vein and disseminated, of which the former is slightly more important economically. Minerals are pitchblende, coffinite and secondary uranium minerals. The associated muscovitization involved transformation or solution of all granitic minerals other than muscovite, with complete alteration of biotite and plagioclase, partial muscovitization of orthoclase, and dissolution of all or most of the quartz. The uranium is considered to have been dissolved from the granite by, and transported in, CO_2-rich non-magmatic solutions, circulating as a result of heating by the lamprophyric dykes; on rising through the granite the solutions entered veins and boiled, precipitating pitchblende.

Emplacement of the mineralized granite, with which pegmatites and quartz veins with tin and tungsten are also associated, took place during the main metamorphism of the Hercynian orogeny, and the granite is considered to be

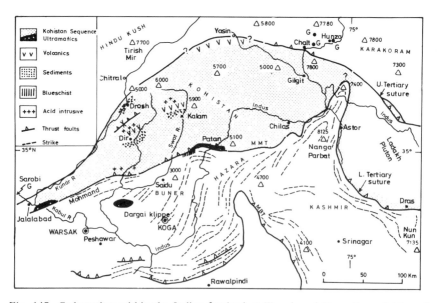

Fig. 147. Carbonatites within the Indian foreland at Warsak and Koga (large dots), and gemstones (G) on the overriding Asian plate, northern Pakistan and adjacent areas; MBT, Main Boundary Thrust, MMT, Main Mantle Thrust (modified from Tahirkheli *et al.*, 1979). By permission of the authors and publisher, Geol. Surv. Pakistan, Quetta.

anatectic. Marignac and Le Roy (unpub. *in* Le Roy, 1978) consider it to be comparable in petrography and tectonic setting to the two-mica granites of the Higher Himalayas, and have suggested that granite generation was related to thrusting analogous to that on the Himalayan Main Central Thrust. Mattauer and Etchecopar (1977) also recognize a major south-directed thrust in the Massif Central which they compare to the Main Central Thrust and interpret in terms of Hercynian continental collision; presumably this was related to the late Devonian collision following northward subduction beneath southern Brittany described by Le Fort (1979).

2 Carbonatites of Pakistan

As discussed above (Ch. 2, II.B.1. (a)) carbonatites are normally restricted to intracontinental hot spots, rift zones and transform faults. However, carbonatite magmas of Pliocene or Quaternary age (Garson, in press) with uranium-bearing pyrochlore are known from the foreland of the Indian plate in Pakistan, in both the Koga area and the Warsak area 100 km to the west (Fig. 147).

Near Koga, where the carbonatites are associated with syenites and granites, a K/Ar age of 50 Ma was obtained on biotite from a nepheline syenite. In the Warsak area, the carbonatites occur within a suite of alkali granite, porphyritic microgranite and meta-basic rocks, and a riebeckite from the alkaline granite yielded a K/Ar age of 41 Ma (Kempe, 1973).

The tectonic setting of the deposits has been discussed by Garson (in press). They lie within the "Swat-Buner Schistose Group" sequence of the Indian foreland, comprising rocks of probable Palaeozoic age which suffered their latest metamorphism in the Tertiary. To the north is the Kohistan region which includes island arc rocks, thrust southwards over the foreland with ophiolitic rocks along the Main Mantle Thrust "suture"; to the south the Main Boundary Thrust forms the northern limit of Siwalik molasse (Tahirkheli *et al.*, 1979). The carbonatites thus lie within the foreland thrust belt of the Indian continent. Emplacement of the carbonatite magmas clearly post-dated the early Tertiary collision between India and the overriding plate of Kohistan to the north (III, B, 1), although it is uncertain why the carbonatites are younger than the associated syenites and alkaline granites. While the absence of evidence for post-collision rifts or graben in the region implies that the alkaline rocks and associated carbonate magmas were emplaced in a collision-related compressional regime, comparison with carbonatites elsewhere would suggest that the Pakistan magmas were generated in a post-collision incipient but aborted episode of rifting which had negligible surface expression.

3 Pegmatite deposits

The South Asian Pegmatite Belt of Rossovsky and Konovalenko (1976) comprises deposits and occurrences of rare metal and gemstone-bearing pegmatites extending from the Hindu Kush through the Himalayas to northern Burma. While the position of this Belt suggests that the pegmatites might all be related in some way to the India–Asia collision, in detail the Belt consists of a number of sub-provinces of distinct age and geological setting, e.g. the ruby deposits of Kashmir and Burma described above (III. B. 1), and situated on what were formerly different plates. However, of the deposits shown by Rossovsky and Konovalenko (1976), those in Nepal and Bhutan lie on a single plate and within a single tectonic setting, the foreland thrust belt of India (see Fig. 130).

In Nepal the pegmatites, with semi-precious polychromic tourmaline, aquamarine, and reportedly sapphire, are restricted to the central gneisses near the base of the Higher Himalayan succession above the Main Central Thrust. The pegmatites clearly cut the gneisses and are generally considered to be related to post-collision movement on the Thrust during the Himalayan orogeny. The pegmatites in Bhutan also contain tourmaline and beryl and reportedly include the lithium pyroxene ore spodumene.

4 The silver–nickel, cobalt arsenide association

It has been suggested (Badham, 1976) that deposits of the hydrothermal vein-type complex Ag- Ni, Co arsenide association are characteristic of magmatic arc plutons emplaced above shallow-dipping subduction zones, with the mid-Proterozoic Great Bear Batholith and the Hercynian granites of Southwest England and the Erzgebirge as examples. In the Hercynian deposits of Europe tin and tungsten deposits are spatially associated with the Ag–Ni–Co ores.

While it is conceivable that formation of the Great Bear mineralization took place in a magmatic arc, the granites associated with the Southwest England and the Erzgebirge deposits are considered here to be collision-related magmas intruded into a foreland thrust belt. Nevertheless, the occurrence in the Cusco area of the eastern Andes of Ag–Ni, Co arsenide deposits of Tertiary age in association with tin mineralization (Badham, 1976) suggests that deposits of this type could also occur in back-arc magmatic belts.

5 Deposits indirectly related to collision: carbonate-hosted lead-zinc mineralization

Many major carbonate-hosted lead–zinc provinces occur in continental shelf rocks of the foreland adjacent to collision belts, although some, e.g. those of Triassic age in the Eastern Alps of Austria and Yugoslavia, are in carbonates of the overriding plate. The presence of lead–zinc mineralization in the foreland does not necessarily indicate a genetic relationship to collision (Ch. 3, I. B. 7), as most ocean basins eventually close, and hence many continental margins are likely to lie on the subducting plate during collision. However, as the mineralization is mostly epigenetic with respect to sedimentation but pre-dates major deformation, it is possible that in some cases mineralizing brines precipitate sulphide in carbonates either within an incipient foreland thrust belt or beneath a foreland basin succession during collision. One example where this may have occurred is in the Missouri province described above (Ch. 5, I.B.5), where a possible source for the mineralizing fluids is the outer arc of the Ouachita flysch. Had plate convergence continued in the Missouri area, the mineralized province would have become a foreland thrust belt and the outer arc would have been largely eroded.

Support for the concept of migration of fluids through foreland sediments during collision is provided by Dickinson's (1974a) suggestion (Ch. 5, I.B.5) that in the Persian Gulf hydrocarbons migrated up-dip from continental margin sediments of the Arabian platform underthrust beneath the Zagros Mts which he interpreted as a foreland thrust belt. The hydrocarbons then accumulated in reservoir rocks partly of Mesozoic age overlying the platform and partly in foreland basin clastic sediments of Tertiary age in the Zagros foothills to the northeast.

Nevertheless in many carbonate-hosted lead–zinc deposits there is little indication of incipient folding or thrusting, and in others the mineralization lies too far from the suture zone for there to be a genetic relationship between collision and mineralization. Without more evidence it therefore appears probable that relatively few deposits of this type are related to collision.

6 Preservation potential

Prolonged erosion in foreland thrust belts results from post-collision isostatic adjustment, which in the Pliocene fold-thrust belt of Taiwan exceeds 5 mm/yr. Consequently, the belts are commonly exposed at a rather deep structural level even in Cenozoic collisions, and in fact are characterized by elevated rocks metamorphosed during the collision. Within the belts the structural vergence and nature of the anatectic syn- to late-tectonic magmatic rocks are still recognizable at even the deepest levels where extensive areas of high-grade mostly Barrovian-type metamorphic rocks are exposed. In a sense therefore foreland thrust belts are well preserved, although not always easily identified. Mineral deposits associated with the apical parts of anatectic collision-related granites are probably usually destroyed by erosion within a few tens of millions of years, but deposits in and beneath the original shelf succession of the foreland will survive syn-collision metamorphism and may be preserved almost indefinitely in steeply inclined thrust sheets.

Deposits of tin and tungsten are mostly formed around the upper levels of granitic plutons and so are not often preserved in Palaeozoic and older collision belts. However, granites in the Glen Lui area of Scotland, interpreted as plutons of a late Cambrian to earliest Ordovician foreland thrust belt (Mitchell, 1978 b), probably carry cassiterite and columbite now found in alluvium, and there are high geochemical backgrounds of tin and niobium in the area. Preservation of uranium mineralization, which can perhaps occur at slightly deeper levels, is indicated by the deposits of Hercynian age in the Massif Central of France, and the Rossing uranium occurs at a deep structural level near to the anatectic source of the mineralized pegmatitic granites. Nevertheless, in many deeply eroded foreland thrust belts syn-tectonic high initial Sr^{87}/Sr^{86} ratio granites within high-grade metamorphic rocks are lacking in both high background and concentrations of tin, tungsten and uranium.

V FORELAND BASINS

A Tectonic Setting

Foreland thrust belts on the subducting continental plate are characteristically bordered on the foreland side by a major sedimentary basin. This develops initially as a result of depression of the continental foreland on entering the subduction zone following closure of the remnant basin, and later in response to isostatic adjustment to the growing mass of the elevated foreland thrust belt. The basins, filled by thick sequences of either clastic continental or marine and continental facies, were recognized long ago by Bertrand (1897), and by Haug (1900) who considered them typical of the late stages of geosynclinal evolution. Both these authors were impressed by what is now the best-known example of a foreland basin, the "molasse" of the Alps. Foreland basins are equivalent to the peripheral basins of Dickinson (1974b), and to some of the exogeosynclines of North America. They belong to the late orogenic evolutionary stage of Aubouin's (1965) European geosynclinal model. Because of their close relationship to continental collision, the successions within foreland basins form an example of the tectofacies of Van Houten (1974), defined as a major lithological facies characteristic of a particular type of tectonic setting.

Foreland basins are by definition situated on continental crust, the surface of which dips beneath the adjacent foreland thrust belt. The basin succession either overlies continental shelf sediments deposited on the passive continental margin of the foreland before collision, or it may lie directly on relatively old foreland "basement". Most Cenozoic foreland basins can be distinguished from hinterland intramontane troughs described later (p. 296) by their position in front of nappes consisting of either outer arc flysch or, more commonly, of foreland thrust belt rocks. Examples of foreland basin successions are the Oligocene to Pliocene "molasse" sediments north and west of the Alps, the Plio-Pleistocene succession of the Po Basin and its continuation into the Adriatic Sea east of the Appenines, the Upper Tertiary Siwaliks and overlying deposits of the Ganges–Brahmaputra river basins south of the Himalayas, the Pliocene–Quaternary sediments beneath the Strait of Taiwan, late Cenozoic sediments of the Northern Australian Shelf south of the Central Highlands in New Guinea, and the Cenozoic sediments of the Persian Gulf. Older examples comprise the Upper Carboniferous successions

Fig. 148. Remnant basin (Carboniferous turbidites) and foreland basins (Appalachian and Illinois Basins) in the Carboniferous Appalachian–Ouachita collision belt (from Graham *et al.*, 1975). By permission of the authors and publisher, *Bull. geol. Soc. Am.* 86, 273–286.

of the Appalachian and Black Warrior Basins on the North American foreland, north of the Ouachita outer arc (Fig. 148), and possibly the Upper Carboniferous Bude Sandstone of Southwest England referred to above (IV.B.I.(d)).

In some foreland basins, for example the sub-Himalayas, the thick predominantly clastic Siwaliks and relatively thin underlying Murees probably overlie marine shelf carbonates, deposited during an immediately preceding marine transgression; in others, e.g. the Alps, shallow marine clastic deposits lie directly on marine flysch. The marine deposits typically pass upwards into deltaic and finally fluviatile facies and in some cases into fanglomerates. Most of the detritus is derived from the adjacent mountain range, and the composition of the clasts or mineral grains is one of the chief sources of information on the age of advancement of nappes and unroofing of plutons and metamorphic rocks in the foreland thrust belt (Fuchtbauer, 1967). In the sub-Himalayan Murees, detritus was derived from the foreland to the south, but in the Siwaliks, the clasts can be matched with the Lower Himalayan metasedimentary succession and overlying klippen of high-grade metamorphic rocks and granites to the north (Stocklin, 1980). Arkoses consisting largely of granitic material have been reported from the Siwaliks, but as in many foreland basins sub-greywackes predominate.

The lower part of the succession in foreland basins is commonly the same age as the main episode of orogeny in the foreland thrust belt, and hence immediately post-collision in age, and the upper part normally post-dates thrusting and elevation. The part of the basin closest to the mountain range normally suffers deformation due to the advance of thrust sheets, and the succession here may be both buried by nappes and itself thrust towards the foreland. Consequently the depositional axis normally migrates away from the mountain range towards the foreland interior during sedimentation. In the Siwaliks, alluvial fan deposits derived from the Higher Himalayas reportedly have been truncated by the Main Boundary Thrust, above which the most recent elevation of the Lower Himalayas took place in the Quaternary (Hagen, 1969).

TABLE XIV

Mineral deposits characteristic of continental collision belts

Tectonic setting	Association	Genesis	Type of deposit/metals	Examples
Remnant basins	Black shale	Biochemical-chemical sedimentary	Phosphorite	Nevada (L Carboniferous)
Suture zones	Obducted ophiolite	Submarine exhalative sedimentary sulphide	Stratiform Cyprus-type Cu, Fe	Betts Cove, Newfoundland (Ordovician)
	Obducted ophiolite	Magmatic	Podiform Cr	Semail, Oman (Cretaceous)
	Metamorphic	Meta-magmatic	Jadeite, nephrite	Burma (pre-Albian)
Hinterland margins	Regional metamorphism, pegmatites or nepheline syenites	Metasomatic	Gemstones	Mogok, Burma, (Tertiary); Hunza, Kashmir?
Foreland thrust belts	Tectonically emplaced shelf rocks		Deposits of continental shelf	
	"S-type" peraluminous granites	Magmatic–meteoric hydrothermal	Sn, W	Higher Himalayas (Tertiary); SW England (L Permian); Main Range Malaysia (L Triassic)

TABLE XIV (Cont'd)

Tectonic setting	Association	Genesis	Type of deposit/metals	Examples
Foreland thrust belts	"S-type" leucogranite	Magmatic–meteoric hydrothermal	U	Massif Central, France (Devonian); Rossing, Namibia (U Proterozoic)
Foreland basins	"Molasse"	Diagenetic or epigenetic	Stratabound sandstone-type U (Cu, V)	Siwaliks, India and Pakistan (U Tertiary)
	"Molasse"	Chemical sedimentary	Evaporites	Ebro Basin, Spain (Tertiary)
Intramontane troughs	Terrigeneous clastics	Epigenetic meteoric hydrothermal	U, Cu	Europe (Permian)
	Lacustrine sediments	Chemical sedimentary	Evaporites	Tarim Basin, Tibet (Quaternary)

Where the direction of subduction and hence polarity of an ancient collision belt or orogen are uncertain, foreland basin successions can be confused with subduction-related back-arc compressive basin sequences on the overriding plate (Ch. 5, V.A), and with overlying successions of post-collision hinterland basins (p. 296). For example, the Mesozoic basins east of the Canadian Cordillera (Eisbacher, 1977) are back-arc with respect to eastward subduction of ocean floor, but are foreland basins with respect to the possible collision which followed inferred westward subduction between North America and island arc complexes on the overriding plate to the west. In general foreland basin successions can be distinguished from those of back-arc compressive and hinterland basins by their proximity (within less than 200 km) to a suture zone indicated by ophiolitic rocks, the common presence of an underlying shelf succession, and the absence of autochthonous magmatic arc rocks between the basin and the suture zone.

B Mineral Deposits

Foreland basins, like other settings in which predominantly clastic terrigeneous sediments without significant evaporites accumulate, are among the less favourable tectonic settings with regard to mineralization.

Alluvial deposits, in particular gold placers, might be expected in fluvial successions derived from rapidly rising land masses, but there appear to be no reports of economically significant gold from any foreland basin. This may reflect either rapid sedimentation with insufficient reworking or the scarcity of primary gold deposits in the thrust belt source area. Coal deposits might also be expected in deltaic facies of foreland basin successions; their scarcity can possibly be explained either by the tectonic instability of the basins, or by the relative lack of recognizable pre-Cenozoic foreland basin successions.

The only economically significant mineralization known from foreland basins is the sandstone-type uranium of the Siwaliks, described below (Table XIV).

1 Stratabound uranium-vanadium mineralization

In the Siwaliks of the Sulaiman Range of Pakistan, the presence of stratabound "sandstone-type" uranium mineralization has been known for several years. The Siwaliks here occupy a tectonic position along the western margin of the Indus Plain continuous with and analogous to that of the Siwaliks south of the Himalayas, and a similar succession of the same age is present.

The mineralization, described by Moghal (1974), is mostly visible as a bright yellow surface coloration. It is restricted to the Middle Siwaliks which here are mostly coarse-grained fluviatile sandstones, and occurs in pebbly subarkose to subgreywacke channel-fill deposits with limonitized wood fragments and logs. Minerals in the non-oxidized zone are mostly uraninite and coffinite, with tyuyamunite in the oxidized zone; surface showings yield up to 0.5% U_3O_8, and minor vanadium is also present. The mineralization is considered to post-date sedimentation and to be related to ground water movement (Basham and Rice, 1974). A possible source is provided by the nearby late Cenozoic uranium-rich carbonatites of the Indian plate, south of the collision zone in Pakistan (Garson, in press; T. Deans, pers. comm. 1978).

To the east of the Pakistan deposits, uranium has also been reported (Udas and Mahadevan, 1974) in the Siwaliks of India, where it is confined to the middle and upper part of the Lower Siwaliks, occurring mostly in greenish sandstones and interbedded mud-pebble conglomerates. The sandstones are subgreywackes with abundant felspar and mica, and calcite cement. Reported U_3O_8 values range from 0.02 to 0.6%.

In the Siwaliks of India, a possible source of uranium is provided by the anatectic granites and leucogranites of the Higher Himalayas to the north, although in any drainage basin the proportion of these rocks is perhaps too low to produce significant concentration in ground waters by leaching.

2 Stratiform copper mineralization in the Ebro Basin

Stratiform copper mineralization within the evaporite-bearing Eocene to Miocene succession of the Ebro Basin, south of the Pyrenees, has been described by Caia (1976). It can be argued that the succession developed in a foreland basin on the Iberian foreland following early Eocene northward subduction beneath and collision with Europe, but in the absence of better evidence for the subduction polarity this remains highly speculative.

3 Preservation potential

The relative scarcity of recognized foreland basin successions bordering ancient orogenic belts compared with those adjacent to Cenozoic collisional orogens suggests that they commonly do not survive the collision by more than a few tens of millions of years. While this is perhaps due to post-collision isostatic rise of the underthrust foreland basement, it could be expected that progressive erosion of rocks elevated above thrusts on the collision side of the basin would expose tectonically underlying foreland basin sediments. Possibly the poor mineral potential of foreland basins, indicated by the scarcity of economic metallic deposits in the successions of Cenozoic age, has resulted in ancient foreland basin successions attracting little attention as exploration targets; this could imply that a large number of basins remain to be identified.

VI HINTERLAND BASINS

Hinterland basins are of minor importance both as tectonic settings diagnostic of plate boundary relationships and with regard to mineral deposits, but are mentioned briefly here for completeness. We define them as former subduction-related back-arc compressive basins of the overriding plate in which sedimentation continued after collision with a continent or inactive island arc on the subducting plate. Under these circumstances thrusting in the former back-arc thrust belt may be maintained or re-activated during and following collision, resulting in continued sedimentation in the basin. There is some evidence that in the Kun Lun Uplift of northern Tibet, south-dipping thrusts antithetic to the ocean floor subduction direction either became or remained active after the India–Eurasia collision, with related deposition of thick "molasse" facies in the hinterland Tarim Basin to the north. Molasse facies north of the Pamir foldbelt may have a similar origin. Underthrusting of continental crust both beneath back-arc thrust belts during ocean floor subduction and beneath hinterland basins during and following collision is termed A-type subduction by Bally (1975).

No mineral deposits are known to us from the Tarim Basin, although because of their thick mostly non-marine clastic successions, derived partly from plutons and uplifted basement in the back-arc thrust belt, hinterland basins in general should be favourable for the formation of stratabound "red bed" copper and sandstone-type uranium deposits, and possibly for evaporites. It is also possible that minor post-collision uplift of the former back-arc basins, which in some cases included shallow marine successions, could result in evaporite deposition.

VII INTRAMONTANE TROUGHS AND GRABEN OF THE FORELAND AND HINTERLAND

A Tectonic Setting

In foreland basins on the subducting continent, clastic sedimentation normally begins during or following collision and continues after development of the foreland thrust belt. Sedimentation in back-arc compressional basin settings on the overriding continental plate also normally continues during and after the syn-collision change to a hinterland basin setting referred to above. In back-arc oceanic basins behind island arcs on the overriding plate which have collided with a continental foreland, for example east of Taiwan, terrigeneous sedimentation accompanies and follows collision, but these deposits are normally neither exposed nor deformed until involved in a later orogeny.

In addition to these foreland and hinterland basins in continental collision belts, a minor but important post-collision setting for sedimentation and volcanism develops within the foreland thrust belt and in some cases within the hinterland adjacent to the suture.

A late Cenozoic example is provided by the Thakkhola valley succession (Fig. 149) in the Nepal Himalayas. The succession, with a mostly faulted eastern boundary and predominant westerly dip, occupies a NE-trending graben or half-graben more than 100 km long and up to 3 km wide which cuts across the

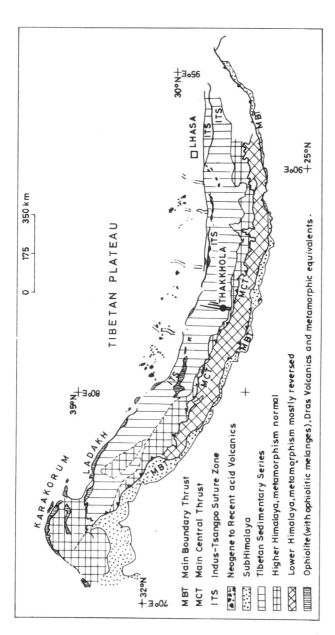

Fig. 149. Main structural features of the Himalaya showing Upper Cenozoic Thakkhola Graben on foreland (after Gansser, 1977, and Bingham and Klootwijk, 1980). Cenozoic graben in hinterland of southern Tibet also shown. ITS, Indus Tsangpo Suture; MBT, Main Boundary Thrust; MCT, Main Central Thrust. By permission of the authors and publishers, Colloque sur l'ecologie et géologie de la Himalaya, No. 268, 181-192, CNRS and *Nature, Lond.* 284, 336-338, Copyright ©1980, Macmillan Journals.

foreland thrust belt of the Himalayas, and is hence clearly post-orogenic. It consists of predominantly alluvial and lacustrine sediments up to 1000 m thick which include widespread polymict conglomerates, locally red in colour. Bordet *et al.* (1971) suggested an Oligo-Miocene age for the Conglomerate and compared it to the Siwaliks foreland basin succession, but an Upper Miocene to Quaternary age seems more probable.

In Tibet, elevation of the volcanic arc, and of the back-arc cratonic basin which evidently occupied much of the area to the north, followed collision with India. Uplift was accompanied over an enormous area in northern Tibet and north-eastern China by movement on high-angle reverse faults, for example the Kun Lun and Altyn Tagh faults, along many of which sinistral movement displaced China eastwards as a result of the collision (Ch. 7, I.A); extensive reactivation of old faults and thrusts has also been interpreted as the response of the overriding plate to the collision (Tapponnier and Molnar, 1977). This suggests that in some older orogens major strike-slip faults, considered by some authors to be the cause of the orogeny, are post-orogenic faults resulting from the collision. Of particular relevance here, in southern Tibet N- to NNE-trending late Cenozoic graben (Fig. 149) can be recognized on topographic maps and satellite imagery, and ring structures, possibly representing late Cenozoic rift-related magmatism, have been identified in some of the graben (K. Burke, pers. comm. 1980).

There is evidence in Tibet of post-collision silicic potassic volcanism which accompanied both predominantly lacustrine sedimentation in fault troughs and hydrothermal activity which continues today. Dewey and Burke (1973) suggested that the Tibetan Plateau is a zone of intense post-collisional basement reactivation with regional metamorphism, emplacement of potassic granites and anorthosites at depth, and generation of migmatites analogous to that in many Precambrian collision belts. However, while post-Palaeogene volcanism in southern Tibet is possibly related to underthrusting of the Indian plate (Bingham and Klootwijk, 1980) it may also be due at least partly to post-collision rifting, and possibly the calc-alkaline and alkaline basalt volcanism in Central Tibet (Bally *et al.*, 1980) is also rift-related. We consider that in most collision belts "basement reactivation" probably takes place in the thrust belt of the fore-land, and in the former back-arc thrust belt of the hinterland, where present, rather than over a very extensive area on the overriding plate as postulated for Tibet. Therefore development of the intramontane basins and volcanism does not necessarily imply basement reactivation at depth.

In the Northern Appenines region early Tertiary continent–arc collision along a suture between Italy and Corsica was followed in the Pliocene by a change to extensional tectonics (see Fig. 136), with development of horst and graben structure and formation of the Tyrrhenian Sea, alkalic volcanism and an increase in the geothermal gradient (Kligfield, 1979). This post-collision volcanism and block faulting corresponds to the post-geosynclinal period of geosynclinal development of Aubouin (1965) based on the late Cenozoic geology of Greece.

Within the Hercynian orogenic belt of Europe, the post-orogenic basins of Stephanian age, in particular in southern Brittany, the Massif Central and Bohemian Massif, were noted by Dewey and Burke (1973). They compared these basins to similar basins of Tertiary age in Tibet, suggesting that the Hercynian basins formed on an overriding southern European continental plate following collision with northern Europe. More recent interpretations suggest that the Hercynian orogeny in the Massif Central took place on a continental foreland following northward subduction and collision with an arc system in northern France (IV.B.I.(g)), and was distinct from and older than the end-Carboniferous orogeny following southward subduction in Southwest England. According to this model, the post-collision Hercynian basins of Europe developed in the foreland thrust belt, rather than on the hinterland as in Tibet.

The European Hercynian basins are mostly narrow linear intramontane troughs, post-dating major collision-related thrusting and related to normal faults. Sedimentation was mostly non-marine and clastic and was accompanied by potassic ignimbrite volcanism. Arthaud and Matte (1977) emphasized the graben or half-graben like features of these and similar early Permian basins beyond the orogen in Britain, and related them to wrench faulting caused by westward movement of Africa relative to Europe to the north.

Late Palaeozoic post-collision intramontane basins in northwestern Africa have been described by Van Houten and Brown (1977). Following collision of Africa with North America and Spain at the end of the Carboniferous, thick clastic sediments accumulated on the African Platform south of the orogen, and sedimentation also took place in small basins on the African plate within the collision belt. Van Houten (1976b) and Van Houten and Brown (1977) described these basins, developed within the mountain range, as being locally fault-bounded with successions up to 2000 m thick, comprising non-marine polymict conglomerates overlain by finer-grained locally red to brown detrital sediments, lignites and thin coals. The successions lie in angular unconformity on older rocks and in places include acidic to intermediate lavas.

The development of major post-collision rifts such as the Rhine graben, formed on the former either overriding or subducting plate, has been considered together with hot-spot related rifts in Ch. 2. Presumably the tectonic processes which result in the major collision-related rifts, of which the Baikal rift is also a possible example, are related to those which form the smaller rifts and intramontane basins such as those of Tibet and the Hercynian orogenic belt of Europe described above.

B Mineral Deposits

Mineral deposits in post-collision intramontane troughs (Table XIV) are apparently largely restricted to low-grade uranium mineralization, lignite and mostly low-rank coals, e.g. in the successions referred to above in France and Northwest Africa, and salt lake deposits on the Tibetan Plateau. It is uncertain whether the diamond occurrences reported from Tibet are related to the late Cenozoic rift zones, in which the only economic deposits known to us are of native sulphur (Bally *et al.*, 1980).

The salt lake deposits of the Tsaidam Basin northeast of the Tibetan Plateau have a rather limited preservation potential due not to their economic importance alone, and hence are unlikely to be typical of ancient post-collision intramontane basins. The lakes contain chlorides of sodium, potassium and magnesium, borate and less abundant salts of lithium, bromine and iodine. The basin is a major producer of the chlorides and borax.

1 Sandstone-type uranium deposits

The common presence of anatectic granites in foreland thrust belts suggests that the intramontane troughs might be a favourable environment for formation of post-tectonic sandstone-type uranium deposits, the uranium being leached from the granites which may have high background uranium values and can include economic deposits. Although none is known from the late Cenozoic Thakkhola Graben in the Himalayas, uranium occurrences in a comparable setting are present in Autunian successions of red and grey terrigeneous sediments in the intramontane basins of the Massif Central of France, where uranium leached from collision-related Hercynian granites was deposited in palaeochannels with carbonaceous material. Highest grades, up to 0.1% U, are mostly restricted to boundaries between red and grey sediments. Small uranium deposits also occur in the post-collision Lower Rotliegende of Germany (Barthel, 1974). Tischler and Finlow-Bates (1980) have suggested that "red bed" sandstone-type uranium and gypsum deposits in the Upper Austro-Alpine unit of the Eastern Alps accumulated in Permian graben developed across the Hercynian collisional orogen (Fig. 150).

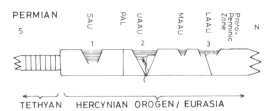

Fig. 150. Permian post-collision graben developed in the Hercynian orogen of the Eastern Alps. 1, Tregiovo Pb–Zn mineralization; 2, Hochfilzen-Gierberbrunn red-bed style uranium and gypsum deposits in the extreme Eastern Alps. The upward movement of mantle-derived material along graben-features is indicated by the presence of the Mitterberg Ni–Co mineralization (3). SAU, South Alpine Unit; PAL, Periadriatic Lineament; UAAU, Upper Austro-Alpine Unit; MAAU, Middle Austro-Alpine Unit; LAAU, Lower Austro-Alpine Unit (from Tischler and Finlow-Bates, 1980). By permission of the authors and publisher, *Mineral. Deposita* **15**, 19–34, Springer-Verlag, Heidelberg.

2 Trachytic volcanic rocks with uranium and fluorite

The uranium and fluorite mineralization associated with Quaternary alkaline volcanic rocks in the Rome Province (Locardi and Mittempergher, 1971) is of stratiform type in a rift setting which post-dates the Cenozoic Appenine collision. The alkaline rocks cover an area of 5000 km^2 between Rome and Orvieto (Fig. 151) and comprise volcanic centres surrounded by trachytic to latitic pyroclastics containing up to ten times the average crustal content of uranium. Large basins between the volcanic centres are infilled with fluvio-lacustrine sediments derived from them; baryte, calcite and very fine-grained fluorite comprise up to 60% of some beds. Secondary concentration of uranium within the volcanic and volcaniclastic rocks by leaching is believed to be related to the continuing exhalation of CO_2 + H_2S gas from faults in the basement (Fig. 152; Kimberley, 1978).

A similar type of mineralization associated with trachytic rocks in a presumed

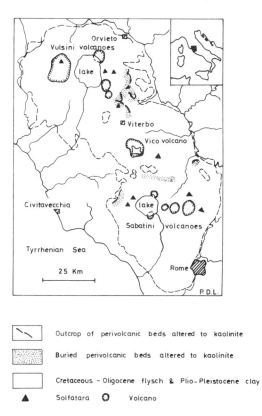

Fig. 151. Uraniferous Quaternary volcanic rocks in Italy in a possible rift-related tectonic setting (from Kimberley, 1978, modified from Locardi and Mittempergher, 1971). By permission of the authors and publishers, *Min. Ass. Can.* **3**, and *Bull. Volcanologique* **35**, 173–184.

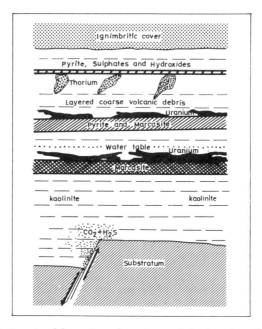

Fig. 152. Postulated mode of formation of uranium and thorium deposits within a Quaternary basin in Italy (from Kimberley, 1978, modified from Locardi and Mittempergher, 1971). Volcanic gas emanates into groundwater from a fracture zone with resultant alteration. Fe, U, Th are leached downwards. Uranium and ferrous sulphides precipitate along the water table. Silicification diminishes downwards and authigenic kaolinite to opal ratio increases. By permission of the authors and publishers, *Min. Ass. Can.* **3**, and *Bull. Volcanologique* **35**, 73–184.

rift setting occurs at Rexspar in British Columbia (Joubin and James, 1956). The stratiform mineralization consists of layers carrying very fine-grained uraninite with thorium and rare earth minerals, together with fluorite, celestite and Fe–Mg mica or vermiculite and pyrite; this is intercalated with layers of tuffaceous fragments with pyrite, fluorite and celestite.

3 Intramontane graben, collision-related rifts, and preservation potential

Of more importance than these examples of intramontane graben with regard to mineralization are the similar and perhaps analogous collision-related intra-continental rift zones described above (Ch. 2, II.A.2), which include the carbonatites of the Rhine graben and possibly the copper deposits of the Kupferschiefer and overlying evaporites. The similarities between intramontane graben and the intracontinental rift successions suggest that stratabound copper deposits could also occur in the graben, although it is arguable that in the absence of major evaporites significant copper mineralization is unlikely to be present. In view of the difficulty in distinguishing pre-Mesozoic collision-related rifts from those generated above hot-spots, comparisons of mineralization in intramontane graben with that in collision and hot-spot-related rifts are not pursued further here.

The preservation potential of deposits in intramontane graben, particularly those within the foreland thrust belt, is limited by continuing uplift and erosion of the orogenic belt long after collision, indicated by the fact that in many collision belts of Mesozoic and Palaeozoic ages or older the foreland thrust belt still forms a mountain range.

CHAPTER 7

Transform Faults and Lineaments in Continental Crust

In this chapter we consider aspects of mineralization in transform faults which intersect continental crust and island arcs and discuss the possible relationship of mineralized lineaments within continents to ancient transforms. As might be expected with the waning in popularity of the fault-controlled epigenetic hydrothermal hypothesis of origin for many types of ore body, relatively few types of mineralization are considered to be associated with transform plate boundaries compared to those in rift, subduction and collision-related settings.

Transform faults form plate boundaries along which motion is dominantly strike–slip with plates sliding past one another with relatively minor convergence or divergence (Ch.4, II.A), in contrast to spreading plate boundaries at which floor is created, and to consuming plate boundaries along which ocean floor or continental crust is lost. While major lateral movement of slices of continental crust commonly takes place at these strike–slip boundaries, they are not the direct cause of extensive magmatism or orogeny.

Non ridge–ridge transform faults in the oceans, and some major transform faults within continental crust, have been referred to as fundamental transforms by Gilliland and Meyer (1976), and contrasted with transforms related to ocean ridges. However, the projected extensions of some ridge–ridge transforms intersect continental crust, and it is more satisfactory here to consider all types of transform fault other than those confined to oceanic crust.

For convenience we divide transform faults into those which are genetically related to oblique subduction of ocean floor, and extensions of transforms related to the start of ocean floor spreading beneath intracontinental rifts. We also discuss briefly the possible relationship between some continental lineaments and mineralization.

I SUBDUCTION-RELATED TRANSFORMS IN CONTINENTAL MARGINS AND ISLAND ARCS

A Tectonic Setting

Transform faults related to subduction are developed in or adjacent to continental margin or island arc systems as a result of oblique subduction, and on former passive continental margins during and following continental collision. The precise tectonic settings are highly variable and best illustrated by a few examples described below, but certain geological features are common to many transforms.

The Sagaing-Namyin Fault in Burma (see Fig. 130) is an example of an active transform fault in a back-arc position linking the small Andaman Sea spreading centre (Curray *et al.*, 1978, 1980) with an intracontinental thrust (Mitchell *et al.*, 1978). The Fault extends northwards for more than 1200 km from the Andaman Sea marginal basin spreading centre to northernmost Burma where it swings northwest and becomes the southwest-directed Miju Thrust. It lies mostly within rocks of Palaeozoic age or older, and offsets of major tectonic belts suggest a dextral displacement of at least 400 km. Much or all of this movement is probably post-early Miocene in age, although in one segment of the fault mid-Tertiary sediments show little apparent displacement. Movement is no doubt related fundamentally to northward movement of the Indian plate with respect to the Asian plate, but is more specifically associated with opening of the Andaman Sea. The fault shows some similarities to the Median Tectonic Line of Japan.

The transform fault investigated in most detail is the San Andreas Fault (see Fig. 120), a ridge–ridge transform extending for 1200 km through continental crust of westernmost North America. The Fault is characterized by a sharp bend near its centre and by numerous splay faults resulting in displaced tectonic slices of crust. Late Cenozoic sinistral movement of at least 300 km is generally accepted, although the total displacement including that along splay faults may be much greater (Crowell, 1979). Although a ridge–ridge transform, the San Andreas Fault's initiation and growth resulted from oblique subduction of Pacific ocean floor beneath the continent followed by impingment of a rise system on the subduction zone. Attempted subduction of the rise-transform system led to northward and southward migration of triple junctions joined by the developing San Andreas transform (Atwater and Molnar, 1973).

The Central Range Fault in Taiwan provides an example of an active transform formed as a result of diachronous and oblique continent–arc collision. Following eastward subduction of the South China Sea beneath the Manila Trench, the related island arc system collided with the Asian mainland to the west in the Pliocene (Biq Chingchang, 1971). Oblique plate convergence was resolved into a component of thrusting perpendicular to the arc in the continental shelf and rise succession forming a foreland thrust belt, and a dextral transform immediately west of the suture. The transform terminates in the eastern end of the Ryukyu Arc to the north, and in the still active zone of oblique subduction to the south (see Fig. 129).

Strike-slip or transform faults developed in the overriding plate following continent-continent collision have been postulated following the recognition on satellite imagery of the major faults in Asia north of the Himalayas (Ch. 6, VII. B). Tapponier and others (e.g. Molnar and Tapponier, 1977) have identified a number of very large predominantly easterly trending faults north of the Himalayas, mostly showing features indicative of strike-slip movement since the Eocene and with some of which large earthquakes are associated. They relate these, for example the Altyn Tagh and Kun Lun Faults, to eastward movement of China since its collision with India in the Eocene, and consider that the Lake Baikal rift and Shansi graben can be interpreted as tensional feature related to the strike-slip faults.

B Mineral Deposits

In general subduction-related transform faults in continental crust show little associated mineralization at the present erosion level, and the clastic successions characteristic of small basins developed as a result of the displacement are also normally unmineralized. For example, the Sagaing–Namyin Fault and its various splay faults all lack mineralization, although the Fault transects older mineral provinces with different metals and of different ages. Similarly, the San Andreas Fault, the Central Range Fault in Taiwan, and the Semangko Fault in Sumatra are unmineralized, and there is little evidence for mineralization genetically related to the Philippines Fault or the Median Tectonic Line of Japan. However the location of the metal-rich Salton Sea within the San Andreas fault system (White, 1968), although not underlain by normal continental crust, suggests some potential for mineralization in subduction-related transform faults, while a carbonatite apparently lacking mineralization at its present erosion level is located on the Alpine Fault in New Zealand (Cooper, 1971).

The only sulphide deposits considered by us to be possibly genetically related to a Cenozoic subduction-related transform are the calcite-quartz-stibnite bodies in Pakistan (Sillitoe, 1978) referred to above (Ch. 5, I. B.3). The mineralization occurs within mid-Tertiary flysch but is localized along faults adjacent to the major Chaman transform. Sillitoe has explained the mineralization by circulation of connate fluids during metamorphism related to fault movement, and has compared it to the major Upper Cenozoic mercury deposits of the Coast Ranges of California that are perhaps associated with the San Andreas transform system (Garson and Mitchell, 1977).

Despite the scarcity of mineral deposits related to Cenozoic subduction-related transforms, it is not impossible that some types of mineralization form

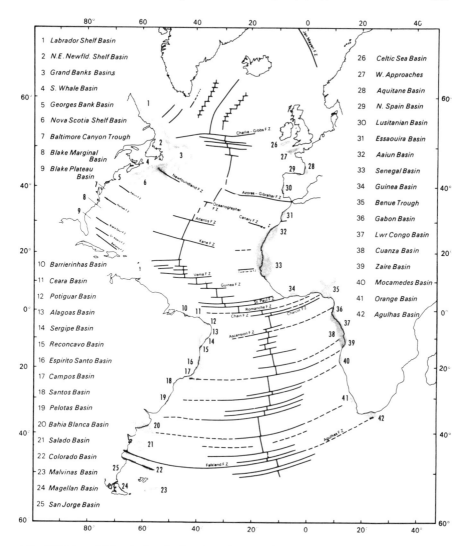

Fig. 153. Map of Atlantic showing location of oceanic fracture zones and continental margin basins (from Wilson and Williams, 1979). Note position of Cuanza Basin (38) tensional rifted margin basin, referred to in Ch. 2. By permission of the authors and publisher, *J. geol. Soc. Lond.* **136**, 311–320.

along the faults at deep structural levels, associated either with ultrabasic intrusions or more probably related to circulation of mineralizing brines for which the faults provided a conduit. This possibility of mineralization along deeply eroded transforms is considered together with continental lineaments discussed later.

II RIDGE-RIDGE TRANSFORM FAULT EXTENSIONS INTO
CONTINENTAL MARGINS

A Tectonic Setting

Transform fault extensions related to incipient development of ocean rises can be divided into those formed in intracontinental rifts prior to emplacement of ocean floor, and those formed with the onset of ocean floor emplacement and now preserved in continental margins. The latter, which are more significant features, comprise the postulated continuation into continents of the major ocean ridge-ridge transforms, and most of the following section is concerned with fault extensions inferred to have developed duirng the early stages of ocean floor spreading.

Wilson's (1965a) concept that offsets along ridge-ridge transform faults can be matched with offsets in the margins of formerly adjacent continents is now generally accepted, in particular with reference to the western Africa-eastern South America coastlines. Less widely accepted is Wilson's (1965a) suggestion that during their initial formation prior to ocean opening, transform faults normally coincide with ancient continental fractures. This concept of transform fault extensions along continental faults is of particular interest because of the frequently postulated relationship of some mineralized lineaments to the continental extension of transform faults and mineralization in North America.

The role of oceanic transform fault traces in controlling development of continental margin sedimentary basins around the Atlantic has been discussed by Wilson and Williams (1979), who recognized three types of basins according to the orientation of their long axes relative to the continental margin and related transform faults (Fig. 153). Tensional rifted margin basins lie parallel to the continental margin and perpendicular to transform faults which form the basin ends; they are typical of passive continental margins. Taphrogenic basins lie perpendicular to the continental margin and parallel to the transform direction, and Wilson and Williams consider the Benue Trough as the best example. Sheared margin basins form parallel to the continental margin and transform fault and are hence best developed in the equatorial region of the Atlantic margins of Africa and South America.

The most direct relationship of modern ridge-ridge transform faults to continental fractures is probably that reported from Nigeria and Cameroun (Wright, 1976) where the Romanche, Chain and 27°S fracture zones of the South Atlantic are more or less continuous with belts of major faulting extending far into the continent and considered to represent deep lineaments (cf. Fig. 155). Dextral displacements of at least 10 km are reported from both the fault zones in Nigeria, and intrusive and extrusive igneous rocks are concentrated in or near the zones. Wright (1976) suggested that the last major displacements on the faults were in the latest Precambrian, although minor vertical movements may have taken place subsequently.

In the Red Sea area, Garson and Krs (1976) have recognized numerous

NE-trending fractures in Egypt and Saudi Arabia, many of which extend seawards into the axial zone of the Red Sea (see Fig. 157). These are interpreted as old (Precambrian) structures which extended seaward during the early Tertiary spreading episode in the Red Sea, presumably linking up with ridge–ridge transforms.

In South America and Angola, Marsh (1973) recognized lineaments defined by alkalic intrusions which he interpreted as continental extensions of Cretaceous transforms which offset the Mid-Atlantic Ridge (Fig. 154). However, Neill (1973) interpreted belts of Mesozoic subvolcanic alkaline ring complexes in the same regions as rifts which became failed arms of trilete systems, the other arms having developed to form the South Atlantic Ocean. More recently, Herz (1977) had suggested that in Brazil the Panama Basalts and alkalic rocks of Lower Cretaceous age were related to a failed rift, and that some of the Upper Cretaceous and younger volcanism and sedimentation has been controlled by drift of the continent over a hot spot now underlying the island of Trinidad. To the east, in the central Brazil shield, Upper Jurassic or lowermost Cretaceous ring complexes of the Velasco alkaline province define a NE-trending lineament considered to be related to the initial rifting of the Africa–South America plate, but not associated with a failed rift or transform extension (Darbyshire and Fletcher, 1979).

The relationship of mineral deposits to postulated transform extensions in these four regions is discussed below.

If ridge–ridge transforms develop preferentially along lines of pre-existing weakness in continental crust, it would be expected that intracontinental rifts such as the East African Rift system should be intersected by easily recognizable deep-seated fractures. While transverse structures within the East African Rift have been identified as favourable sites for rift magmatism, there is little evidence for major reactivation of old faults extending beyond the Rift.

B Mineral Deposits

In recent years an extensive literature has accumulated on the relationship of several types of mineral deposit to on-land extensions of oceanic transforms. These deposits include carbonatites, kimberlites, and basic and ultrabasic intrusions with Cu, Ni, Pt, Au and Ti on or landward of passive continental margins, metalliferous brines in the Red Sea deeps, volcanogenic deposits in the East Japan Arc, and mineralized pegmatites (Table XV). In view of this extensive documentation of mineralization associated with Upper Mesozoic and Cenozoic transform extensions, it is surprising that older deposits distributed along similar faults are not more numerous.

It may be noted that in the transform fault-bounded Atlantic margin basins described above there is some relationship between the faults and mineral deposits within the basins, although this relationship is likely to be most direct

TABLE XV

Mineral deposits characteristic of continental transform faults.

Tectonic setting	Association	Genesis	Type of deposit/metals	Examples
Subduction-related transform faults	Calcite–quartz veins and pods	Connate hydrothermal	Sb	Chaman transform, Pakistan (U Tertiary)
	Carbonatites	Magmatic–metasomatic	Alkaline igneous complexes	Angola, Namibia, Brazil (Cretaceous–L Tertiary)
	Kimberlites	Magmatic	Diamonds	W Africa, Brazil, Australia (mostly Cretaceous)
Ridge–ridge transform extensions	Basic–ultrabasic intrusions	Magmatic	Cu, Ni, Pt, Au, Ti	Freetown, Sierra Leone (Jurassic); SE Desert Egypt
	Brine pools and basalt	Submarine exhalative	Metalliferous sediments	Red Sea (Recent)
	Pegmatites	Magmatic–hydrothermal	Sn, Li	Peninsular Thailand (U Mesozoic)

in the case of petroleum. The Cuanza Basin (see Fig. 153) with stratiform copper deposits overlain by evaporites (Ch. 2, II.B.2) is an example of the tensional ritted margin basin of Wilson and Williams (1979), and the northern and southern terminations of the basin were clearly related either to transform fault extensions or to hot spot tracks as discussed above.

1 Carbonatites in Namibia, Angola, Brazil and Uruguay

In Angola, Namibia, Brazil and Uruguay, most of the alkaline igneous complexes and carbonatites occur either along distinct lineaments (Marsh, 1973), or at intersections of these lineaments with fractures trending roughly parallel to the Mid-Atlantic Ridge (Rodrigues, 1972). In Brazil, several alkaline igneous complexes and important carbonatite plutons, e.g. Araxa, Tapira and Catalao I and II, occur within a narrow rift-zone 700 km long (Alves, 1960) trending NNW roughly parallel to the spreading ridge. The lineaments lie along small circles centred on the early Cretaceous (120-135 Ma) and 80 Ma to present poles of rotation for the south Atlantic (Marsh, 1973), which can be correlated with unique transform faults off-setting the Mid-Atlantic Ridge (Fig. 154).

Most of the complexes were emplaced during the initial rifting of the Africa-South America plate in the early Cretaceous. Three 80 Ma old complexes in the Brazil Southern Lineament and complexes in the Brazil Northern Lineament (dated at 51-83 Ma), and possibly several undated complexes were emplaced when the South Atlantic pole of rotation shifted northwards to the present position. Marsh considered that consequent stresses, propagated along the transform faults and their continental extensions, resulted in renewal of alkaline activity. The problem remains why continental alkaline complexes are associated with only some transform extensions and not others. Possibly the transform fault positions were influenced by pre-rift continental fracture zones as is apparent for the localization of transform faults in the Red Sea area (Garson and Krs, 1976) referred to below.

Mineralization of economic interest associated with carbonatites in the continental transform fractures in Angola (Lapido-Loureiro, 1973), Namibia and Brazil (Deans, 1966), includes pyrochlore in residual soils, large tonnages of apatite in eluvial residual material at Araxa, Tapira and Jacupiranga in Brazil, and fluorite deposits of 7-10 million tonnes carrying more than 35% CaF_2 at Okorusu in Namibia.

2 Kimberlites in western Africa, Brazil and Australia

The kimberlite pipes of the Lucapa graben in Angola are situated on the extreme north-eastern elongation of one of the major NE-trending continental transform fault lineaments (Rodriques, 1972) referred to in the previous section (see Fig. 154). Most of the other kimberlitic centres fall on parallel lineaments. In general diamondiferous diatremes form kimberlitic provinces at the intersection

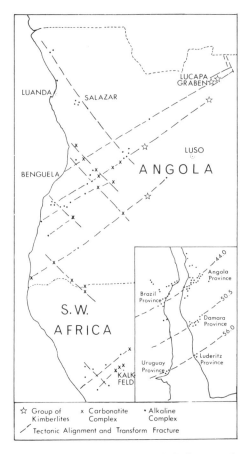

Fig. 154. Distribution of kimberlites, carbonatites and alkaline complexes in Angola and SW Africa, with (inset) complexes centred on extensions of oceanic transforms forming small circles centred on Cretaceous pole of rotation, distances from pole in degrees (from Marsh, 1973). By permission of the author and publisher, *Earth planet. Sci. Lett.* **18**, 317-323.

of faults running NE and NNW- trending faults (Reis, 1970) within the Congo-Angolan craton, while alkaline complexes and carbonatites characterize the more highly fractured coastal sections. This tectonic setting is comparable to that of the diamondiferous kimberlites in the Tanzanian craton and the carbonatite complexes of east Africa situated near or within the main rift zones (see Fig. 29). Bailey (1980) has postulated that kimberlites are the limiting case of cratonic magmatism with different modes of alkaline-carbonatite magmatic activity produced by volatile fluxing along steeper geotherms away from the cool cratonic nucleii.

In Brazil, kimberlites have been located only during the past decade, although diamonds have been known since the 1700s. Published information so far is scanty but in western Minas Gerais district, kimberlite pipes of probable Upper

Cretaceous age in the Coromandel–Monte Carmelo area are known to occur on the flanks of the rift-like fracture zone with which are associated the Tapira, Araxa, Serra Negra and Catalao carbonatites (Svisero *et al.,* 1976). The kimberlites and the belt of carbonatites are situated within a NW-trending zone of Precambrian rocks separating the Parana and Alto San–Franciscana Basins. The distribution of diamonds in Brazil suggests that this zone may extend into the Mato Gosso district, i.e. a distance of over 1000 km; on the same trend some hundreds of kilometres inland, diamonds and a possible kimberlitic pipe have been located in Rondônia (Svisero *et al.,* 1976). Northwest of Rio de Janeiro the zone is intersected by a NE-trending zone of diamond localities adjacent to the 44° continental transform lineament, along which numerous alkaline complexes are distributed in Brazil, and alkaline complexes, carbonatites and diamondiferous kimberlites are located in Angola (Marsh, 1973). The presence of kimberlites within both these intersecting fracture zones suggests that the transform fault extension tapped the NW-trending zone, believed to be a reactivated Precambrian fracture, and that in each zone similar tectonic conditions allowed rapid access from the 200 km depth considered necessary for formation and preservation of diamonds in kimberlite pipes.

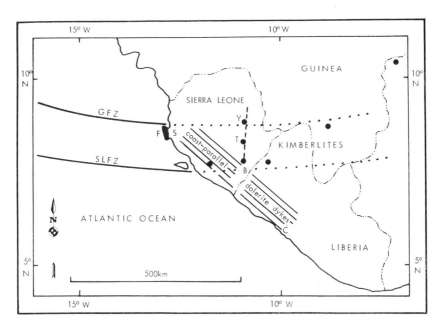

Fig. 155. Sketch map showing the relationship between magmatism and regional structure in Sierra Leone and adjacent countries. B, Baghe; C, Cape Mount; F, Freetown Complex; S, Songo; T, Tongo; Y, Yengema; GFZ, Guinea Fracture Zone; SLFZ, Sierra Leone Fracture Zone; solid circles, kimberlite; dotted lines, hypothetical continental continuations of fracture zones; dashed line, major fault controlling kimberlite magmatism (from Culver and Williams, 1979). By permission of the authors and publisher, *J. geol. Soc. Lond.* **136,** 605–618.

In the northwestern bulge of Africa, kimberlites are also considered to be situated along continental extensions of transform faults in Guinea, Mali, Sierra Leone, Liberia, the Ivory Coast and Ghana (Williams and Williams, 1977). The relevant oceanic transforms are the Vema, Guinea and Sierra Leone fracture zones (Fig. 155), which are believed to pass inland into brittle fractures within cratonic crust from 2700–2000 Ma in age. There is some evidence that these continental fractures are reactivated pre-rift features. The magmatic events in the western Africa-eastern South America region may be summarized as:(a) intrusion of dolerite dykes associated with tensional stresses prior to the break-up of this part of Gondwanaland from 180–140 Ma;(b) eruption of basaltic lava flows at 130–120 Ma during incipient drift and formation of oceanic crust followed by; (c) alkali volcanism and carbonatite eruptions from 120–50 Ma during active drift; (d) soon after the initial carbonatite activity, kimberlites were intruded in continental extensions of transform faults at about 90 Ma.

Wyllie (1967) has noted that within this part of Africa there are kimberlites of at least four different ages (2300–2000 Ma, 1150 Ma, 700 Ma and 100–80 Ma). Similarly in South Africa, Precambrian diamondiferous kimberlites occur in the same general area as Cretaceous kimberlites. Relevant tectonic features related to kimberlitic activity of different ages in the same area were probably an already thick cratonic crust in the Proterozoic, and the formation of cracks reactivated during periods of stress as the result of rifting and transform faulting.

Within southeastern Australia and Tasmania there are fourteen areas of kimberlitic rocks of Permian, Jurassic and Tertiary age; according to Stracke *et al.* (1979) these occur along continental extensions of transform faults, related to both the Antarctic and Tasman Sea Ridges (Fig. 156). Some

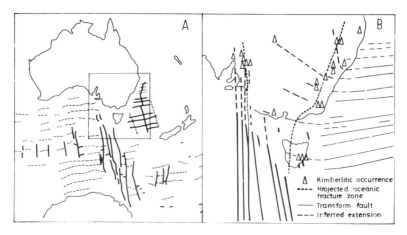

Fig. 156. (A) Tectonic map of the ocean floor in the Australian/Antarctica area; (B) location of kimberlites in south-eastern Australia in relation to transform faults and inferred continental extensions (from Anon. 1978). By permission of the Dept National Development, Canberra.

kimberlitic areas lie at the intersection of the projected continental transforms extending from the two spreading centres while many of the kimberlitic rocks are situated on pre-rifting fractures which later became sites of transform fault extensions.

No diamonds have been located in any of these kimberlitic rocks, some of which have chemical characteristics of normal South African-type kimberlites while others have more close chemical affinities with nephelinites (Ferguson and Sheraton, 1979). The lack of diamonds can be attributed to the absence of cold cratonic crust of age greater than 1500 Ma and the situation of the kimberlitic rocks within a wide region of high heat flow stretching diagonally northwest across Australia (Cull and Denham, 1979). The only diamonds so far confirmed in Australia are situated in the Ellendale area of Western Australia at the margin of the Kimberley craton. There is an apparent association with a SE-trending lineament (Garlick, 1979) perhaps related to an oceanic fracture.

3 Basic/ultrabasic intrusions with Cu, Ni, Pt, Au, Ti

Intrusions of layered basic/ultrabasic rocks within fracture zones considered to be related to transform faulting have been described from Sierra Leone and the South-Eastern Desert of Egypt.

(a) The Freetown layered basic complex, Sierra Leone

A sheet-like layered basic/ultrabasic complex at Freetown within a rift-like structure consists of a 6 km thick series of gabbroic cumulate rocks comprising layers of dunite, troctolite, olivine gabbro, gabbro, leucogabbro and anorthosite (Wells, 1962). Ilmenomagnetite-rich troctolite layers occur at the bottom of rhythmic sequences, and sulphides are ubiquitous as immiscible droplets and late-stage hydrothermal veins and replacements (Bowles, 1978). Platinum occurs in placer deposits in the area, and Cu, Ni, Pt and Au are associated with sulphides in one part of the complex (Stumpfl, 1966). The Freetown complex is about 193 Ma old (Beckinsale *et al.,* 1977), of similar age to an adjacent swarm of basic dykes paralleling the coast which are considered to be extensional features developed during early Atlantic opening (Culver and Williams, 1979). Williams and Williams (1977) have suggested that the complex occurs at the intersection of the Guinea Fracture Zone and the Atlantic protorift (see Fig. 155) i.e. its position is related both to early rifting and to the continental extension of a transform fault.

(b) Basic/ultrabasic layered complexes, South-Eastern Desert of Egypt

In southeastern Egypt, dominantly NNW-trending folded belts of ophiolitic rocks of Proterozoic age are offset by a series of N60° -trending, deep-seated fracture zones, first recognized from geophysical evidence (Garson and Krs, 1976). These fractures are locally intruded by layered ultramafic and basic rocks interpreted as continental extensions of transverse faults, later rejuvenated during sea-floor

spreading in the Red Sea (Garson and Shalaby, 1977). Areas of metavolcanics
with Kuroko-type mineralization are also apparently centred on some of these
fractures.

The largest intrusion (3 km x 10 km) occurs at Gabbro Akarem, 130 km
due east of Aswan. Two intrusive phases have been identified, a lopolithic

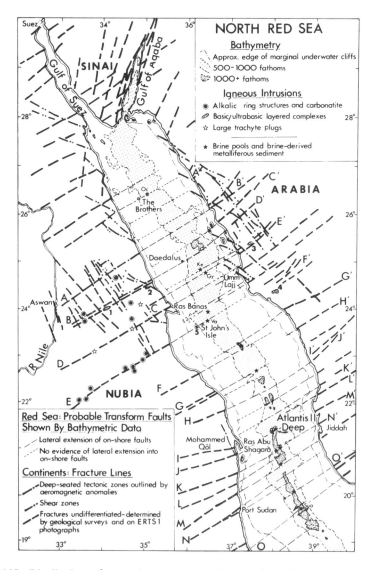

Fig. 157. Distribution of tectonic zones, probable transform faults, related igneous
intrusions, brine pools and brine-derived metalliferous sediment in the northern Red Sea
area (from Garson and Krs, 1976). By permission of the authors and publisher, *Bull. geol.
Soc. Am.* 87, 169–181.

unmineralized norite with mafic layers and a later suite of steeply dipping mineralized dykes and plugs of peridotite, plagioclase–peridotite, melanorite and pyroxenite. The mineralization occurs as disseminated and massive sulphides in magmatic segregations up to 15 m thick consisting of pyrrhotite, mackinawite, pentlandite, chalcopyrite and cubanite.

Titanium–iron mineralization occurs in gabbroic intrusions, also probably in transverse tectonic structures parallel to the major zone through Gabbro Akarem. The largest deposit is at Abu Ghalaga where reserves of more than 3 million tonnes of ilmenite average 36% TiO_2. The steeply-inclined lenses of ore show a distinct near-horizontal layering of massive ilmenite, ilmenite/chalcopyrite and gabbro.

4 Base metal deposits in the Red Sea area

A study in the Red Sea area of the distribution of transform faults, and of faults presumed from bathymetry and the on-shore fracture pattern to be of transform type (Fig. 157), shows that the Recent brine pools and metalliferous sediments are almost certainly located at or very near to the intersection of transform faults with the Red Sea spreading ridge (Mitchell and Garson, 1976). Similarly most of the Miocene base metal deposits in the coastal areas are located on extensions of transform or presumed transform faults. Deposits include extensive manganese mineralization in Sinai and at Erba (associated with baryte) and relatively minor occurrences elsewhere. At Ras Banas, Miocene sandstones are impregnated with sub-economic secondary copper minerals. The ancient working at Umm Gheig in Middle Miocene lime-grits and calcareous sandstones consists of secondary zinc and lead minerals with occasional primary galena and several percent of strontianite and baryte. In general in these

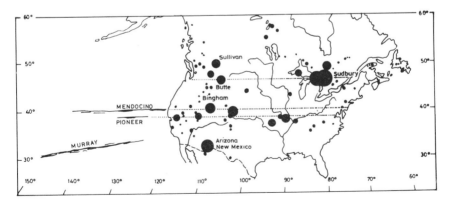

Fig. 158. Distribution of the main hydrothermal ore deposits of USA and adjacent parts of Canada. Note broad belt of large ore deposits extending eastwards along landward projection of Mendocino and Pioneer fracture zones (from Kutina, 1980a). By permission of the author and publisher, *Global Tectonics Metallogeny* **1**, 134-193.

deposits the content of zinc exceeds lead and that of strontianite exceeds baryte.

The Miocene Zn, Pb, Cu and Mn deposits evidently provided a readily available source of metals for the younger metalliferous deposits of the Red Sea deeps, although it is arguable whether a source other than ridge basalt is necessary (Ch. 4, I.B.1. (b). (ii)). Bignell (1975) has postulated that even a volcanic heat source is unnecessary in the Red Sea deeps to produce circulation, the geothermal gradient alone being adequate to heat the brines and produce movement of the solutions through Miocene evaporites, with subsequent discharge of metalliferous brines at the intersection of transform faults and the spreading ridge. A similar process may have operated in Miocene times, the metals having been scavenged from Proterozoic rocks including both ophiolites and island arc rocks with Kuroko-type deposits.

5 Hydrothermal ore deposits landward of the Mendocino Fracture Zone

Kutina (1974, 1980a) has developed the hypothesis of concentration of ore deposits along equally spaced east–west continental lineaments, most of which he interprets as the on-land extension of oceanic transform faults. In particular he has emphasized how in North America a broad E-trending belt of hydrothermal ore deposits extends landward of the Mendocino and Pioneer Fracture Zones of the East Pacific (Fig. 158). These deposits mostly either pre- or post-date the Laramide orogeny, and the fracture zones are considered to be reactivated deep-seated faults. Along the projected on-land extension of the Mendocino Fracture Zone, major ore deposits are concentrated near the intersections of

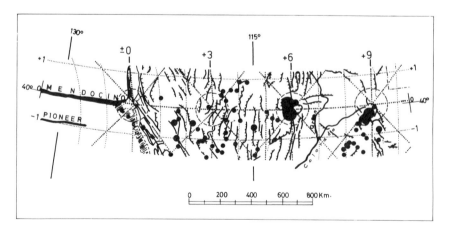

Fig. 159. Part of "empirical prospecting net for hydrothermal ore deposits of the western United States". Broken circles indicate areas of intersection of N, NE, and NW-trending major deep-seated fractures with projection of E-trending Mendocino fracture zone. Major ore deposits shown by solid circles (from Kutina, 1980a). By permission of the author and publisher, *Global Tectonics Metallogeny* **1**, 134–193.

NW- and NE-trending deep-seated fractures, e.g. the Bingham Mine in Utah and the Climax Mine of the Colorado Mineral Belt (Fig. 159). In the Nevada intersection, a number of mineral deposits, mostly of precious metals, occur in the area of intersection of the fracture zones, but there is no obvious concentration within this area compared with the surrounding region.

Gilluly (1976) has emphasized that there is no suggestion in the exposed geology of the USA of any on-land extension of the Mendocino Fracture Zone. He has pointed out moreover that Kutina's pattern of lineaments, although very abundant, fails to include many of the largest ore deposits of the western states including Butte, Tintic, Yerington, Ely, Globe-Miami, Bisbee, Ajo and many others.

6 Tin–lithium pegmatite mineralization in peninsular Thailand

In the Phuket–Phangnga area of peninsular Thailand a series of parallel tin–lepidolite–albite pegmatites (Fig. 99B) and tin–tourmaline pegmatites, both worked extensively as tin ores, are intruded into the Phangnga Fault system (Garson *et al.*, 1975). This forms part of the major NE-trending trans-current Khlong Marui Fault zone which, it has been suggested, offsets the Peninsula with a sinistral displacement of about 150 km (Garson and Mitchell, 1970). Related lineated tin-bearing tourmaline–biotite granites occupy a graben within the fault zone, tentatively interpreted as the continental extension of a late Mesozoic transform fault.

The lepidolite pegmatites are relatively fine-grained unzoned bodies up to 20 m wide and 1 km long with accessory topaz, cassiterite and ilmenite and smaller amounts of microlite, tantalite and xenotime (Garson *et al.*, 1969). The tourmaline pegmatites worked on Phuket Island are up to 10 m across and carry cassiterite, wolframite and columbite. The tin-bearing lineated biotite granites in the Khlong Marui graben have been affected by movements in this fault system during and after emplacement. They are characterized by perthitic feldspar and abundant tourmaline, apatite and sphene with accessory cassiterite, yttrotantalite and orthite.

Tin–lepidolite pegmatites are also associated with the Ranong Fault system to the north which is possibly also a continental expression of a transform fault.

7 Upper Cenozoic sulphur and gold–silver deposits in eastern Japan

Upper Cenozoic ore deposits including the Kuroko ores referred to above (Ch. 5, II. B. 2) are abundant in the Green Tuff magmatic arc of eastern Japan, mainly within 50 km of the volcanic front. A concentration of several of these deposits has been noted at intersections of the volcanic front with transverse faults, which are postulated extensions of oceanic transform faults on the subducting plate, drawn by Carr *et al.* (1973) on the basis of

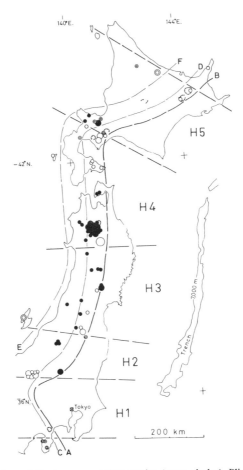

Fig. 160. Distribution of Quaternary sulphur–pyrite (open circles), Pliocene–Miocene gold–silver (double circles), and Miocene base metal deposits (solid circles) in relation to transverse faults (from Ishihara *et al.*, 1974). Circle size category: 1, more than 5×10^6 tons (only one sulphur deposit); 2, 5×10^6 to 5×10^5 tons and 5×10^5 to 5×10^4 tons for the sulphur and base metal deposits; 3, more than 50 tons and 50 tons to 5 tons for the gold deposits (silver omitted). Segments H1 to H5 and transverse faults after Carr *et al.*, (1973). Lines A–B, C–D, and E–F represent volcanic front, tholeiite, high alumina, and alkali basalt provinces. Circles with hachure, Krakatoa-type caldera larger than 10 km in diameter. By permission of the authors and publisher, *Bull. geol. Soc. Am.* **85**, 292–294; **84**, 2917–2930.

seismic evidence (Fig. 160). In particular a remarkable coincidence is reported between the transverse faults and the concentration of sulphur deposits. The Miocene–Pliocene gold–silver quartz vein deposits, although distributed more sporadically, are also located close to the transverse faults.

Kutina's (1974) proposed "Forty-north fracture zone", a continuation of the Mendocino and Pioneer Fracture Zones of the East Pacific, passes through the Kosaka area of Honshu Island, Japan, with extensive Kuroko deposits, and continues across the Sikhote-Alin Peninsula near Vladivostok, into north China where there are alignments of endogenic deposits. However, the Japanese Kuroko deposits are distributed along the length of the East Japan Arc and are not significantly associated with transform fault directions (e.g. Ishihara *et al.,* 1974), while the sulphur and gold–silver deposits are not confined to the Forty-north fracture zone through Kosaka.

III CONTINENTAL LINEAMENTS AND MINERALIZATION

The concept of control of mineralization by lineaments, rectilinear or arcuate topographic features related to faults or fault zones, has attracted some geologists throughout this century. Recently attempts to demonstrate the value of satellite imagery in locating ore bodies have relied heavily on the concept that mineralization can occur preferentially along lineaments and in particular at intersections of lineaments of preferred trends (e.g. Norman, 1980). These ideas have been developed in particular in North America by Kutina, referred to above, and a number of other workers, and are especially fashionable in the Precambrian shield areas of USSR.

Lineaments other than those either related to oceanic transform extensions or associated with carbonatites and kimberlites have been postulated as controls on for example the location of porphyry copper, Mississippi Valley, vein-type mineralization, and different types of deposit of various ages considered to lie on a single trend. The suggested lineaments range from less than 100 to several thousand km in length, and up to 200 km in width. Examples are the Texas lineament in North America, crustal failures or lineaments in Nigeria (Fig. 161), and postulated lineaments in Burma which Crawford (1980) considers control several types of mineralization of different ages including the Cambrian Kuroko-type Ag–Pb–Zn mine at Bawdwin, the (Tertiary?) ruby deposits of Mogok and of jadeite of Kachin State. These deposits are of different ages, as Crawford points out, and the postulated major lineament crosses at least one Mesozoic suture zone; control of mineral localities by a long-buried deep-seated fracture is hence difficult to visualize (see Fig. 130).

In Nevada and Utah Rowley *et al.* (1978) described the E-trending Blue Ribbon Lineament, a zone about 25 km wide and possibly 360 km long, defined by range terminations, E-trending valleys, alignment of Upper Tertiary alkalic rhyolite eruptive centres and mineralized and hydrothermally altered rocks, magnetic highs and E-trending basin and range faults. Mineralization, mostly within middle Miocene to Pliocene host rocks, consists largely of fluorine, uranium and tungsten, and numerous warm springs are present. The Lineament crosses at a low angle the Pioche Mineral Belt and is parallel to sub-parallel to a number of other lineaments of similar age in Nevada and Utah. Rowley *et al.* (1978) interpret the Blue Ribbon Lineament as a fault zone, either related to an E-trending warp in the underlying subducting plate, or part of an intracontinental transform fault active during extensional rifting in the eastern Great Basin.

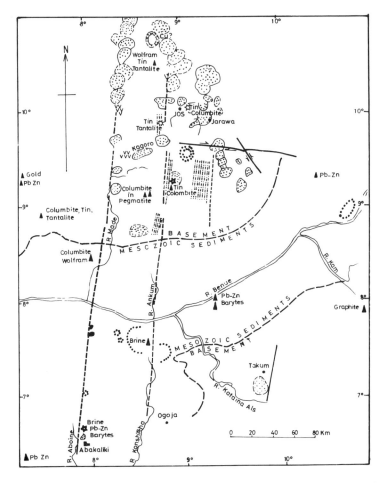

Fig. 161. Selected crustal failures in Nigeria interpreted from Landsat imagery, aligning Jurassic tin-bearing granites, older igneous intrusions, lead–zinc occurrences and warm brine springs (from Norman, 1980). By permission of the author and publisher, *Trans. Instn Min. Metall.* **89**, B63–72.

Recently, Noble (1980) has produced detailed metallogenic maps for North America which he claims show linear patterns transverse to the edge of the continent. In particular a plot of "First Order Deposits" (200 million dollars or more at time of production or present value of reserves) indicates four latitudinal belts of deposits of all ages across North America, separated by almost barren zones. The belts are (1) between N35° and N41°, i.e. broadly the Forty-north fracture zone of Kutina (1974), (2) between N44° and N53°, (3) between latitude N54° and N60° and (4) a very wide belt covering the southern states and Mexico which with more information may be sub-divided (Fig. 162). In addition he describes three other belts of deposits, the first being near and about parallel to the Pacific Coast, the second transverse to these and the third comprising linear NE-, ENE- and E-trending zones within the latitudinal belts and partly overlapping the coastal type. Noble has concluded that the metals must have been segregated into the latitudinal belts when the earth's axis of rotation was very close to the present axis, and that if the continent has drifted it has retained or regained its approximate position relative to the axis of rotation.

Noble's ideas, developed in detail over a considerable length of time, cannot be dismissed lightly. If his latitudinal belts of ore-concentration are valid and not over-selective, as apparently are some of Kutina's fracture

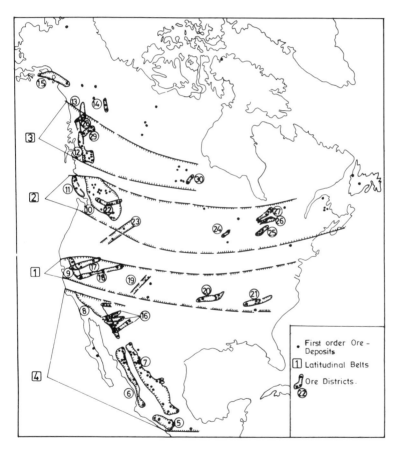

Fig. 162. A metallogenic map of the North America (simplified from Noble, 1980). Filled circles are major ore occurrences of First Order (200 million dollars recorded production plus reserves). Solid lines outline contiguous major occurrences plus adjacent areas of significant ore occurrences. Noble has outlined 26 First Order ore districts of which all but 2 lie within 4 main latitudinal belts, the southernmost (No. 4) being very wide and lacking the presumed latitudinal aspects of the others. By permission of the author and publisher, *Geol. Rdsch.* **69**, 594–608.

zones with ore metals, then the overall patterns must be explicable. One possibility may have been that primitive heterogeneous enrichments of metals resulted from rotation of lithosphere over wide Archaen plumes. Younger ore deposits in these belts presumably formed by normal plate tectonic processes, deposition perhaps having benefited from the early concentration of metals.

Gilluly (1976), in a review of lineament-controlled mineralization in western North America, argued that most lineaments are not linear, and that most topographic features associated with abundant ore bodies are broad zones, many of which are arcuate and controlled by tectonic processes unrelated to major strike–slip faults. An example is the mineralization belt of Billingsly and Locke (1935), which is sinuous and coincides with the back-arc thrust belt indirectly related to eastward subduction of ocean floor. While in a few cases mineralization of a particular type may be related to a linear feature such as the surface trace of a hot spot, or a shear zone, e.g. the Mother Lode of California described by Gilluly, it is barely conceivable that different types of deposit of different ages, some of which are syngenetic with their host rocks, can be related in any way to a topographic feature or even a major fault which happens to pass through or near them. Nevertheless, Gilluly's conclusion that the value of lineaments as guides to ore is virtually negligible has been disputed by Kutina (1980b) with particular reference to ore fields and lineaments in the Bohemian Massif.

The above discussion of transform faults demonstrates that Cenozoic transforms developed at obliquely converging plate margins are rarely mineralized at the present level of erosion, but that some ancient continental fractures, possibly reactivated during development of ridge–ridge transforms associated with initial stages of continental rupture and ocean floor spreading, are important controls on carbonatites and kimberlites, magmatic rocks originating below the crust and including specific types of economic deposit. This implies that in general major fractures and hence most lineaments may be preferred sites for mineralization only when there is a long history of intermittent displacement, and that only deposits which are epigenetic with respect to non-intrusive host rocks are likely to be present.

Confirmation of the inhomogeneous distribution of metal in the lithosphere, implicit in the concepts of Kutina and Noble, will presumably require either recognition of high trace metal contents in igneous rocks or successful prediction of preferentially mineralized belts. Pending this confirmation, the postulated systematic mineralized lineament patterns remain at least somewhat speculative.

CHAPTER 8

Mineral Deposits and the Wilson Orogenic Cycle

In the preceding chapters we have discussed the generation and emplacement of most types of post-Archean mineral deposit in terms of major modern and analogous ancient tectonic settings. We have also shown in the later chapters how the present position of the deposits and their host rocks in the ancient settings can be explained in terms of the orogenic cycle of ocean opening and closing proposed by Wilson (1968), and of the more specific processes of the plate tectonic hypothesis. Here we summarize the relationships of mineral deposits to tectonic settings by treating the various types of deposit as part of their enclosing host rocks and describing a sequence of mineral formation and emplacement events in terms of the orogenic cycle. An example of earlier attempts to relate formation of deposits to the orogenic cycle (e.g. Mitchell, 1978a, Table 2) is shown in Fig. 11.

As there are differences in the type of deposit formed on ocean margins facing east from those facing west, and as many oceans close by subduction at only one margin, not all the types of mineral deposit discussed can be shown in an opening and closing cycle involving only one ocean. Therefore in the following idealized model we consider two oceans which open as a result of spreading on N-trending rises, and which close following subduction, in one case eastward and in the other predominantly westward.

It must be emphasized that this discussion of mineral deposits in terms of the largely mechanical orogenic cycle is not directly related to any postulated geochemical cycle. Ideally the formation, generation, exposure and erosion of metallic mineral deposits could perhaps be related to the orogenic cycle by considering the path of metals through a sequence of ocean floor spreading, subduction and eventual collision. For example it is sometimes suggested that copper is extracted from the upper mantle at ocean rises, concentrated in submarine exhalative sulphide deposits and hydrogeneous ferromanganese nodules on the ocean floor, subducted and remobilized to form porphyry and stratiform ores in volcanic arcs, and incorporated into orogenic belts during continental collision, from which it is leached and deposited as stratabound syngenetic or epigenetic sulphides in post-collision rifts. However, this approach

325

is at present qualitative and largely speculative because of the widespread uncertainty as to how much ocean floor sediment is subducted rather than tectonically accreted to outer arcs, whether any metals in volcanic arcs are derived from subducted rocks, and whether metalliferous deposits in rifts are derived from older orogenic rocks or from syn-rifting volcanics. In the case of tin, there is some evidence that the metal may be scavenged from pre-existing concentrations at depth during magmatic activity in continental margin arcs, remobilized and concentrated to form workable primary deposits during subsequent collision-related generation of anatectic granites, and eroded and deposited as placer deposits which are buried to provide a source for repetition of the cycle. This model requires an initial concentration of tin, sometimes postulated to reflect local inhomegeneities in upper mantle metal distribution either formed early in the earth's history (e.g. Noble, 1970) or, as recently suggested by Norman (1980), resulting from accretion of large meteorites mostly during the Archean. In the absence of compelling evidence for these long-lived mantle inhomegeneities we also omit this inheritance concept from the discussion.

I HOT SPOT MAGMATISM AND INTRACONTINENTAL RIFTING

Initiation of the 'Wilson' orogenic cycle can conveniently be considered to start within a major continent, α in Fig. 163A, stationary with respect to underlying asthenosphere. Mantle plumes (1, 2 and 3), arbitrarily considered to be fixed relative to each other, heat the overlying continental lithosphere resulting in 'hot spots', with consequent crustal doming and mostly alkaline magmatism, predominantly silicic but with some basalts, concentrated in pre-existing lines of crustal weakness. Intrusive rocks include carbonatite with associated apatite, magnetite, vermiculite and pyrochlore mineralization, per-aluminous granite with tin, tungsten, and in some cases niobium and base metals, and uranium in alkaline or per-alkaline granites which commonly form ring complexes.

With continued doming above hot spots a radiating system of rifts develops, arms from adjacent rifts eventually linking up to form a continuous system (Fig. 163B). Rift systems also originate in or adjacent to orogenic belts following continent–continent or continent–arc collision, and while they are not associated with pre-rift doming their subsequent development appears to be similar to that of hot spot-related rifts. Rifts of both origins are equivalent to the taphro-geosynclines of geosynclinal terminology.

Magmatic activity associated with rifts largely comprises undersaturated alkaline and per-alkaline rocks including carbonatites with apatite, pyrochlore, rare earths, strontianite, copper, uranium and baddeleyite mineralization; diamondiferous kimberlites; and probably ultrabasic–basic intrusions at depth along the rift axis with Cr–Ni–Pt–Cu ores. Porphyry molybdenum mineralization is associated with some alkali granites in rift zones.

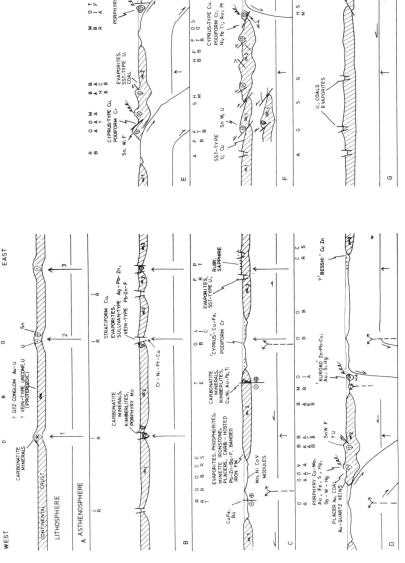

Fig. 163. Mineralization in the Wilson orogenic cycle. Variations in thickness of lithosphere omitted. A, aulacogen; BACB, back-arc compressive basin; BAEB, back-arc extensional basin; BAMB, back-arc magmatic belt; CR, continental rise; CS, continental shelf; FB, foreland basin; FR, failed rift; FTB, foreland thrust belt; G, troughs and graben; HB, hinterland basin; HD, hot spot dome; HM, hinterland margin; IB, intracontinental basin; IR, intracontinental rift; LIC, linear island chain; MA, magmatic arc; MB, marginal basin; OA, outer arc; OAT, outer arc trough; OB, ocean basin; OIA, oceanic island arc; OR, ocean rise; PT, continental plume trace; RB, remnant basin; RRT, ridge-ridge transform; S, suture zone; TE, transform extension; TF, transform fault. Pluton symbols: dots, alkaline; * carbonatite; crosses, per-aluminous granite or leucogranite; oblique crosses, calc-alkaline diorite-granodiorite. Thick short bars, ophiolite; broken arrows, spreading lithosphere; arrows, plumes; dots, oceanic island chains and continental plume-trace basalts.

Predominantly clastic sediments deposited in the developing rifts become host rocks to stratabound and sometimes stratiform sulphides, mostly epigenetic or diagenetic with respect to the host rocks. These include major copper and copper–cobalt deposits, and probably some 'Sullivan type' Ag–Pb–Zn ore bodies. In failed rifts evaporites may accumulate, and sandstone-type uranium mineralization can form in terrigeneous mostly non-marine sediments. Economic deposits of sodium and potassium salts, magnesite and phosphate can also form in lacustrine rift environments. Some rift successions include vein-type lead–zinc and fluorite mineralization.

Early Proterozoic quartz-pebble conglomerate gold–uranium deposits may have accumulated in aulacogens, but it is also possible that these and Proterozoic unconformity vein-type uranium deposits formed in continental basins either between hot spot domes or possibly in back-arc compressive basins referred to later.

II OCEAN FLOOR SPREADING

With emplacement of oceanic crust in the rift axis continental separation accelerates and a spreading ocean rise flanked by ocean basins develops (Fig. 163 C). As a result of ocean floor spreading at two centres, continent α splits into three, α 1, α 2 and α 3; we arbitrarily assume that continent α 3 moves east slowly with respect to underlying asthenosphere. Consequently a continental hot spot track of alkaline basaltic rocks forms above plume 3, possibly with associated gem-quality corundum deposits. Continent α 2 moves west rapidly with respect to asthenosphere and plume-related volcanism is extinguished. The spreading ridge between α 2 and α 3 is also forced to move west, away from plume 3; continuing volcanism above this plume forms a linear island chain or ridge progressively older towards continent α 3; the ridge rocks are normally unmineralized, although theoretically carbonatites and associated mineralization could be expected.

On the spreading oceanic rise sulphide-rich muds and brines accumulate in the axial rift around zones of hydrothermal discharge located on topographic highs, probably forming stratiform copper-rich deposits. Hydrothermal iron–manganese deposits also form on the spreading rise. Baryte and possibly stratiform sulphides accumulate on or adjacent to oceanic ridge–ridge transform faults. Hydrogeneous manganese nodules, with high contents of Cu, Ni and Co, are precipitated in areas of low sediment accumulation on the ridge flanks, in the ocean basins and on topographic highs. Near or immediately beneath the Moho magmatic podiform chromite accumulates within dunite, and rare economic deposits of Ni, Fe, Ti, Au and Pt can develop in the crust or uppermost mantle rocks.

On the rifted or passive margins of the separating continents, transgressive evaporitic sequences accumulate in low latitudes, perhaps providing mineralizing brines which result in stratabound copper mineralization in the underlying clastic rift sediments. With continued separation of the rift flanks deposition of miogeosynclinal continental shelf successions takes place, locally interrupted by deltas in failed rifts. During marine transgression in low latitudes phosphorites are deposited preferentially on shelves of west- rather than east-facing margins, and metal-rich black shales accumulate in deeper water. Consequently the upward sequence stratabound copper–evaporites–phosphorites is not uncommon. Placer deposits of titanium minerals and very rarely diamonds accumulate on the coastal plain and on beaches, where suitable source rocks are present inland. Deeply buried shelf successions are potential host rocks for epigenetic but mostly pre-orogenic carbonate-hosted stratabound lead–zinc and baryte–fluorite deposits, in some cases above older hot spot-related granites, in which radiogenic heating causes convective circulation of mineralizing brines. Near-shore environments on passive margins are favourable environments for deposition of minette-type ironstones, and in the early Proterozoic deposition of most banded iron formations probably took place on continental shelves.

Continental rise successions oceanward of the shelf are normally unmineralized, although possibly Besshi-type deposits could form where hot spot-related volcanism intersects the sedimentary prism.

Oceanic ridge–ridge transform faults formed at the start of spreading may extend into continents as zones of displacement coinciding with pre-existing lines of weakness; along or adjacent to these diamondiferous kimberlites, carbon-carbonatites, and possibly Cu, Ni, Ti, Au and Pt associated with basic and ultrabasic intrusions may be emplaced, as shown near the eastern margin of continent α 2 (Fig. 163C).

III OCEAN FLOOR SUBDUCTION AND BACK-ARC SPREADING

1 West-facing arc systems

Subduction of the spreading ocean floor, related at least partly to its cooling and increase in density with distance from the spreading ridge, is often initiated at a continental margin. Eastward subduction beneath the westward-moving continent α 2 (Fig. 163D) results in a shallow-dipping Benioff zone and formation of a continental margin arc system under lateral compression. In the calc-alkaline volcanic arc, magmas of intermediate and silicic composition predominate and porphyry copper–molybdenum deposits form around or just above the upper levels of mostly granodioritic plutons. Other mineralization, characteristic of continental margin magmatic arcs but less common and of less economic importance, includes tin and tungsten associated with granitic plutons, magnetite–haematite–apatite lava flows, and possibly stratabound Sb–W–Hg deposits. Mercury, native sulphur and primary gold mineralization occur in both continental margin and oceanic magmatic arcs.

Landward of the west-facing volcanic arc a back-arc thrust and magmatic belt commonly forms, with anatectic granites related to tectonic thickening of the continental crust. Tin and associated metals are commonly deposited either around granite cupolas or in sub-volcanic settings, and other types of magmatic hydrothermal deposit may also form in this setting.

East of the back-arc thrust belt, depression of the continental crust results in a compressive basin with predominantly terrigeneous sedimentation and a depositional axis which migrates continentward in advance of the thrusts. Sandstone-type uranium deposits are characteristic of these back-arc basins, the metal being derived partly from granites in the magmatic belt, and evaporites and coals may also accumulate. The basins are possibly equivalent to some of the zeugogeosynclines of American authors.

2 Outer arcs

During subduction, deposition of submarine fans on the ocean floor seaward of the continent may take place, perhaps due to the start of diachronous collision of continent α 1 with α 2 to the south or north, outside the plane of the cross-section in Fig. 163. The sediments are tectonically accreted above the subduction zone to form an outer arc, in which auriferous quartz veins may develop during low-grade metamorphism. Ophiolitic rocks representing either a trapped segment of oceanic crust east of the trench, or an oceanic topographic high tectonically accreted to the overriding plate at an early stage of subduction, are elevated landward of the outer arc imbricate flysch (Fig. 163E). The ophiolites may contain associated Cyprus-type sulphides, podiform chromite or rarely manganese nodules, and asbestos, talc and magnesite formed subsequently in the ultrabasic rocks. In the outer arc trough landward of the outer arc, gold placers and potential coal deposits may accumulate.

The ocean floor, outer arc and magmatic arc are now equivalent to the eugeosyncline in the early orogenic stage of development (Table II) of Aubouin's (1965) model, with the continental hinterland of α 2 to the east and the miogeo-syncline and foreland of α 3 to the west.

With continued subduction the spreading ridge enters the trench and is lost beneath the outer arc, resulting in partial melting of the outer arc rocks at depth to form granitic plutons of variable initial Sr^{87}/Sr^{86} isotope ratios; tin mineralization, and possibly porphyry copper deposits, may be associated with these plutons.

3 East-facing arc systems

Westward subduction beneath the westward-drifting continent α 2 (Fig. 163D) takes place along a steeply-dipping Benioff zone and results in tensional stresses in and behind the initial continental margin volcanic arc. Rifting between the arc and continent to the west is followed by back-arc spreading and development of a marginal basin with mineral deposits probably identical to those in oceanic crust and uppermost mantle, although arguably there may be differences in the trace metal contents of the crustal host rocks. In or on the flanks of the eastward-migrating volcanic island arc, which includes a fragment of rifted continent (α 4), Kuroko-type Zn–Pb–Cu (Au–Ag) stratiform sulphides are deposited from submarine hydrothermal exhalations associated with rhyolitic volcanism; this mineralization may be restricted to short-lived but widespread episodes of rifting within the volcanic arc. The oceanic volcanic arc may include deposits of mercury, sulphur in basaltic rocks sometimes associated with rhyolites, and gold both associated with dioritic plutons and forming epigenetic deposits in thick successions of otherwise barren basaltic andesites.

Events accompanying incipient rifting of the arc from the continent are not well documented for east-facing arcs and are inferred largely from the Miocene events which followed termination of eastward subduction beneath south-western North America. This indicates that extensional block faulting, crustal thinning, a high geothermal gradient and bimodal basalt–rhyolite volcanism can be accompanied by epithermal gold–silver mineralization, emplacement of porphyry molybdenum deposits and possibly deposition of stratiform sulphides.

During westward subduction beneath the island arc α 4 the spreading ridge to the east enters the trench, and if plate convergence is oblique subduction may be replaced by strike–slip motion on a transform fault (Fig. 163E). Subduction-related transform faults are not normally mineralized, with the possible exception of stibnite associated with calcite and quartz in serpentinites. Subsequently a change in plate motion may lead to eastward subduction of the marginal basin, with mineralization including porphyry copper–gold and perhaps porphyry gold deposits in the magmatic arc. The marginal basin spreading system in the marginal basin west of α 4 will enter the trench, and in the absence of a major outer arc, ophiolitic rocks of the spreading ridge will be tectonically accreted to the overriding plate to the east.

IV CONTINENT-CONTINENT AND CONTINENT-ARC COLLISION

Continued eastward subduction beneath the oceanic arc α 4 and closure of the remnant basin to the west results in collision of the arc with continent α 2 (Fig. 163E, F), corresponding to the late orogenic stage of Aubouin (1965). In the absence of a major outer arc of imbricate flysch, collision is accompanied by thrusting of the previously accreted ophiolitic rocks onto the continental fore-land to form an ophiolite sheet with associated mineralization. West or landward of the ophiolite and suture a foreland thrust belt develops, with crustal thickening and generation of anatectic high initial Sr^{87}/Sr^{86} ratio granites with tin, tungsten and uranium mineralization. Rarely deposits of the silver–nickel, cobalt arsenide association may also be generated. West of the foreland thrust belt the developing foreland basin or exogeosyncline succession may become host to stratabound sandstone-type uranium–vanadium–copper mineralization related to the granitic source rocks to the east, and possibly stratabound copper deposits.

Similarly, continued eastward subduction beneath continent α 2 results in closure of the remnant ocean basin to the west and collision with the continental foreland α 1. The major outer arc of α 2 will inhibit accretion and hence subsequent obduction of ophiolites, but may be largely destroyed by erosion, exposing jadeite and nephrite in the suture zone; alternatively, it may be overridden either by the magmatic arc of the overriding plate or, less commonly, by rocks of the foreland. The continental margin volcanic arc on the western margin of α 2 is eroded, exposing gemstone corundum in metamorphosed marbles and in pegmatites. In embayments in the continental margins where syn- and post-collision convergence is limited, mineralizing brines may be squeezed from the outer arc accretionary prism to form epigenetic carbonate-hosted lead–zinc deposits in the more deeply buried continental shelf rocks of α 1.

As in the α 2 – α 4 collision, a foreland thrust belt and foreland basin develop on α 1 which formerly extended to the ocean to the east now terminates in the Tectonic slices of the continental shelf and underlying rift succession are elevated, including carbonate-hosted lead–zinc deposits, ironstones, stratabound copper, possibly phosphorites, and evaporites which may form preferred horizons for thrusting. Included in the thrust slices may be basement rocks and associated mineralization of a previous orogenic cycle. On α 2, the former back-arc compressive basin becomes a hinterland basin within which sedimentation and adjacent thrusting may continue. As a result of the collision, the failed rift on α 1 which formally extended to the ocean to the east now terminates in the collision belt and hence becomes an aulacogen.

Following collision of oceanic arc α 4 with continent α 2, continuing convergence may result in a further Benioff zone flip, with consequent westward subduction of ocean floor and eventual collision between continents α 2 and α 3. The location of foreland thrusting in α 3 may be determined by the position of the failed rift generated in Fig. 163B.

Continents α 1, α 2 and α 3 and oceanic arc α 4 are now recombined to form a single continent. Graben formation within the former foreland and hinterland, corresponding to the post-geosynclinal stage of Aubouin (1965), commonly follows collision and cessation of plate convergence in the foreland thrust belt, the graben sediments becoming host to sandstone-type uranium and stratabound uranium within volcanic rocks; evaporites and potential coal beds may also accumulate.

V POST-COLLISION RIFTS AND THE OROGENIC CYCLE

We have referred in Ch. 6 to the development in Cenozoic collision belts of post-collision rifts within perhaps five to 20 Ma of continent–arc and continent–continent collisions, and to the difficulty in determining whether some pre-Cenozoic rifts developed above hot spots or following collision was discussed briefly in Chapter 2. It is clear from the post-collision geological history of the Appenine region that graben formed across or within continent–arc collision belts can develop into oceanic spreading centres without evidence for the presence of causal hot spots, but whether rifts in continent–continent collision belts can also become oceanic spreading centres is less certain. If continent-continent collisions can rift in the absence of underlying hot spots, then the orogenic cycle of ocean opening and closing, once initiated by hot spot-related rifting within a continent, could continue indefinitely, and the presence of ancient sutures situated within continents known to have been stable since the last collision in the Proterozoic or Palaeozoic would be anomalous.

This suggests that while orogenies rarely cease with continent–arc collisions, they are normally stabilized by continent–continent collision, and that the post-collision rifts in the latter do not become sites of ocean floor spreading, unless the collision belt happens to be at rest with respect to underlying mantle plumes. Thus in Fig. 163 we show the overall orogenic cycle, which involved opening and closing of oceanic and marginal basins, terminating in a continent–continent collision, the resulting single continent drifting west with respect to the asthenosphere. If the continent should subsequently come to rest with respect to underlying asthenosphere, the collision belts and presumably the post-collision rifts would be favourable locations for penetration of the continental lithosphere by mantle plumes, leading to hot spot doming and magmatism and the initiation of a new orogenic cycle. Since collisions take place in zones of ocean floor subduction, it is difficult to envisage any direct relationship between post-collision rifting and models of upper mantle convection in which mantle material descends beneath subduction zones, for example that recently discussed by O'Nions *et al.* (1980).

This interpretation implies that orogeny is completed in the Himalayas, but that in the Mediterranean and Indonesian regions, each with a late Mesozoic and Cenozoic history of arc–continent collision, orogeny will continue until Africa and Europe are welded together and until Australia has collided with mainland Southeast Asia. We may safely predict that in these future collisions many of the mineral deposits in the Mediterranean region and the abundant deposits in Southeast Asia will be either eroded or overridden by thrust sheets of deeper and hence less mineralized structural levels.

CHAPTER 9

Plate Tectonics as a Guide to Mineral Exploration

The tectonic settings we have described are as far as possible defined in terms of their lithology or style of deformation, position relative to other settings, and in some cases their morphology. Nevertheless, recognition of many of the pre-late Cenozoic examples, now in settings different from those in which they formed, depends on acceptance of Wilson's concept of ocean opening and closing which the plate tectonic hypothesis alone can satisfactorily explain. For example, without continental drift it would be impossible to interpret the now intracontinental calc-alkaline rocks of southern Tibet as an ancient analogue of active magmatic arcs, however great the lithological similarity.

The significance to mineral exploration of the relationship between plate tectonics and mineralization has been emphasized by a number of authors and at least one journal is devoted entirely to this topic. Since the ultimate aim of attempts to understand the genesis of mineral deposits is to find new or additional ore bodies, it is necessary to consider briefly whether the plate tectonic hypotheses can assist in selection of areas for exploration on a regional scale.

Areas with abundant mineral occurrences, none of which are necessarily economic, are often considered to be potentially favourable for ore bodies. This is either because of the observation that ore bodies are often associated with sub-economic deposits and occurrences, or because the occurrences are considered to reflect anomalous concentrations of metals in the upper mantle or lower crust as suggested by, for example, Noble (1970) and Norman (1980). In either case, selection of area is then based on either proximity to or density of major occurrences, and geological considerations are not necessarily involved. More commonly, geological considerations play at least some role in selection of areas, and assessment of the mineral potential is usually based on the geology of the area considered together with the relationship of known ore bodies or mineral occurrences to areas of similar geology elsewhere. The geological guides to the mineral potential of an area, however sophisticated in detail, all fall within one of three main concepts, involving lineaments, stratigraphic correlation or analogy.

334

I LINEAMENTS AND SELECTION OF AREA

An exploration programme is sometimes based on the presence of a lineament extending from an adjacent area where it is known to be mineralized. The role of lineaments in controlling the position of mineral deposits has been discussed above (Ch. 7, III) and may involve on-land lineaments only or on-shore extensions of oceanic fracture zones. While the location of some types of deposit, in particular kimberlites and carbonatites, is in some cases clearly controlled by lineaments interpreted as continental fracture zones, there is a relationship to plate tectonics only where it can be shown that the lineament's origin was related to the relative movement of plates. Many other postulated lineaments, e.g. the 38th parallel lineament in USA, have no apparent relationship to past or present plate boundaries, and the existence of a world-wide systematic lineament system or regmatic shear pattern of fixed orientation, favoured by some geologists, is incompatible with the concept of continental drift. Hence whether or not these lineaments are a guide to ore bodies there is no demonstrable relationship to plate tectonics.

II STRATIGRAPHIC CORRELATION AND SELECTION OF AREA

Selection of exploration area is often based on simple stratigraphic or rock unit correlation, on scales ranging from a horizon to a supergroup and including plutonic as well as sedimentary and metasedimentary rocks. This approach depends on the recognition within the area of possible interest of a rock unit either physically continuous with, or considered to have been formerly continuous with, a unit of the same age elsewhere which contains either economic mineralization or mineral occurrences.

Where the stratigraphic unit can be traced from the area where it is mineralized to the possible exploration area, there is clearly no necessity to involve plate tectonic or any other global tectonic scheme in selection of area. However, in many cases there is evidence that a formerly single mineralized rock unit or metal province has been split into two or more fragments which became widely separated as a result of continental drift. This concept is particularly applicable to Upper Palaeozoic and Mesozoic mineral provinces, because the continents and continental fragments which split off as Gondwanaland broke up can in most cases be reassembled with increasing accuracy as more data become available on their pre-drift position. Clearly any extensive mineralized rock unit of pre-Mesozoic or pre-Upper Moseozoic age known from one continent may have originally extended into adjacent continents then in Gondwanaland, and recognition within these continents of the same rock unit indicates that similar associated mineralization might be present. Truncation of a mineralized rock unit by a continental margin naturally suggests that its continuation may be expected in the originally contiguous continent.

This approach was used by Schuiling (1967) in what was perhaps the earliest discussion of the relationship of plate tectonics to mineral provinces. Schuiling demonstrated how a previously single tin province in Africa and South America had been split by the opening of the Atlantic Ocean, and emphasized how the continuation of the truncated province in one continent could be located in another. Similarly Reid (1974) suggested that a single pre-drift source could explain the presence of detrital diamonds in Guiana and Sierra Leone, Burton (1970) suggested that the original source of the diamonds of the Andaman Sea coast in Thailand lay in India, and Ragland and Rogers (1980) considered "Pan-African" uranium provinces in their pre-drift position (Fig. 164). However, in each of these examples the known distribution of minerals was related to continental drift, rather than new provinces being predicted from the inferred original position of the continents.

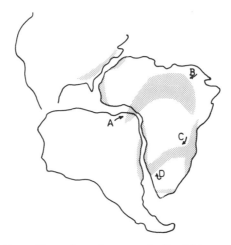

Fig. 164. Approximate positions of uranium areas in Pan-African zones, North America, South America, and Africa reconstructed in predrift positions. (A) Currais Novos, Brazil; (B) Younger Granites, Red Sea Hills, Egypt; (C) Katanga area, Zaire; (D) Rossing, Namibia. The Appalachian crystalline terrane appears to be an orogenic belt along the northwestern margin of the Archaen craton of western Africa (after Ragland and Rogers, 1980). By permission of the authors and publisher, *J. geophys. Explor.* **13**, 181–199.

There are evidently numerous instances where attempts to predict mineral provinces from the re-assembly of continents could be applied. For example, Lock (1980) suggested that the Cape Fold Belt was a back-arc magmatic belt extending from South Africa into South America in the west and Antarctic and Australia in the east (Ch. 5, V. B. I (c)). It follows that deposits of uranium analogous to those in the back-arc basin of the Karroo Supergroup of Southern Africa, to the north of the Cape Fold Belt, might be expected in similar sediments, if present, in the other continents north of the projected fold belt. Similarly recognition of any mineralization in the fold belt itself would imply that comparable mineralization could occur in the east–west extensions of the belt.

Another example is in the Himalayan-Southeast Asia region (Mitchell, 1976b), where there is evidence that western Southeast Asia formerly lay adjacent to northern India in Gondwanaland. This would suggest that pre-drift stratabound mineralization similar to that in Southeast Asia, for example carbonate-hosted lead–zinc in Ordovician limestones of Burma and Malaysia, and less probably Kuroko-type Zn–Pb–Cu ores in Cambrian volcanic rocks in Burma, could be expected in the Higher Himalayas of the Indian plate, south of the Indus suture. Extensive Lower Ordovician carbonates are known from the Higher Himalayas, and there are recent reports of lead–zinc mineralization within this unit in Bhutan.

III GEOLOGICAL ANALOGY AND SELECTION OF AREA

In the analogy approach, selection of a prospective area for exploration is based on recognition of geological features comparable to those of post-Archean age elsewhere with which economic deposits are associated.

The geological features may be identifiable on air photographs or satellite imagery: e.g. a fold and fault pattern used successfully in petroleum exploration in Wyoming (Fussell, 1980), erosional features resembling those of ring complexes or circular depressions indicative of kimberlites. The features could also be simply lithological; an example might be the identification of volcanic rocks similar to those which elsewhere contain economic sulphides. Interpretations of regional geophysical or geochemical anomalies also make use of the analogy approach: e.g. an airborne gravity anomaly indicating ultrabasic rocks implies a Cr, Cu or Ni potential, while the regional geochemical analogy is based simply on the observation that known ore bodies are often surrounded by rocks with above-background metal content.

By far the most significant and widely used application of the analogy approach is the tectonic analogy, or recognition of an overall tectonic setting analogous to that known elsewhere to contain ore bodies. This can be applied at various scales, but is usually concerned with tectonic settings on the scale of those described in Chs 2-7, with an area of at least 1000 square kilometres. Examples might include identification of an ancient volcanic arc with potential for either porphyry copper–molybdenum or stratiform submarine exhalative deposits, based on analogy with the young mineralized arc systems of Chile or Japan; identification of ancient fore-arc basin rocks analogous to those of modern basins with coal deposits; or recognition of an on-land ophiolite sequence closely comparable to either a mineralized ophiolite elsewhere, e.g. in Cyprus, or to an ocean floor area with mineralization, e.g. part of the East Pacific Rise. A further example, involving a combination of tectonic analogy and pre-drift of continents, is the recognition of black shale–carbonate–chert sequences deposited on west-facing continental margins as favourable for phosphorite deposits. It is a fundamental aim of exploration geologists to refine these analogies so that progressively smaller targets are identified before exploration work is started.

IV THE INDIRECT VALUE OF PLATE TECTONICS IN SELECTION OF AREA

Most accounts of the value of the plate tectonic hypothesis to exploration have considered either simple lithological or tectonic analogies, especially those of subduction-related settings and, less commonly, of intracontinental rifts. They consider, for example, that Cenozoic magmatic arcs formed on overriding plates are usually strongly mineralized with copper and that therefore recognition of an ancient magmatic arc will provide a large-scale exploration target. However, the position of an ancient magmatic arc can rarely be predicted from, for example, the identification of a collision belt or suture, as this requires knowledge of the palaeo-subduction polarity, which in geologically poorly known areas can usually be determined only from the recognition of the magmatic arc itself. Recognition of the magmatic arc requires identification of calc-alkaline volcanic rocks or plutons, which have long been known to exploration geologists as a favourable indicator of porphyry copper mineralization regardless of the former existence of an overriding plate margin and underlying Benioff zone. Even where the magmatic arc rocks are recognized only in one locality, prediction of the arc's continuation or trend is based on the trend of adjacent belts of, for example, flysch or serpentinites, and is not dependent on these having been emplaced at a former plate boundary. Similarly in the case of Cyprus-type sulphides; the association between ores of this type and basaltic pillow lavas has long been known, and is independent of theories of origin of the host rocks.

It is also sometimes argued that recognition of a zonal distribution of metals near a plate boundary, e.g. the presence of a tin belt landward of copper deposits in the Peru-Bolivia segment of the Andes, provides a guide to metal distribution in other arc systems based on plate tectonic theory. However, here also observation of the distribution of metals in the Andes is independent of any relationship to plate boundaries. The Peruvian segment of the Andean arc and the Burma–Thailand segment of the Sunda Arc are both west-facing continental margin arcs and are both bordered on their continental side by a tin-bearing granitic belt. By analogy the Sumatran section of the Sunda Arc might be expected to include a tin-bearing back-arc magmatic belt, regardless of whether subduction has taken or is taking place beneath it.

It can therefore be seen that the *direct* application of the plate tectonic hypothesis to exploration strategies is somewhat limited and largely confined to providing evidence for continental drift, and reassembly of continents in their pre-drift position; it does not relate to the interpretation of rocks in terms of a tectonic, or specifically plate tectonic theory. However, it is clear from consideration of the focus of exploration activities in the last decade that the plate tectonic hypothesis has had a profound *indirect* effect on exploration strategies based on the tectonic analogy approach. This can be explained by the increase in confidence, consequent on understanding and acceptance of the hypothesis, with which geologists now recognize ancient tectonic settings as analogous to the young mineralized settings which can be related to present plate boundaries. For example, a decision to search for porphyry copper deposits within the Triassic volcanic rocks in northern Thailand could be based on the similarities of these rocks to those in modern mineralized magmatic arcs or, slightly more elaborately, on similarities between the position of the Triassic arc with an ophiolite-flysch belt to the west, and the relative position of the submarine trench and magmatic arc in the porphyry copper-bearing west-facing Andean arc. However, the plate tectonic interpretation immensely strengthens this simple tectonic analogy which prior to the plate tectonic hypothesis was not sufficiently convincing to most exploration geologists.

In the future increasingly sophisticated tectonic analogies, indirectly supported by refinements in the plate tectonic hypothesis, will assist in exploration in two ways. First, by differentiating between two apparently similar areas only one of which is extensively mineralized, e.g. the west-facing continental magmatic arcs of the Andes and Sumatra, the former with numerous porphyry copper ore bodies and the latter with none, or the African Rift region with diamondiferous kimberlites south of Lake Victoria but barren kimberlites adjacent to the Eastern Rift to the east and southeast. Secondly, by identifying tectonic settings of progressively smaller size, thereby eventually indicating where, within for example a back-arc magmatic belt or shelf sequence, a specific type of mineral deposit is most likely to occur. In addition, evidence that some mineralization associated with non-magmatic rocks. (e.g. copper in terrigeneous sediments and carbonate-hosted lead-zinc ores) formed much later than their host rocks, will require that detailed understanding of the post-host rock geological history of prospective areas be applied to prediction of the location of ore bodies.

Eventually, increasing understanding of plate tectonic processes based on late Cenozoic settings will facilitate recognition of older tectonic settings, each of well-defined area and age and each with a characteristic mineral potential. One result will be realistic metallogenic maps based on more detailed tectonic and mineral genesis concepts presenting in condensed form a mass of information of value to the exploration geologist.

References

ALABASTER, T., PEARCE, J. A., MALLICK, D. I. J. and ELBOUSHI, J. M. (1980). The volcanic stratigraphy and location of massive-sulfide deposits in the Oman ophiolite. *In* "Ophiolites, Proceedings International Ophiolite Symposium" Cyprus, 1979 (A. Parayoitou, ed.), 751-757. *Geol. Surv. Dept Nicosia, Cyprus.*

ALSTINE, R. E. VAN. (1976). Continental rifts and lineaments associated with major fluorspar districts. *Econ. Geol.* **71**, 977-987.

ALVES, B. P. (1960). Distro nióbio-titanifero de Tapira. *Brasil Dept, Nac. Fomento Prod. Mineral Bol.* **103**.

ANDREWS, J. E. (1980). Morphologic evidence for reorientation of sea-floor spreading in the West Philippine Basin. *Geology* **8**, 140-143.

ANDRIEUX, J., BRUNEL, M. and HAMET, J. (1977). Metamorphism, granitisation and relations with the Main Central Thrust in Central Nepal: RB[88]/Sr[87] age determination and discussion. *In* "Colloque sur l'ecologie et géologie de l'Himalaya No. 268", 31–40. Centre National de la Recherche Scientifique, Paris.

ANNELS, A. E. (1979). Mufulira greywackes and their associated sulphides. *Trans. Instn Min. Metall.* **88**, B15-23.

ANON. (1978). "Year Book of the Bureau of Mineral Resources, Geology and Geophysics." Dept National Development, Canberra.

ARGAND, E. (1924). La tectonique de l'Asia. 13th Internat. Geol. Cong., Belgium 1922, Comptes Rendus pt 1, 171–372.

ARMSTRONG, R. L., TAUBENECK, W. H. and HALES, P. O. (1977). Rb-Sr and K-Ar geochronometry of Mesozoic granitic rocks and their Sr isotopic compositions, Oregon, Washington and Idaho. *Bull. geol. Soc. Am.* **88**, 397–411.

ARSENYEV, A. A. (1962). Laws of the distribution of kimberlites in the eastern part of the Siberian platform. *Dokl. Akad. Nauk SSR Earth Sci. Sect.* **137**, 355-357.

ARTHAUD, F. and MATTE, P. (1977). Late Palaeozoic strike-slip faulting in southern Europe and northern Africa: result of a right-lateral shear zone between the Appalachians and the Urals. *Bull. geol. Soc. Am.* **88**, 1305-1320.

ATWATER, T. and MACDONALD. K. C. (1977). Are spreading centres perpendicular to their transform faults? *Nature, Lond.* **270**, 715-719.

ATWATER, T. and MOLNAR, P. (1973). Relative motion of the Pacific and North American plates deduced from sea-floor spreading in the Atlantic. Indian and South Pacific Oceans. *Stanford Univ. Publs. geol. Sci.* **13**, 136-148.

AUBOUIN, J. (1965). "Geosynclines". Elsevier, Amsterdam.

AUDLEY-CHARLES, M. G., CARTER, D. J., BARBER, A. J., NORVICK, M. S. and TJOKROSAPOETRO, S. (1979). Reinterpretation of the geology of Seram: implications for the Banda Arcs and northern Australia. *J. geol. Soc. Lond.* **136**, 547–568.

AVEDIKO, G. P. (1971). Evolution of geosynclines of Kamchatka. *Pacif. Geol.* **3**, 1–13.

BÄCKER, H. and SCHOELL, M. (1974). Anreicherungen von Elementen zu Rohstoffen in mariner Bereich. *Chemker Zeitung* **98**, 299–305.

BACKSTROM, J. W. VON (1974). Uranium deposits in the Karroo Supergroup near Beaufort West, Cape Province, South Africa. *In* "Formation of Uranium Ore deposits", 419–424. IAEA, Vienna.

BADHAM, J. P. N. (1976). Orogenesis and metallogenesis with reference to the silver-nickel, cobalt arsenide ore association. *Spec. Pap. geol. Ass. Can.* **14**, 559-571.

BADHAM, J. P. N. (1978). Slumped sulphide deposits at Avoca, Ireland, and their significance. *Trans. Instn. Min. Metall.* **87**, B21–26.

BADHAM, J. P. N. and HALLS, C. (1975). Microplate tectonics, oblique collisions, and evolution of the Hercynian orogenic systems. *Geology* **3**, 373-376.

BALLARD, R. D., HOLCOMB, R. T. and VAN ANDEL, T. N. (1979). The Galapagos Rift at 86° W: 3. Sheet flows, collapse pits and lava lakes of the rift valley. *J. Geophys, Res.* **84**, 5407–5422.

BALLARD, R. D. and VAN ANDEL, T. H. (1977). Morphology and tectonics of the inner rift valley at lat. 36° 50′ N on the Mid-Atlantic Ridge. *Bull. geol. Soc. Am.* **88**, 507–530.

BAILEY, D. K. (1960). Carbonatites of the Rufunsa Valley, Feira District. *Bull. N. Rhod. geol. Surv.* **5**.

BAILEY, D. K. (1977). Lithosphere control of continental rift magmatism. *J. geol. Soc. Lond.* **133**, 103-106.

BAILEY, D. K. (1980). Volatile flux, geotherms, and the generation of the kimberlite-carbonatite-alkaline magma spectrum. *Min Mag.* **43**,695–699.

BAKER, B. H., MOHR, P. A. and WILLIAMS, L. A. J. (1972). Geology of the eastern rift system of Africa. *Spec. Pap. geol. Soc. Am.* **136**.

BALDOCK, J. W. (1969). Geochemical dispersion of copper and other elements at the Bukusu carbonatite complex, Uganda. *Trans. Instn. Min Metall.* **78**, B12-28.

BALLY, A. W. (1975). "A Geodynamic Scenario for Hydrocarbon Occurrences." Proc. 9th World Petroleum Cong. No. 2, 33–44. Applied Science, Essex.

BALLY, A. W. and SNELSON, S. (1980). Realms of subsidence. *In* "Facts and Principles of World Petroleum Occurrence". Can. Soc. Petroleum Geologists Mem. No. 6.

BALLY, A. W., ALLEN, C. R., GEYER, R. B., HAMILTON, W. B., HOPSON, C. A., MOLNAR, P. H., OLIVER, J. E., OPDYKE, N. D., PLAFKER, G. and WU, F. T. (1980). Notes on the geology of Tibet and adjacent areas—report of the American plate tectonics delegation to the People's Republic of China. U.S. Geol. Surv. Open File Rept, 80-501.

BARR, S. M. and MACDONALD, A. S. (1978). Geochemistry and petrogenesis of Late Cenozoic alkaline basalts of Thailand. *Bull. geol. Soc. Malaysia* **10**, 25–52.

BARR, S. M. and MACDONALD, A. S. (1979). Palaeomagnetism, age and geochemistry of the Denchai basalt, northern Thailand. *Earth planet. Sci. Lett.* **46**, 113-124.

BARSUKOV, V. L. and KURIL'CHIKOVA, G. Y. (1966). On the forms in which tin is transported in hydrothermal solutions. *Geochem. Internat.* **3**, 759–764.

BARTHEL, F. H. (1974). Review of uranium occurrences in Permian sediments in Europe, with special reference to uranium mineralizations in Permian sandstone. *In* "Formation of Uranium Ore Deposits", 277–289. IAEA, Vienna.

BASHAM, I. R. and RICE, C. M. (1974). Uranium mineralization in Siwalik Sandstones from Pakistan. *In* "Formation of Uranium Ore Deposits", 405–417. IAEA, Vienna.

BATCHELOR, B. C. (1979). Geological characteristics of certain coastal and offshore placers as essential guides for tin exploration in Sundaland Southeast Asia. *Bull. geol. Soc. Malaysia* **11**, 283–313.

BATEMAN, A. M. (1950). "Economic Mineral Deposits". Wiley, New York.

BAUMANN, L. (1970). Tin deposits of the Erzgebirge. *Trans. Instn. Min. Metall.* **79**, B68–75.

BECK, R. H. (1972). The oceans, the new frontier in exploration. *Aust. Petroleum Explor. Ass. J.* **12**, 7–28.

BECKINSALE, R. D. and MITCHELL, A. H. G. (1981). Ore deposits associated with subduction. In "Economic Geology and Geotectonics" (D. H. Tarling, ed.), 135–147. Blackwell Scientific, Oxford.

BECKINSALE, R. D., BOWLES, H. F.W., PANKHURST, R. J. and WELLS, M. K. (1977). Rubidium-strontium age studies and geochemistry of acid veins in the Freetown Complex, Sierra Leone. *Min. Mag.* **41**, 501–512.

BECKINSALE, R. D., SUENSILPONG. S., NAKAPADUNCRAT, S. and WALSH, J. N. (1979). Geochronology and Geochemistry of granite magmatism in Thailand in relation to a plate tectonic model. *J. geol. Soc. Lond.* **136**, 529-537.

BELOUSSOV, V. V. (1968). "The Earth's Crust and Upper Mantle of the Oceans". Nauka, Moscow.

BELYAYEA, K. V. and UYAD'YEV, L. I. (1978). Palaeozoic dike complexes of the Kola peninsula and Northern Karelia. *Int. Geol. Rev.* **20**, 273–280.

BEMMELEN, R. W. VAN (1949). "The Geology of Indonesia. IA, General Geology". Mart Nijhoff, The Hague.

BEMMELEN, R. W. VAN. (1978). The present formulation of the undation theory. *Z. Geol. Wiss. Berlin* **6**, S23–40.

BEN AVRAHAM, Z. B. and NUR, A. (1976). Slip rates and morphology of continental collision belts. *Geology* **4**, 661–664.

BENNETT, E. H. (1980). Granitic rocks of Tertiary age in the Idaho Batholith and their relation to mineralization. *Econ. Geol.* **75**, 278–288.

BERANGÉ, J. P. and JOBBINS, E. A. (1975). The geology, gemmology, mining methods and economic potential of the Pailin ruby and sapphire gemfield, Khmer Republic. Inst. geol. Sci., Overseas Div. Rep. No. 35.

BERGSTØL, S. (1972). The jacupirangite at Kodal, Vestfold, Norway. *Mineral. Deposita* **7**, 233–246.

BERNING, J., COOKE, R., HEIMSTRA, S. A. and HOFFMAN, U. (1976). The Rossing uranium deposit, South West Africa. *Econ. Geol.* **71**, 351–368.

BERTRAND, M. (1897). Structure des Alpes francaises et recurrence de certains facies sedimentaires. 5th Internat. Geol. Cong., Zurich, 161-177.

BEST, M. G., ARMSTRONG, R. L. GRAUSTEIN, W. C., EMBREE, G. F. and AHLBORN, R. C. (1974). Mica granites of the Kern Mountains Pluton, Eastern White County, Nevada: remobilized basement of the Cordilleran Miogeocline? *Bull. geol. Soc. Am.* **85**, 1277–1286.

BICHAN, R. (1970). The evolution and structural setting of the Great Dyke, Rhodesia. *In* "African Magmatism and Tectonics" (T. N. Clifford and I. G. Gass, eds), 51–71. Oliver and Boyd, Edinburgh.

BIGNELL, R. D. (1975). Timing, distribution and origin of submarine mineralization in the Red Sea. *Trans. Instn Min. Metall.* **84**, B1–6.

BILIBIN, Y. A. (1948). On geochemical types of orogenic zones. 18th Internat. Geol. Congr., London, pt 2, 22-28.

BILIBIN, Y. Al (1955). "Metallogenetic Provinces and Epochs". Gosgeolte-khizdat, Moscow.

BILIBIN, Y. A. (1968). "Metallogenic Provinces and Metallogenic Epochs", 1–35. Queens College Press, Flushing, New York.

BILLINGSLEY, R. P. and LOCKE, A. (1935). Tectonic position of ore districts in the Rocky Mountain regions. *Trans. Am. Inst. Min. metall. Petrol. Engrs* **115**, 59-68.

BINGHAM, D. K. and KLOOTWIJK, C. T. (1980). Palaeomagnetic constraints on Greater India's underthrusting of the Tibetan Plateau. *Nature, Lond.* **284**, 336–338.

BIQ CHINGCHANG (1971). Dual-trench structure in the Taiwan–Luzon region. *Proc. geol. Soc. China* **15**, 65–75.

BIQ CHINGCHANG (1978). Taiwan vis-a-vis New Guinea: a comparison of their continent-arc collisions. *Acta Oceanographica Taiwanica* **8**, 22–42.

BLONDEL, F. (1936). "La géologie et les mines des vieilles platformes". Bureau d'Études Géologiques et Minères Coloniales, Paris.

BOCCALETTI, M., MANETTI, P., PECCERILLO, A. and PELTZ, S. (1973). Young volcanism in the Calimani-Harghita Mountains (East Carpathians): evidence of a palaeoseismic zone. *Tectonophysics* **19**, 299–313.

BOCCALETTI, M., HORVATH, F., LODDO, M., MONGELLI, F. and STEGENA, L. (1976). The Tyrrhenian and Pannonian basins: a comparison of two Mediterranean interarc basins. *Tectonophysics* **35**, 45–69.

BOCKSTROM, A. A. (1977). The magnetite deposits of El Romeval, Chile. *Econ. Geol.* **72**, 1101–1130.

BOCTOR, N. Z. and BOYD, F. R. (1979). Oxide minerals in layered kimberlite-carbonate sills from Benfontein, South Africa. Ann. Rep. Director Geophys. Lab. 1978-1979, 493-496. Carnegie Inst, New York.

BONATTI, E. and HARRISON, C. G. A. (1976). Hot lines in the earth's mantle. *Nature, Lond.* **263**, 402–404.

BONATTI, E., FISHER, D. E., JOENSUU, O., RYDELL, H. S. and BEYTH, M. (1972). Iron-manganese-barium deposit from the northern Afar rift (Ethiopia). *Econ. Geol.* **67**, 717–730.

BONATTI, E., HONNOREZ-GUERSTEIN, M. B., HONNOREZ, J. and STERN, C. (1976a). Hydrothermal pyrite concretions from the Romanche Trench (equatorial Atlantic): metallogenesis in oceanic fracture zones. *Earth planet. Sci. Lett.* **32**, 1–10.

BONATTI, E., ZERBI, M., KAY, R. and RYDELL, H. S. (1976b). Metalliferous deposits from the Apennine ophiolites: Mesozoic equivalents of modern deposits from oceanic spreading centers. *Bull. geol. Soc. Am.* **87**, 83-94.

BORCHERT, H. (1960). Genesis of marine sedimentary iron ores. *Trans. Instn Min. Metall.* **69**, B261–279.

BORCHERT, H. (1964). Principles of the genesis and enrichment of chromite ore deposits. Paris Org. Econ. Coop. Dev. 175–202.

BORDET, P., COLCHEN, M., KRUMMENACHER, D., LE FORT, P., MOUTERDE, R. and REMY, M. (1971). "Recherches geologiques dans l'Himalaya du Nepal, Region de la Thakkhola". Centre National de la Recherche Scientifique, Paris.

BOR-MING JAHN, CHEN, P. Y. and YEN, T. P. (1976). Rb–Sr ages of granitic rocks in southeastern China and their tectonic significance. *Bull. geol.*

Soc. Am. **86**, 763-776.

BORODIN, L. S., GOPAL, V., MORALEV, V. M., SUBRAMANIAM, V. and PONIKAROV, V. (1971). Precambrian carbonatites of Tamil Nadu, South India. *J geol. Soc. India* **12**, 101-112.

BORTOLOTTI, V., LAPIERRE, H. and PICCARDO, G. B. (1976). Tectonics of the Troodos Massif (Cyprus): preliminary results. *Tectonophysics* **35**, T1-T5.

BOSTROM, K., RYDELL, H. and JOENSUU, O. (1979). Långban-an exhalative sedimentary deposit. *Econ. Geol.* **74**, 1002-1011.

BOSTROM, R. C. (1971). Westward displacement of lithosphere. *Nature, Lond.* **234**, 536-538.

BOWDEN, P. and KINNAIRD, J. A. (1978). Younger granites of Nigeria—a zinc-rich tin province. *Trans. Instn Min. Metall.* **87**, B66-69.

BOWDEN, P., VAN BREEMEN, O., HUTCHINSON, J. and TURNER, D. C. (1976). Palaeozoic and Mesozoic age trends for some ring complexes in Niger and Nigeria. *Nature, Lond.* **259**, 297-299.

BOWIE, S. H. U., SIMPSON, P. R. and RICE, C. M. (1973). In "Geochemical exploration 1972", Proc. 4th Internat. Geochem. Explor. Sym. (M. J. Jones, ed.), 359-372. Instn Min. Metall., London.

BOWLES, J. F. W. (1978). The geochemical role of copper-sulphur mineralisation in the crystallisation of the Freetown (Sierra Leone) layered gabbro. *Min. Mag.* **42**, 111-116.

BRECKE, E. A. (1979). A hydrothermal system in the midcontinent region. *Econ. Geol.* **74**, 1327-1335.

BREEMAN, O., VAN, AFTALION, M. and JOHNSON, M. R. W. Age of the Loch Borolan Complex and late movements along the Moine Thrust. *J. geol. Soc. Lond.* **136**, 489-495.

BREY, G. P. and GREEN, D. H. (1976). Solubility of CO_2 in olivine melilitite at high pressures and role of CO_2 in the earth's upper mantle. *Contr. Mineral. Petrol.* **55**, 217-230.

BROMLEY, A. V. (1976). Granites in mobile belts—the tectonic setting of the Cornubian batholith. *J. Cambourne School Mines* **76**, 40-47.

BROMLEY, A. V. (1978). Mineralization associated with acid magmatism and other aspects of ore geology. *J. geol. Soc. Lond.* **135**, 253-257.

BROOKFIELD, M. E. (1977). The emplacement of giant ophiolite nappes. I. Mesozoic-Cenozoic examples. *Tectonophysics* **37**, 247-303.

BROWN, A. C. (1978). Stratiform copper deposits—evidence for their post-sedimentary origin. *Minerals Sci. Engng* **10**, 172-181.

BROWN, G. C., CASSIDY, J., TINDLE, A. G. and HUGHES, D. J. (1979), The Loch Doon granite: an example of granite petrogenesis in the British Caledonides. *J. geol. Soc. Lond.* **136**, 745-753.

BROWN, M. A. (1980). Textural and geochemical evidence for the origin of some chromite deposits in the Oman ophiolite. *In* "Ophiolites, Proceedings International Ophiolite Symposium", Cyprus, 1979 (A. Panayiotou, ed.), 714-721. Geol. Surv. Dept. Nicosia, Cyprus.

BRYNER, L. (1969). Ore deposits of the Philippines-an introduction to their geology. *Econ. Geol.* **64**, 644-666.

BURKE, K. C. (1975). Atlantic evaporites formed by evaporation of water spilled from Pacific, Tethyan, and Southern oceans. *Geology* **3**, 613-616.

BURKE, K. C. (1976). The Chad Basin: an active intra-continental basin. *Tectonophysics* **36**, 197-206.

BURKE, K. C. (1977). Aulacogens and continental break-up. *A. Rev. Earth planet. Sci.* **5**, 371-396.

BURKE, K. C. and DEWEY, J. F. (1973). Plume-generated triple junctions: key indicators in applying plate tectonics to old rocks. *J. Geol.* **81**, 406–433.

BURKE, K. C. and DEWEY, J. F. (1974). Two plates in Africa during the Cretaceous. *Nature, Lond.* **249**, 313-316.

BURKE, K. C. and WHITEMAN, A. J. (1973). Uplift, rifting and the break-up of Africa. *In* "Implications of Continental Drift to Earth Sciences", (D. H. Tarling and S. K. Runcorn, eds), 735–755. Academic Press, London and New York.

BURKE, K. C. and WILSON, J. T. (1976). Hot spots on the earth's surface. *Sci. Am.* **238**, 46–57.

BURKE, K. C., KIDD, W. S. F. and WILSON, J. T. (1973). Plumes and convective plume traces of the Eurasian plate. *Nature, Lond.* **241**, 128–129.

BURNS, R. G. and FYFE, W. S. (1964). Site preference energy and selective uptake of transition-metal ions from a magma. *Science* **144**, 1001-1003.

BURRI, C. (1960). Petrochemie der Capvervden und Vergleich des Capverrdischen Vulkanisms mit demjenigen des Rheinlandes. *Schweiz. Mineral. Petrog. Mitt.* **40**, 115-161.

BURTON, C. K. (1970). The palaeotectonic status of the Malay Peninsula. *Palaeogeog. Palaeoclimatol. Palaeoecol.* **7**, 51–60.

BUTTON, A. (1976). Iron-formation as an end-member in carbonate sedimentary cycles in the Transvaal Supergroup, South Africa. *Econ. Geol.* **71**, 193-201.

CADY, W. W., McKELVEY, V. M. and WELLS, F. G. (1950). Geotectonic relationships of mineral deposits. (Abs.), *Bull. geol. Soc. Am.* **61**, 1447.

CAHEN, L. and SNELLING, N. J. (1966). "The geochronology of equatorial Africa". North-Holland, Amsterdam.

CAIA, J. (1976). Palaeogeographical and sedimentological controls of copper, lead and zinc mineralizations in the Lower Cretaceous sandstones of Africa. *Econ. Geol.* **71**, 409–422.

CALLAHAN, W. H. (1967). Some spatial and temporal aspects of the localization of Mississippi Valley-Appalachian type ore deposits. *Econ. Geol. Mon.* **3**, 14–19.

CARNEY, J. N. and MACFARLANE, A. (1979). Geology of Tanna, Aneityum, Futuna and Aniwa New. Hebrides Govt Geol. Sur. Reg. Rept.

CARPENTER, A. B., TROUT, M. L. and PICKETT, E. E. (1974). Preliminary report on the origin and chemical evolution of lead- and zinc-rich oil field brines in Central Mississippi. *Econ. Geol.* **69**, 1191-1206.

CARR, M. J., STOIBER, R. E. and DRAKE, C. L. (1973). Discontinuities in the deep seismic zones under the Japanese arcs. *Bull. geol. Soc. Am.* **84**, 2917-2930.

CARTER, D. J., AUDLEY-CHARLES, M. G. and BARBER, A. J. (1976). Stratigraphical analysis of island arc – continental margin collision in eastern Indonesia. *J. geol. Soc. Lond.* **132**, 179–198.

CARTER, W.D. (1960). Origin of "manto-type" copper deposits of the Cabildo mining district, central Chile. 21st Internat. Geol. Congr. Rep., Copenhagen, pt 16, 17–28.

CHANG CHENG-FA and CHENG HSI-LAN (1973). Some tectonic features of the Mt. Jolmo Lungma area, Southern Tibet, China. *Scientia Sinica* **16**, 257–265.

CHAPPELL, B. W. and WHITE, A. J. R. (1974). Two contrasting granite types. *Pacif. Geol.* **8**, 173-174.

CHARRIER, R., LINARES, E., NIEMEYER, H. and SKARMETA, J (1979). K-Ar ages of basalt flows of the Meseta Buenos Aires in southern Chile and their relation to the southeast Pacific triple junction. *Geology* **7**, 436-439.

CHURKIN, M. Jr. (1974). Palaeozoic marginal ocean basin volcanic arc systems in the Cordilleran foldbelt. *In* "Modern and ancient geosynclinal sedimentation" (R. H. Dott Jr and R. H. Shaver, eds). *Spec. Publs Soc. econ. Palaeont. Miner.* **19**, 174-192.

CLAGUE, D. A. and STRALEY, P. F. (1977). Petrologic nature of the oceanic Moho. *Geology* **5**, 133-136.

COATS, R. R. (1962). Magma type and crustal structure in the Aleutian arc. *Aust. Geophys. Union, Geophys. Mon.* **6**, 92-109.

COBBING, E. J. and PITCHER, W. S. (1972). Plate tectonics and the Peruvian Andes. *Nature Phys. Sci.* **240**, 51-53.

COETZEE, G. L. (1963). The origin of the Sangu carbonatite complex and associated rocks, Karema depression, Tanganyika Territory, East Africa. PhD thesis, Univ. of Wisconsin, USA.

COLEMAN, R. G. (1971). Plate tectonic emplacement of upper mantle peridotites along continental edges. *J. geophys. Res.* **76**, 1212-1222.

COLEMAN, R. G. (1977). "Ophiolites: Ancient Oceanic Lithosphere?" Minerals and Rocks, Vol. 12. Springer-Verlag, Berlin.

COLLETTE, B. J. (1974). Thermal contraction joints in a spreading sea-floor as origin of fracture zones. *Nature, Lond.* **251**, 299-300.

COLLEY, H. (1976). Classification and exploration guide for Kuroko-type deposits based on occurrences in Fiji. *Trans. Instn Min. Metall.* **85**, B190-199.

CONEY, P. J. (1976). Plate tectonics and the Laramide orogeny. *In* "Tectonics and Mineral Resources of Southwestern North America" (L. A. Woodward and S. A. Northrop, eds), 5-10. New. Mex. Geol. Soc., Spec. Publ. 6.

CONSTANTINOU, G. and GOVETT, G. J. S. (1972). Genesis of sulphide deposits, ochre and umber of Cyprus. *Instn. Min. Metall.* **81**, B34-46.

COOK, F.A., ALBAUGH, D. S., BROWN, L. D., KAUFMAN, S., OLIVER, J. E. and HATCHER, R. D. Jr. (1979). Thin-skinned tectonics in the crystalline southern Appalachians; COCORP seismic-reflection profiling of the Blue Ridge and Piedmont. *Geology* **7**, 563-567.

COOK, P. J. and McELHINNY, M. W. (1979). A re-evaluation of the spatial and temporal distribution of sedimentary phosphate deposits in the light of plate tectonics. *Econ. Geol.* **74**, 315-330.

COOPER, A. F. (1971). Carbonatites and fenitization associated with a lamprophyric dike-swarm intruded into schists of the New Zealand Geosyncline. *Bull. geol. Soc. Am.* **82**, 1327-1340.

CORLISS, J. B. (1971). The origin of metal-bearing submarine hydrothermal solutions. *J. geophys. Res.* **76**, 8128-8138.

CORLISS, J. B., DYMOND, J., GORDON, L., EDMOND, J. M., VON HERZEN, R. P., BALLARD, R. D., GREEN, K., WILLIAMS, D., BAINBRIDGE, A., CRANE, K. and ANDEL, T.H. VAN (1979). Submarine thermal springs on the Galapagos Rift. *Science, N.Y.* **203**, 1073-1083.

COUSINS, C. A. (1959). The structure of the mafic portion of the Bushveld Igneous Complex. *Trans. geol. Soc. S. Afr.* **62**, 179-189.

COX, K. G. (1978). Flood basalts, subduction and the break-up of Gondwanaland. *Nature, Lond.* **274**, 47-49.

COX, K. G. (1980). Kimberlite and carbonatite magmas. *Nature, Lond.* **238**, 616-617.

CRAWFORD, A.R. (1980). Burma: NNW-SSE lineaments and relation to drainage, mineralization and tectonics. *J. geol. Soc. India* **21**, 273-285.

CRONAN, D. S. and TOOMS, J. S. (1969). The geochemistry of manganese nodules and associated pelagic deposits from the Pacific and Indian Oceans. *Deep-Sea Res.* **16**, 335-359.

CROSS, T. A. and PILGER, R. H. Jr. (1978). Constraints on absolute plate motion and plate interaction inferred from Cenozoic igneous activity in the western United States. *Am. J. Sci.* **278**, 865–902.

CROWELL, J. C. (1979). The San Andreas fault system through time. *J. geol. Soc. Lond.* **136**, 293–302.

CULL, J. P. and DENHAM, D. (1979). Regional variations in Australian heat flow. *Bur. Mineral. Resour. J. Aust. Geol. Geophys.* **4**, 1–13.

CULVER, S. J. and WILLIAMS, H. R. (1979). Late Precambrian and Phanerozoic geology of Sierra Leone. *J. geol. Soc. Lond.* **136**, 605–618.

CURRAY, J. R., MOORE, D. G., LAWVER, L. A., EMMEL, F. J., RAITT, R. W., HENRY, M. and KIECKHEFER, R. (1978). Tectonics of the Andaman Sea and Burma. *In* "Geological and Geophysical Investigations of Continental Margins". 189-198. Am. Ass. Petroleum Geologists Mem. 29.

CURRAY, J. R., EMMEL, F. J. and MOORE, D. G. (1980). Structure, tectonics and geological history of the northeastern Indian Ocean. *In* "The Ocean Basins and Margins", Vol. 6. The Indian Ocean (A. E. M. Nairn and F. G. Stehli, eds). Plenum Press, New York.

DAHL, A. R. and HAGMAIER, J. L. (1974). Genesis and characteristics of the Southern Powder River Basin uranium deposits, Wyoming, U.S.A. *In* "Formation of uranium ore deposits", 201-216. IAEA, Vienna.

DAHLKAMP, F. J. (1978). Geologic appraisal of the Key Lake U-Ni Deposits, Northern Saskatchewan. *Econ. Geol.* **73**, 1430–1449.

DAHLKAMP, F. J. (1980). The time-related occurrence of uranium deposits. *Mineral. Deposita* **15**, 69–79.

DANA, J. D. (1873). On some results of the earth's contraction from cooling, including a discussion of the origin of mountains and the nature of the earth's interior. *Am. J. Sci.* **5**, 432–434, **6**, 6–14, 104–115, 161–171.

DARBYSHIRE, D. P. F. and FLETCHER, C. J. N. (1979). A Mesozoic alkaline province in eastern Bolivia. *Geology* **7**, 545–548.

DAVIES, F. B. and WINDLEY, B. F. (1976). Significance of major Proterozoic high-grade linear belts in continental evolution. *Nature, Lond.* **263**, 383–385.

DAWSON, J. B. (1962). The geology of Oldoinyo Lengai. *Bull. Volcanologique.* **24**, 349–387.

DAWSON, J. B. (1967). A review of the geology of kimberlite. *In* "Ultramafic and Related Rocks" (P. J. Wyllie, ed.), 269–278. Interscience, New York.

DAWSON, J. B. (1970). The structural setting of African kimberlite magmatism. *In* "African Magmatism and Tectonics" (T. N. Clifford and I. G. Gass, eds), 321–335. Oliver and Boyd, Edinburgh.

DEANS, T. (1966). Economic geology of African carbonatites. *In* "Carbonatites" (O. F. Tuttle and J. Gittins, eds), 385–413. Interscience, New York.

DEANS, T. and POWELL, J. L. (1968). Trace elements and strontium isotopes in carbonatites, fluorites and limestones from India and Pakistan. *Nature, Lond.* **218**, 750–752.

DEGENS, E. T. and ROSS, D. A. (eds) (1969). "Hot Brines and Recent Heavy Metal Deposits in the Red Sea: A Geochemical and Geophysical Account." Springer-Verlag, New York.

DE JAGER, D. H. and FOURIE, P. J. (1978). A review of some new aspects on phosphate mineralization in the Palabora Igneous Complex (South Africa). Preprint Am. Instn. Min. Metall. Eng. Meeting, Denver, Colorado, Feb/March, 1978.

DELONG, S. E. and SCHWARZ, W. M. (1979). Thermal effects of ridge subduction. *Earth planet. Sci. Lett.* **44**, 239–246.

DELONG, S. E., DEWEY, J. F. and FOX, P. J. (1977). Displacement history of oceanic fracture zones. *Geology* **5**, 199–202.

DEN, N. and HOTTA, H. (1973). Seismic refraction and reflection evidence supporting plate tectonics in Hokkaido. *Pap. Met. Geophys.* **24**,

31–54.

DENHOLM, L. S. (1967). Geological exploration for gold in the Tavua Basin, Vitu Levu, Fiji. *N.Z. J. Geol. Geophys.* **10**, 1185–1186.

DERKMANN, K. and KLEMM, D. D. (1977). Strata-bound kies-ore deposits in ophiolitic rocks of the "Tauernfenster" (Eastern Alps, Austria/Italy). *In* "Time and Stratabound Ore Deposits" (D. D. Klemm and H. J. Schneider, eds), 305-313. Springer-Verlag, Berlin.

DE RUITER, P. A. C. (1979). The Gabon and Congo Basins salt deposits. *Econ. Geol.* **74**, 419–431.

DEWEY, J. F. and BIRD, J. M. (1970). Mountain belts and the new global tectonics. *J. geophys. Res.* **75**, 2625–47.

DEWEY, J. F. and BIRD, F. M. (1971). Origin and emplacement of the ophiolite suite: Appalachian ophiolites in Newfoundland. *J. geophys. Res.* **76**, 3179-3206.

DEWEY, J. F. and BURKE, K. C. (1973). Tibetan, Variscan and Precambrian basement reactivation: products of continental collision. *J. Geol.* **81**, 683–692.

DEWEY, J. F. and KIDD, W. S. F. (1977). Geometry of plate accretion. *Bull. geol. Soc. Am.* **88**, 960–968.

DICKEY, J. S. (1975). A hypothesis of origin for podiform chromite deposits. *Geochim. cosmochim. Acta* **39**, 1061–1074.

DICKINSON, W. R. (1971). Clastic sedimentary sequences deposited in shelf, slope and trough settings between magmatic arcs and associated trenches. *Pacif. Geol.* **3**, 15–30.

DICKINSON, W. R. (1974a). Subduction and oil migration. *Geology* **2**, 421–424.

DICKINSON, W. R. (1974b). Plate tectonics and sedimentation. *Spec. Publs. Soc. econ. Palaeont. Miner.* **22**, 1–27.

DICKINSON, W. R. (1976). Sedimentary basins developed during evolution of Mesozoic-Cenozoic arc-trench system in western North America. *Can. J. Earth Sci.* **13**, 1268–1287.

DICKINSON, W. R. (1977). Subduction tectonics in Japan. *Eos* **58**, 948–952.

DICKINSON, W. R. (1978). Plate tectonic evolution of North Pacific rim. *J. Phys. Earth* **26**, Suppl. S1–S19.

DICKINSON, W. R. and SNYDER, W. S. (1979). Geometry of triple junctions related to San Andreas transform. *J. geophys. Res.* **84**, 609–628.

DIETRICH, V. J. (1976). Evolution of the Eastern Alps: a plate tectonics working hypothesis. *Geology* **4**, 147–152.

DIETZ, R. S. (1961). Vredefort ring structure: meteorite impact scar? *J. Geol.* **69**, 499–516.

DIETZ, R. S. (1964). Sudbury structure as an astrobleme. *J. Geol.* **72**, 412–434.

DIETZ, R. S. and HOLDEN, J. C. (1966). Miogeoclines (miogeosynclines) in space and time. *J. Geol.* **74**, 566–585.

DIETZ, R. S. and SPROLL, W. P. (1970). East Canary Islands as a microcontinent within Africa-North America continental drift fit. *Nature, Lond.* **226**, 1043.

DIMROTH, E. (1972). The Labrador geosyncline revisited. *Am. J. Sci.* **272**, 487–506.

DIXON, C. J. (1979). "Atlas of Economic Mineral Deposits". Chapman and Hall, London.

DOE, B. R. and STACEY, J. S. (1974). The application of lead isotopes to the problems of ore genesis and ore prospect evaluation. *Econ. Geol.* **69**, 757–776.

DOIG, R. (1970). An alkaline rock province linking Europe and North America. *Can. J. Earth Sci.* **7**, 22–28.

DRAKE, C. L., EWING, M. and SUTTON, G. H. (1959). Continental margins and geosynclines: the east coast of North America, north of Cape Hatteras. *In* "Physics and Chemistry of the Earth" (L. H. Ahrens, F. Press, S. K. Runcorn and H. C. Urey, eds). Vol. 3, 110–198. Pergamon Press, Oxford.

DREVER, J. I. (1974). Geochemical model for the origin of Precambrian banded iron formations. *Bull. geol. Soc. Am.* **85**, 1099–1106.

DRONOV, V. I., KARARSKIY, A. K., DENIKAEV, S. S., SALAH, A. S., SONIN, I. I., CHMYRIOV, V. M. and ABDULLAH, J. (1973). Mineral resources of Afghanistan. *In* "Geology and Mineral Resources of Afghanistan", 44–85. Geol. Surv. Dept, Afghanistan.

DUKE, N. A. and HUTCHINSON, R. W. (1974). Geological relationships between massive sulfide bodies and ophiolitic volcanic rocks near York Harbour, Newfoundland. *Can. J. Earth Sci.* **11**, 53–69.

DUNCAN, R. A. and GREEN, D. H. (1980). Role of multistage melting in the formation of oceanic crust. *Geology* **8**, 22–26.

DUNNET, D. (1976). Some aspects of the Panantarctic cratonic margin in Australia. *Phil. Trans. R. Soc. Lond.* **280A**, 641–654.

DUNSMORE, H. E. (1973). Diogenetic processes of lead-zinc emplacement in carbonates. *Trans. Instn. Min. Metall.* **82**, B168–173.

DU TOIT, A. L. (1937). "Our Wandering Continents". Oliver and Boyd, Edinburgh.

ECKERMANN, H. VON (1948). The alkaline district of Alnö Island. *Sver. geol. unders.* **36**, Stockholm.

EDEN, J. G. VAN (1978). Stratiform copper and zinc mineralization in the Cretaceous of Angola. *Econ. Geol.* **73**, 1154–1161.

EGANOV, E. A. (1979). The role of cyclic sedimentation in the formation of phosphorite deposits. *In* "Proterozoic–Cambrian Phosphorites" (P. J. Cook and J. H. Shergold, eds), 22–25. IGCP Project 156, Canberra.

EGUCHI, T., UYEDA, S. and MAKI, T. (1979). Seismotectonics and tectonic history of the Andaman Sea. *Tectonophysics* **57**, 35–51.

EHRENBURG, H., PILGER, A. and SCHRODER, G. (1954). Das Schwerfelkies-Zinkblende-Schwerspatlager von Meggen (Westfalen). Niedersachsisches Landesanst fur Bodenforschung, Monograph der Deutschen Blei-Zlink Erzlagerstatten No. 7, Hanover.

EISBACHER, G. H. (1974). Evolution of successor basins in the Canadian Cordillera. *In* "Modern and Ancient Geosynclinal Sedimentation" (R. H. Dott Jr and R. H. Shaver, eds), *Spec. Publs Soc. econ. Palaeont. Miner.* **19**, 274–291.

EISBACHER, G. H. (1977). Mesozoic-Tertiary basin models for the Canadian Cordillera and their geological constraints. *Can. J. Earth Sci.* **14**, 2414–2421.

ERDOSH, G. (1979). The Ontario carbonatite province and its phosphate potential. *Econ. Geol.* **74**, 331–338.

EUGSTER, H. P, and CHOU, I. M. (1973). The depositional environments of Precambrian banded iron-formations. *Econ. Geol.* **68**, 1144–1168.

EVERNDEN, J. F., KRIZ, S. J. and CHERRONI, M. C. (1977). Potassium-argon ages of some Bolivian rocks. *Econ. Geol.* **72**, 1042–1061.

FALCON, R. M. S. (1977). Coal in South Africa—Part 1: the quality of South African coal in relation to its uses and world energy resources. *Minerals. Sci. Engng* **9**, 198–217.

FALVEY, D. A. (1974). The development of continental margins in plate tectonic theory. *Aust. Petroleum Explor. Ass. J.* **10**, 95–106.

FAUCONNIER, D. and SLANSKY, M. (1979). The possible role of dinoflagellates in phosphate sedimentation. *In* "Proterozoic-Cambrian Phosphorites" (P. J. Cook and J. H. Shergold, eds), 93–101. IGCP Project 156, Canberra.

FEISS, P. G. (1978). Magmatic sources of copper in porphyry copper deposits. *Econ. Geol.* **73**, 397–404.

FEATHER, C. E. and KOEN, G. M. (1975). The mineralogy of the Witwatersrand reefs. *Minerals Sci. Engng* **7**, 189–224.

FERGUSON, J. and SHERATON, J. W. (1979). Petrogenesis of kimberlitic rocks and associated xenoliths of southeastern Australia. *In* "Kimberlites, Diatremes and Diamonds: Their Geology, Petrology and Geochemistry" (F. R. Boyd and H. O. A. Meyer, eds), 140–160. Proc. Second Internat. Kimberlite Conf. 1. Am. Geophys. Union, Washington.

FERGUSON, S. A. (1971). Columbium (Niobium) deposits of Ontario. *Ontario Dept Mines, Mineral Resour. Cir.* **14.**

FERNANDEZ, H. E. and DAMASCO, F. V. (1980). Gold deposition in the Baguio Gold District and its relationship to regional geology. *Econ. Geol.* **74**, 1852–1868.

FINCH, W. J. (1980). Uranium and thorium. *Geotimes* **25**, no. 2, 46–47.

FINLOW–BATES, T. and LARGE, D. E. (1978). Water depth as a major control on the formation of submarine exhalative ore deposits. *Geol. Jb. D.* **36**, 27–39.

FLEET, A. J. and ROBINSON, A. H. F. (1978). Metalliferous sediments of the Semail Nappe, Oman. *Geol. Soc. Lond. Newsletter* **7**, Nov. 1978.

FLEET, A. J. and ROBINSON, A. H. F. (1980). Ocean-ridge metalliferous and pelagic sediments of the Semail Nappe, Oman. *J. geol. Soc. Lond.* **137**, 403–422.

FLEISCHER, V. D., GARLICK, W. G. and HALDANE, R. (1976). Geology of the Zambian Copper belt. *In* "Handbook of Stratabound and Stratiform Ore Deposits", Vol. 6" (K. H. Wolf, ed.), 223–352. Elsevier, New York.

FRANCHETEAU, J., NEEDHAM, H. D., CHOUKROUNE, P., JUTEAU, T., SEGURET, M., BALLARD, R. D., FOX, P. J., NORMARK, W., CARRANZA, A., CORDOBA, D., GUERRERO, J., RANGIN, C., BOUGAULT, H., CAMBON, P. and HEKINIAN, R. (1979). Massive deep-sea sulphide ore deposits discovered on the East Pacific Rise, *Nature, Lond.* **277**, 523–528.

FRANCIS, P. W. (1974). A new interpretation of the 1968 Fernandina Caldera collapse and its implications for the Mid-Oceanic Ridges. *Geophys. J.* **39**, 301–318.

FRANCIS, P. W., MOORBATH, S. and THORPE, R. S. (1977). Strontium isotope data for recent andesites in Ecuador and North Chile. *Earth planet. Sci. Lett.* **37**, 197–202.

FRIETSCH, R., PAPUNEN, H. P. and VOKES, F. M. (1979). The ore deposits in Finland, Norway and Sweden - a review. *Econ. Geol.* **74**, 975–1001.

FRONDEL, C. and BAUM, J. L. (1974). Structure and mineralogy of the Franklin zinc-iron-manganese deposit, New Jersey. *Econ. Geol.* **69**, 157–180.

FRUTOS, J. and OYARZÚN, J. (1975). Tectonic and geochemical evidence concerning the genesis of El Laco magnetite lava flow deposits, Chile. *Econ. Geol.* **70**, 988–990.

FUCHTBAUER, H. (1967). Die Sandsteine in der Molasse Nordlich der Alpen. *Geol. Rdsch.* **56**, 266–300.

FULLER, A. O. (1979). Phosphate occurrences on the western and southern coastal areas and continental shelves of Southern Africa. *Econ. Geol.* **74**, 221–231.

FUSSELL, J. (1980). Lineaments lead to oil in Wyoming. *Geotimes* **25**, no. 2, 19–20.

FYFE, W. S. (1978). The evolution of the earth's crust: modern plate tectonics to ancient hot spot tectonics. *Chem. Geol.* **23**, 89–114.

GANSSER, A. (1964). "Geology of the Himalayas". Wiley Interscience, London.

GANSSER, A. (1977). The great suture zone between Himalaya and Tibet. A preliminary account. *In* "Colloque sur l'ecologie et géologie de la Himalaya No. 268", 181–192. Centre National de la Recherche Scientifique, Paris.

GARLICK, H. J. (1979). Australian diamond prospects – the story so far. *Ind. Miner.* 17–29.

GARLICK, W. G. (1961). The syngenetic theory. *In* "The Geology of the Northern Rhodesian Copper Belt" (F. Mendelssohn, ed.) 146–165. Macdonald, London.

GARRELS, R. M., PERRY, E. A. and MACKENZIE, F. T. (1973). Genesis of Precambrian iron-formations and development of atmospheric oxygen. *Econ. Geol.* **68**,1173–1179.

GARSON, M. S. (1965a). Carbonatites in southern Malawi. *Bull. geol. Surv. Malawi* **15**.

GARSON, M. S. (1965b). Carbonatite and agglomerate vents of the western Shire Valley. *Geol. Surv. Malawi Mem.* **3**.

GARSON, M. S. (in press). Relationship of carbonatites to plate tectonics. Sukheswala. St. Xaviers College. Bombay.

GARSON, M. S. and KRS, M. (1976). Geophysical and geological evidence of Red Sea transverse tectonics to ancient fractures. *Bull. geol. Soc. Am.* **87**, 169–181.

GARSON, M. S. and MITCHELL, A. H. G. (1970). Transform faulting in the Thai Peninsula. *Nature, Lond.* **228**, 45–47.

GARSON, M. S. and MITCHELL, A. H. G. (1977). Mineralization at destructive plate boundaries: a brief review. *In* "Volcanic Processes in Ore Genesis", 81–97. Instn. Min. Metall. and Geol. Soc. Lond. Spec. Publ. No. 7.

GARSON, M. S. and MITCHELL, A. H. G. (1981). Precambrian ore deposits and plate tectonics. *In* "Precambrian Metallogeny and Plate Tectonics" (A. Kröner. ed.) 689–731. Elsevier Scientific, Amsterdam. In press.

GARSON, M. S. and SHALABY, I. B. (1976). Precambrian -Lower Palaeozoic plate tectonics and metallogenesis in the Red Sea Region. *Spec. Pap. Geol. Ass. Can.* **14**, 573–596.

GARSON, M. S., BRADSHAW, N. and RATTAWONG, S. (1969). Lepidolite pegmatites in the Phangnga area of peninsular Thailand. *In* "2nd Technical Conference on Tin", 325–350. Intern. Tin Council and Dept. Min. Resources, Bangkok, Thailand.

GARSON, M. S., YOUNG, B., MITCHELL, A. H. G. and TAIT, B. A. R. (1975). The geology of the tin belt in peninsular Thailand around Phuket, Phangnga and Takua Pa. Inst. Geol. Sci., Overseas Div. Mem. No. 1, London.

GEORGE, R. P. (1975). The internal structure of the Troodos Ultramafic Complex, Cyprus. PhD thesis (unpub.), New York State Univ.

GEORGE, U. (1979). An ocean is born. "Geo" Vol. 1, 93–112. Gruner and Jahr, New York.

GERASIMOVSKI, V. I., VOLKOV, V. P., KOGARKO, L. N. and POLYAKOV, A. I. (1974). Kola Peninsula. *In* "The Alkaline Rocks" (H. Sørensen, ed.) 206–221. Wiley, London.

GEYTI, A. and SCHØNWANDT, H. K. (1979). Bordvika – a possible porphyry molybdenum occurrence within the Oslo Rift, Norway. *Econ. Geol.* **74**, 1211–1220.

GILLILAND, W. N. and MEYER, G. P. (1976). Two classes of transform faults. *Bull. geol. Soc. Am.* **87**, 1127–1130.

GILLULY, J. (1976). Lineaments – ineffective guides to ore deposits. *Econ. Geol.* **71**, 1507–1514.

GILMOUR, P. P. (1971). Strata-bound massive pyritic sulphide deposits – a review. *Econ. Geol.* **66**, 1239–1244.

GINZBURG, A. I. (1962). "Geology of Rare-Element Deposits. 17. Geological Structure and Mineralogical Characteristics of Rare Metal Carbonatites". Gosgeoltechizdat, Moscow.

GITTINS, J., MacINTYRE, R. M. and YORK, D. (1967). The ages of carbonatite complexes in eastern Canada. *Can. J. Earth Sci.* **4**, 651–655.

GLIKSON, A. Y., DERRICK, G. M., WILSON, I. H. and HILL, R. M. (1974). Structural evolution and geotectonic nature of the Middle Proterozoic Mount Isa Fault Trough, northwestern Queensland. *Bur. Mineral Resour., Geol. Geophys. Rec.* **14**.

GLIKSON, A. Y. (1976). Archaen to early Proterozoic shield elements: relevance of plate tectonics. *Spec. Pap. geol. Ass. Can.* **14**, 489–516.

GNIBIDENKO, H. S. (1973). Crustal structure and evolution in the north-western part of the Pacific belt. *In* "The Western Pacific: Island Arcs, Marginal Seas, Geochemistry" (P. J. Coleman, ed.), 435–449. University Western Australia Press, Nedlands.

GNIBIDENKO, H. S. and SHASHKIN, K. S. (1970). Basic principles of the geosynclinal theory. *Tectonophysics* **9**, 5–13.

GOLD, D. P., VALLEE, M. and CHARETTE, J-P. (1967). Economic geology and geophysics of the Oka alkaline complex, Quebec. *Bull. Can. Instn. Min. Metall.* **60**, 1131–1144.

GOOSSENS, P. J. (1976). Lithologic, geochemical and metallogenic belts in the Northern Andes, and their structural relationships. *Trans. Soc. Min. Eng. Am. Instn. Min. Metall. Engng* **260**, 59–67.

GOOSSENS, P. J. and HOLLISTER, V. P. (1973). Structural control and hydrothermal alteration pattern of Chaucha porphyry copper, Ecuador. *Mineral. Deposita* **8**, 321–331.

GRAHAM, S. A., DICKINSON, W. R. and INGERSOLL, R. V. (1975). Himalayan-Bengal model for flysch dispersal in the Appalachian-Ouachita system. *Bull. geol. Soc. Am.* **86**, 273–286.

GRANT, J. N., HALLS, C., AVILA, W. and AVILA, G. (1977). Igneous geology and the evolution of hydrothermal systems in sone sub-volcanic tin deposits. *In* "Volcanic Processes in Ore Genesis", 117–126. Instn Min. Metall. and Geol. Soc. Lond. Spec. Publ. No. 7.

GRANT, J. N., HALLS, C., SALINAS, W. S. and SNELLING, N. J. (1979). K-Ar ages of igneous rocks and mineralization in part of the Bolivian tin belt. *Econ. Geol.* **74**, 838–851.

GREENBAUM, D. (1977). The chromitiferous rocks of the Troodos ophiolite complex, Cyprus. *Econ. Geol.* **72**, 1175–1194.

GREENE, H. G., DALRYMPLE, G. B. and CLAGUE, D. A. (1978). Evidence for northward movement of the Emperor Seamounts. *Geology* **6**, 70–74.

GREENWOOD, W. R., HADLEY, D. G., ANDERSON, R. E., FLECK, R. J. and SCHMIDT, D. L. (1976). Late Proterozoic cratonization in the southwestern Saudi Arabia. *Phil. Trans. R. Soc. Lond.* **280A**, 517–527.

GRUENEWALDT, G. VON (1977). The mineral resources of the Bushveld Complex. *Minerals Sci. Engng* **9**, 83–95.

GRESENS, R. L. (1978). Evaporites as precursors of massif anorthosite. *Geology* **6**, 46–50.

GUILD, P. W. (1971). Metallogeny: a key to exploration. *Min. Engng* **23**, 69–72.

GUILD, P. W. (1972). Massive sulphides vs porphyry deposits in their global tectonic settings. Print No. G13, MMIJ-AIME Joint Meeting, Tokyo. 1–12.

GUILD, P. W. (1974a). Distribution of metallogenic provinces in relation to major earth structures. *In* "Metallogenetische und geochemische Provinzen", (W. E. Petrascheck, ed.), 10–24. Springer-Verlag, Vienna.

GUILD, P. W. (1974b). Application of global tectonic theory to metallogenic studies. *In* "Symposium on Ore Deposits of the Tethys Region in the Context of Global Tectonics", CTOD-IAGOD. Varna, Bulgaria (preprint).

GUILD, P. W. (1978a). Metallogenic maps: principles and progress. *Global Tectonics Metallogeny* 1, 10–15.

GUILD, P. W. (1978b). Metallogenesis in the western United States. *J. geol. Soc. Lond.* 135, 355–376.

GUSTAFSON, L. B. (1978). Some major factors of porphyry copper genesis. *Econ. Geol.* 73, 600–607.

HACKETT, J. P. and BISCHOFF, J. L. (1973). New data on the stratigraphy, extent and geologic history of the Red Sea geothermal deposits. *Econ. Geol.* 68, 533–564.

HAGEN, T. (1969). Report on the geological survey of Nepal, vol. 1, preliminary reconnaissance. *Denkschr. Schweiz. Naturff. Gessell. Bd.* 86, Hfl.

HALL, J. (1859). Description and figures of the organic remains of the lower Helderberg Group and the Oriskany Sandstone. *In* "Natural History of New York" Palaeontology, New York 3, 1–544.

HALLAM, A. and BRADSHAW, M. J. (1979). Bituminous shales and oolitic ironstones as indicators of transgressions and regressions. *J. geol. Soc. Lond.* 136, 157–164.

HALLIDAY, A. N., STEPHENS, P. H. and HARMON, R. S. (1980). Rb-Sr and O isotopic relationships in 3 zoned Caledonian granitic plutons, Southern Uplands, Scotland: evidence for varied sources and hybridization of magmas. *J. geol. Soc. Lond.* 137, 329–348.

HALLS, C., REINSBAKKEN, A., FERRIDAY, I., HAUGEN, A. and RANKIN, A. (1977). Geological setting of the Skorovas orebody within the allochthonous volcanic stratigraphy of the Gjersvik Nappe, central Norway. *In* "Volcanic Processes in Ore Genesis", 128–151. Inst. Min. Metall. and Geol. Soc. Lond. Spec. Publ. No. 7.

HAMET, J. and ALLÈGRE, C. J. (1976). Rb-Sr systematics in granite from central Nepal (Manaslu): significance of the Oligocene age and high Sr^{87}/Sr^{86} ratios in Himalayan orogeny. *Geology* 4, 470–472.

HAMILTON, W. (1970). Bushveld complex—product of impacts? *Spec. Publs Geol. Soc. S. Afr.* 1, 367.

HAMILTON, W. (1973). Tectonics of the Indonesian region. *Bull. geol. Soc. Malaysia* 6, 3–10.

HAMILTON, W. (1979). Tectonics of the Indonesian region. US Geol. Surv. Prof. Paper 1978. Govt Printing Office, Washington.

HANEKOM, H. J. VAN, STADEN, C. M. V. H., SMIT, P. J. and PIKE, D. R. (1965). The geology of the Palabora Igneous Complex. *S. Afr. geol. Surv. Mem.* 54.

HANSON, R. E. and AL SHAIEB, Z. (1980). Voluminous sub-alkaline silicic magmas related to intracontinental rifting in the southern Oklahoma aulacogen. *Geology* 8, 180–184.

HARGRAVES, R. B. (1970). Palaeomagnetic evidence relevant to the origin of the Vredefort Ring. *J. Geol.* 78, 254–263.

HARLEY, D. N. (1979). A mineralized Ordovician resurgent caldera complex in the Bathurst-Newcastle Mining District, New Brunswick, Canada. *Econ.*

Geol. **74**, 786–796.

HARRIS, J. F. (1961). Summary of the geology of Tanganyika. Pt. IV: Economic Geology. *Mem. Tanganyika geol. Surv.* **1**.

HASSAN, M. A. and AL-SULAIMI, J. S. (1979). Copper mineralization in the northern part of Oman Mountains near Al Fujairah, United Arab Emirates. *Econ. Geol.* **74**, 919–924.

HAUG, E. (1900). Les geosynclinaux et les aires continentales. Contribution a l'étude des regressions et des transgressions marines. *Bull. geol. Soc. France* **28**, 617–711.

HAWKESWORTH, C. J., NORRY, M. J., RODDICK, J. C., BAKER, P. E., FRANCIS, P. W. and THORPE, R. S. (1979). ^{143}Nd/^{144}Nd, ^{87}Sr/^{86}Sr and incompatible element variations in calc-alkaline andesites and plateau lavas from South America. *Earth planet. Sci. Lett.* **42**, 45–57.

HEATON, T. H. E. and SHEPPARD, S. M. F. (1977). Hydrogen and oxygen isotope evidence for sea water–hydrothermal alteration and ore deposition, Troodos complex, Cyprus. *In* "Volcanic Processes in Ore Genesis", 42–57. Instn. Min. Metall. Lond. and Geol. Soc. Lond. Spec. Publ. No. 7.

HEGGE, M. R. and ROWNTREE, J. C. (1978). Geologic setting and concepts on the origin of uranium deposits in the East Alligator River, N. T., Australia. *Econ. Geol.* **73**, 1420–1429.

HEINRICH, E. W. (1966). "The Geology of Carbonatites". Rand McNally, Chicago.

HENLEY, R. W. and ADAMS, J. (1979). On the evolution of giant gold placers. *Trans. Instn Min. Metall.* **88**, B41–50.

HENLEY, R. W. and McNABB, A. (1978). Magmatic vapor plumes and ground-water interaction in porphyry copper emplacement. *Econ. Geol.* **73**, 1–20.

HERZ, N. (1977). Timing of spreading in the South Atlantic: information from Brazilian alkalic rocks. *Bull. geol. Soc. Am.* **88**, 101–112.

HITE, R. J. and JAPAKASETR, T. (1979). Potash deposits of the Khorat Plateau, Thailand and Laos. *Econ. Geol.* **74**, 448–458.

HOFFMAN, P. F. (1973). Evolution of an early Proterozoic continental margin: the Coronation Geosyncline, and associated aulacogens of the N.W. Canadian Shield. *Phil. Trans. R. Soc. Lond.* **273A**, 547–581.

HOFFMAN, P. F., DEWEY, J. F. and BURKE, K. (1974). Aulacogens and their genetic relation to geosynclines, with a Proterozoic example from Great Slave Lake, Canada. *In* "Modern and Ancient Geosynclinal Deposits" (R. H. Dott and R. H. Shaver, eds), *Spec. Publs Soc. econ. Palaeont.. Miner.* **19** 38–55.

HOLL, R. (1977). Early Palaeozoic ore deposits of the Sb–W–Hg formation in the Eastern Alps and their genetic interpretation. *In* "Time and Strata-Bound Ore Deposits" (D. D. Klemm and H-J. Schneider, eds), 169–198. Springer-Verlag, Berlin.

HOLL, R. and MAUCHER, A. (1976). The strata-bound ore deposits in the Eastern Alps. *In* "Handbook of Strata-Bound and Stratiform Ore Deposits", Vol. 5 (K. H. Wolf, ed.), 1–36. Elsevier, Amsterdam.

HOLLAND, H. D. (1979). Metals in black shales—a reassessment. *Econ. Geol.* **74**, 1676–1680.

HOLLISTER, V. F. (1975). An appraisal of the nature and source of porphyry copper deposits. *Minerals Sci. Engng* **7**, 225–233.

HOLLISTER, V. F. and BERNSTEIN, M. (1975). Copaquire, Chile: its geologic setting and porphry copper deposit. *Trans. Soc. Min. Eng. Am. Instn Min. Metall. Engng* **258**, 137–142.

HOLLISTER, V. F. and SIRVAS, E. B. (1974). The Michiquillay porphyry copper deposit. *Mineral Deposita* 9, 261–269.

HOLLISTER, V. F., ALLEN, J. M., ANZALONE, S. A. and SERAPHIM, R. H. (1975). Structural evolution of porphyry mineralization at Highland Valley, British Columbia. *Can. J. Earth Sci.* 12, 807–820.

HOLMQUIST, P. J. (1900). En geologisk profil ofver fjellomradena emellan kvikkjok och norska kustem. *Geol. Foren. Forhandl.* 22, 72–104, 151–177.

HORIKOSHI, E. (1969). Volcanic activity related to the formation of the Kuroko-type deposits in the Kosaka District, Japan. *Mineral. Deposita* 4, 321–345.

HORIKOSHI, E. (1976). Development of late Cenozoic petrogenetic provinces and metallogeny in Northeast Japan. *Spec. Pap. geol. Ass. Can.* 14, 121–142.

HOSKING, K. F. G. (1973). The primary tin mineralization patterns of West Malaysia. *Bull. geol. Soc. Malaysia* 6, 297–308.

HOSKING, K. F. G. (1977). Known relationships between the 'hard-rock' tin deposits and the granites of Southeast Asia. *Bull. geol. Soc. Malaysia* 9, 141–157.

HOUTEN, F. B. VAN (1974). Northern Alpine molasse and similar Cenozoic sequences of Southern Europe. *In* "Modern and Ancient Geosynclinal Sedimentation" (R. H. Dott, Jr and R. H. Shaver, eds). *Spec. Publs Soc. econ. Palaeont. Miner.* 19, 260–273.

HOUTEN, F. B. VAN(1976a). Late Cenozoic volcaniclastic deposits, Anean foredeep, Columbia. *Bull. geol. Soc. Am.* 87, 481-495.

HOUTEN, F. B. VAN (1976b). Late Variscan nonmarine basin deposits, Northwest Africa: record of Hercynotype orogeny. *In* "The Continental Permain in Central, West and South Europe" (H. Falke, ed.), 215–224. Reidel, Holland.

HOUTEN, F. B. VAN and BROWN, R. H. (1977). Latest Palaeozoic-Early Mesozoic palaeogeography, northwestern Africa. *J. Geol.* 85, 143–156.

HUDSON, T. (1979). Mesozoic plutonic belts of southern Alaska. *Geology* 7, 230–234.

HUDSON, T., PLAFKER, G. and PETERMAN, Z. E. (1979). Palaeogene anatexis along the Gulf of Alaska margin. *Geology* 7, 573–577.

HUGHES, J. M., STOIBER, E. R. and CARR, M. J. (1980). Segmentation of the Cascade volcanic chain. *Geology* 8, 15–17.

HUNTER, D. R. (1973). The localization of tin mineralization with reference to southern Africa. *Minerals Sci. Engng* 5, 53–57.

HUSSONG, D. M., UYEDA, S., et al. (1978). Near the Philippines—Leg 60 ends in Guam. *Geotimes* 23, no. 10, 19–22.

HUTCHISON, R. W. (1973). Volcanogenic sulfide deposits and their metallogenic significance. *Econ. Geol.* 68, 1223-1246.

HUTCHISON, R. W. and ENGELS, S. G. (1970). Tectonic significance of regional geology and evaporite lithofacies in northeastern Ethiopia. *Phil. Trans. R. Soc. Lond.* 267A, 313–329.

HUTCHINSON, C. S. (1978). Southeast Asian tin granites of contrasting tectonic setting. *J. Phys. Earth* 26, Suppl. S211–232.

HUTCHINSON, C. S. and TAYLOR, D. (1978). Metallogenesis in SE Asia. *J. geol. Soc. Lond.* 135, 407–428.

ILLIES, J. H. and GREINER, G. (1978). Rhinegraben and the Alpine System. *Bull. geol. Soc. Am.* 89, 770-782.

INGHAM, F. T. (1959). Economic geology. *Cyprus geol. Surv. Mem.* 1, 137–175.

IRVINE, T. N. (1965). Chromian spinel as a petrogenetic indicator. Pt 1.

Theory. *Can. J. Earth Sci.* **2**, 648–672.

ISHIHARA, S. (1973). Molybdenum and tungsten provinces in the Japanese Islands and North American Cordillera: an example of asymmetrical metal zoning in Pacific type orogeny. *Bull. Bur. Miner. Resour. Geol. Geophys. Aust.* **141**, 173-189.

ISHIHARA, S. (1974). Magmatism of the Green Tuff tectonic belt, northeast Japan. *In* "Geology of Kuroko Deposits" (S. Ishihara *et al.*, eds), 235–249. Min. Geol. Spec. Issue No. 6.

ISHIHARA, S. (1977). The magnetite-series and ilmenite-series granitic rocks. *Min. Geol.* **27**, 293–305.

ISHIHARA, S. (1978). Metallogenesis in the Japanese island arc system. *J. geol. Soc. Lond.* **135**, 389–406.

ISHIHARA, S., IGARASHI, T. and NISHIWAKI, C. (1974). A re-examination of the regional distribution of the late Cenozoic ore deposits in the East Japan Arc. *Bull. geol. Soc. Am.* **85**, 292–294.

ITSIKSON, M. I. and KRASNYY, L. I. (1970). Aspects of the geotectonics and metallogeny of the eastern USSR. *Geotectonics* **2**, 132–141.

IVANOV, T., SHALABY, I. M. and HUSSEIN, A. A. A. (1973). Metallogenic characteristics of South-Eastern Desert of Egypt. *Ann. geol. Surv. Egypt* **3**, 139–166.

IVANOVA, T. N. (1963). "Apatite Deposits of the Khibiny Tundra." Gosgeoltechizdat, Moscow.

JACKSON, E. D. and THAYER, T. P. (1972). Some criteria for distinguishing between stratiform, concentric and alpine peridotite-gabbro complexes. 24th Internat. Geol. Congr. Proc., Montreal, Sect. 2, 289–296.

JACOB, R. E. (1978). Granite genesis and associated mineralization in part of the Central Damara Belt. *In* "Mineralization in Metamorphic Terranes" (W. J. Verwoerd, ed.), 417–432. Van Schaik, Pretoria.

JACOBSEN, J. B. E. (1975). Copper deposits in time and space. *Minerals Sci. Engng* **7**, 337–371.

JAKES, P and WHITE, A. J. R. (1971). Composition of islands arcs and continental growth. *Earth planet. Sci. Lett.* **12**, 224-230.

JAMES, D. E. (1978). Subduction of the Nazca plate beneath central Peru. *Geology* **6**, 174–178.

JAMES, T. C. (1958). Carbonatite investigation: a progress report. *Rec. Tanganyika geol. Surv.* **6**, 45.

JENKYNS, H. C. (1978). Pelagic environments. *In* "Sedimentary Environments and Facies" (H. G. Reading, ed.), 314–371. Blackwell Scientific, Oxford.

JENKYNS, H. C. (1980). Cretaceous anoxic events from continents to oceans. *J. geol. Soc. Lond.* **137**, 171–188.

JENSEN, M. L. (1971). Provenance of Cordilleran intrusives and associated · metals. *Econ. Geol.* **66**, 34–42.

JOHNSON, R. L. (1962). The geology of the Dorowa and Shawa carbonatite complexes, Southern Rhodesia. *Trans. geol. Soc. S. Afr.* **61** (1961), 101–145.

JOHNSON, T. L. (1979). Alternative model for emplacement of the Papuan ophiolite, Papua New Guinea. *Geology* **7**, 495–498.

JONES, M. T., REED, B. L., DOE, B. R. and LANPHERE, M. A. (1977). Age of tin mineralization and plumbotectonics, Belitung, Indonesia. *Econ. Geol.* **72**, 745–752.

JONES, O. T. (1954). The characteristics of some Lower Palaeozoic marine sediments. *Proc. R. Soc. Lond.* **222A**, 327–333.

JOUBIN, F. R. and JAMES, D. H. (1956). Rexspar uranium deposits, Canada. *Min. J.* **77**, 59–60.

KAADEN, G. VAN DER (1970). Chromite-bearing ultramafic and related gabbroic rocks and their relationship to "ophiolitic" extrusive basic rocks and diabases in Turkey. *Spec. Publs. Geol. Soc. S. Afr.* 1, 511-531.

KÁNEHIRA, K. and TATSUMI, T. (1970). Bedded cupriferous deposits in Japan, a review. *In* "Volcanism and Ore Genesis" (T. Tatsumi, ed.), 51-76. University of Tokyo Press, Tokyo.

KARIG, D. E. (1970). Ridges and basins of the Tonga-Kermadec island arc. *J. geophys. Res.* 75, 239-155.

KARIG, D. E. (1971). Origin and development of marginal basins in the western Pacific. *J. geophys. Res.* 76, 2542-2561.

KARIG, D. E. (1975). Basin genesis in the Philippine Sea. *In* "Initial Reports Deep Sea Drilling Programme" (J. G. Ingle and D. E. Karig *et al.*, eds), No. 31. US Govt. Printing Office, Washington.

KARIG, D. E. (1980). Initiation of subduction zones: implications for arc evolution and ophiolite development. *In* "Trench and Fore-Arc Sedimentation and Tectonics in Modern and Ancient Subduction Zones (Summary)", 14-15. Geol. Soc. Lond.

KARIG, D. E. and MOORE, G. F. (1975). Tectonically controlled sedimentation in marginal basins. *Earth planet. Sci. Lett.* 26, 233-238.

KARIG, D. E. and SHARMAN, G. F. III (1975). Subduction and accretion in trenches. *Bull. geol. Soc. Am.* 86, 377-389.

KARIG, D. E., LAWRENCE, M. B., MOORE, G. F. and CURRAY, J. R. (1980). Structural framework of the fore-arc basin, NW Sumatra. *J. geol. Soc. Lond.* 137, 77-91.

KARSON, J. and DEWEY, J. F. (1978). Coastal Complex, western Newfoundland: an early Ordovician oceanic fracture zone. *Bull. geol. Soc. Am.* 89, 1037-1049.

KAY, M. (1947). Geosynclinal nomenclature and the craton. *Bull. Am. Ass. Petrol. Geol.* 31, 1289-1293.

KAY, M. (1951). North American geosynclines. *Mem. geol. Soc. Am.* 48.

KEAREY, P. (1976). A regional structural model of the Labrador Trough, northern Quebec, from gravity studies, and its relevance to continental collision. *Earth planet. Sci. Lett.* 28, 371-378.

KEAYS, R. R. and SCOTT, R. B. (1976). Precious metals in ocean-ridge basalts: implications for basalts as source rocks for gold mineralization. *Econ. Geol.* 71, 705-720.

KELLY, W. C. and RYE, R. O. (1979). Geologic, fluid inclusion and stable isotope studies of the tin-tungsten deposits of Panasqueira, Portugal. *Econ. Geol.* 74, 1721-1822.

KEMPE, D. R. C. (1973). The petrology of the Warsak alkaline granites, Pakistan, and their relationship to other alkaline rocks of the region. *Geol. Mag.* 110, 385-404.

KENNEDY, J. M. and DeGRACE, J. F. (1972). Structural sequence and its relationship to sulphide mineralization in the Ordovician Lushs Bight Group of Western Notre Dame Bay, Newfoundland. *Bull. Can. Instn. Min. Metall.* 65, 300-308.

KERR, J. W. (1977). Cornwallis Lead-Zinc District, Mississippi Valley-type deposits controlled by stratigraphy and tectonics. *Can. J. Earth Sci.* 14, 1402-1426.

KESLER, S. E., JONES, L. M. and WALKER, R. L. (1975). Intrusive rocks associated with porphyry copper mineralization in island arc areas. *Econ. Geol.* 70, 515-526.

KIMBERLEY, M. M. (1978). Origin of stratiform uranium deposits in sandstone conglomerate and pyroclastic rock. *In* "Uranium Deposits, their Mineralogy and Origin" (M. M. Kimberley, ed.), 339–381. Min. Ass. Can. No. 3. Univ. Toronto Press, Toronto.

KING, B. C. (1970). Vulcanicity and rift tectonics in East Africa. *In* "African Magmatism and Tectonics" (T. N. Clifford and I. G. Gass, eds), 263–283. Oliver and Boyd, Edinburgh.

KING, B. C. and SUTHERLAND, D. S. (1960). Alkaline rocks of eastern and southern Africa, pts 1-111. *Sci. Prog.* 48, 298-321, 504-524, 709-720.

KING, B. C. and SUTHERLAND, D. S. (1966). The carbonatite complexes of eastern Uganda. *In* "Carbonatites" (O. F. Tuttle and J. Gittins, eds), 73–126. Interscience, New York.

KING, H. F. and THOMPSON, B. P. (1953). Geology of the Broken Hill District. *In* "Geology of Australian Ore Deposits" (A. B. Edwards, ed.), 533–577. 5th Empire Min. Metall. Congr., Australia and New Zealand.

KIRBY, G. A. (1979). The Lizard Complex as an ophiolite. *Nature, Lond.* 282, 58–61.

KISVARZANYI, E. B. (1980). Granitic ring complexes and Precambrian hot-spot activity in the St. François terrane, Midcontinent region, United States. *Geology* 8, 43–47.

KLEIN, G. and BRICKER, C. P. (1977). Some aspects of the sedimentary and diagenetic environment of Proterozoic banded iron-formation. *Econ. Geol.* 72, 1457-1470.

KLEMM, D. D. (1979). A biogenetic model of the formation of the banded iron formation in the Transvaal Supergroup/South Africa. *Mineral. Deposita* 14, 381–385.

KLIGFIELD, R. (1979). The northern Apennines as a collisional orogen. *Am. J. Sci.* 279, 676–691.

KLOOSTERMAN, J. B. (1967). Ring-structure in the Oriente and Massangana granite complexes, Rondônia, Brazil. *Eng. Min. Met.* 45, No. 266, 72-77.

KNIGHT, C. L. (1957). Ore genesis – the source bed concept. *Econ. Geol.* 52, 808–819.

KNORRING, O. VON and DUBOIS, C. G. B. (1961). Carbonatitic lava from Fort Portal in western Uganda. *Nature, Lond.* 192, 1063-1064.

KOPECKY, L. (1971). Relationship between fenitization, alkaline magmatism, barite-fluorite mineralization and deep-fault tectonics in the Bohemian Massif. Upper Mantle Project Programme in Czechoslovakia 1962–1970, Geology Final Report, Academia Nakla Cesk. Akad. Ved. Prague 73–95.

KUMARAPELI, P. S. (19760. The St Lawrence rift system, related metallogeny, and plate tectonic models of Appalachian evolution. *Spec. Pap. Geol. Ass. Can.* 14, 301–320.

KUNO, H. (1966). Lateral variation of basaltic magma across continental margins and island arcs. *In* "Continental Margins and Island Arcs" (W. H. Poole, ed.), 317–336, Can. Geol. Surv. Paper 66-15.

KUTINA, J. (1974). Structural control of volcanic ore deposits in the context of global tectonics. *Bull. Volcanologique* 38, 1039-1069.

KUTINA, J. (1980a). Regularities in the distribution of ore deposits along the 'Mendocino Latitude', western United States. *Global Tectonics Metallogeny* 1, 134–193.

KUTINA, J. (1980b). Are lineaments ineffective guides to ore deposits? *Global Tectonics Metallogeny* 1, 200–205.

KUZNETSOV, V. A. (1977). Deposits of mercury. *In* "Ore deposits of the USSR", Vol. 2 (V. I. Smirnov, ed.), 298–348. Pitman, London.

LAMBERT, I. B. (1977). Notes on exploration guides for stratiform lead-zinc ores, geochemical and geobiological evolution in the Precambrian, and massive sulphide deposits of the Noranda area. *Min. Res. Lab. Tech. Commun.* **61**, CSIRO.

LAMBERT, I. B. and SATO, T. (1974). The Kuroko and associated ore deposits of Japan: a review of their features and metallogenesis. *Econ. Geol.* **69**, 1215–1236.

LAMBERT, I. B., DONNELLY, T. H. and ROWLANDS, N. J. (1980). Genesis of Upper Proterozoic stratabound copper mineralization, Kapunda, South Australia. *Mineral. Deposita* **15**, 1–18.

LANDIS, C. A. and BISHOP, D. G. (1972). Plate tectonics and regional stratigraphic-metamorphic relations in the southern part of the New Zealand geosyncline. *Bull. geol. Soc. Am.* **83**, 2267–2284.

LANGE, I. M. and MURRAY, R. C. (1977). Evaporite brine reflux as a mechanism for moving deep warm brines upward in the formation of Mississippi Valley-type base metal deposits. *Econ. Geol.* **72**, 107–109.

LANGFORD, F. F. (1978a). Uranium deposits in Australia. *In* "Uranium Deposits, their Mineralogy and Origin" (M. M. Kimberley, ed.), 205–216. Min. Ass. Can. No. 3. Univ. Toronto Press, Toronto.

LANGFORD, F. F. (1978b). Origin of unconformity-type pitchblende deposits in the Athabasca Basin of Saskatchewan. *In* "Uranium deposits, Their Mineralogy and Origin" (M. M. Kimberley, ed.), 485–499. Min. Ass. Can. No. 3. Univ. Toronto Press, Toronto.

LAPIDO-LOUREIRO, F. W. (1973). Carbonatites de Angola. *Mem. Trabal. Inst. Invest. Cient. Angola* **2**.

LARGE, D. (1979). Proximal and distal stratabound ore deposits: A discussion of the paper by I. R. Plimer, *Mineral Deposita* **13**, 345–353 (1978). *Mineral Deposita* **14**, 123–124.

LAWRENCE, L. J. (1978). Porphyry-type gold mineralization in shoshonite at Vundu, Fiji. *Aust. Inst. Min. Metall.* **268**, 21–31.

LAZNICKA, P. (1976). Porphyry copper and molybdenum deposits of the U.S.S.R. and their plate tectonic settings. *Trans. Instn Min. Metall.* **85**, B14–32.

LAZNICKA, P. and WILSON, H. D. B. (1972). The significance of a copper-lead line in metallogeny. 24th Internat. geol. Congr., Canada, sect 4, 25–36.

LEACH, D. (1973). A study of the barite-zinc-lead deposits of central Missouri and related mineral deposits in the Ozark region. *In* J. M. Sharp, PhD thesis (unpub.) 1978, Univ. Missouri, Columbia.

LEAKE, R. C. and BROWN, M. J. (1980). Porphyry-style copper mineralization at Black Stockarton Moor, southwest Scotland. *Trans. Instn Min. Metall.* **88**, B177–181.

Le FORT, P. (1979). Iberian-Armorican arc and Hercynian orogeny in western Europe. *Geology* **7**, 384–388.

Le FORT, P. (1973). Les leucogranites à tourmaline de la Himalaya sur l'example du granite du Manaslu (Nepal central). *Bull. geol. Soc. France* **15**, no. 5-6, 556–561.

Le FORT, P. (1975). Himalayas: the collided range, present knowledge of the continental arc. *Am. J. Sci.* **275A**, 1-44.

Le FORT, P., DEBON, F. and SONET, J. (1980). The "Lesser Himalayan" cordierite granite belt: typology and age of the pluton of Manserah (Pakistan). Proc. Int. Geodynamic Conf., Geol. Bull. Univ. Peshawar. Pakistan (in press).

LEGGETT, J. K., McKERROW, W. S. and CASEY, D. M. (1980). Anatomy of a Lower Palaeozoic accretionary prism: the Southern Uplands of Scotland. *In* "Trench and Fore-Arc Sedimentation and Tectonics in Modern and Ancient Subduction Zones. (Summary)", 15–16. Geol. Soc. Lond.

LeROY, J. (1978). The Margnac and Fanay uranium deposits of the La Crouzille District (Western Massif Central, France): geologic and fluid inclusion studies. *Econ. Geol.* **73**, 1611–1634.

LEYDEN, R., ARMUS, H., ZEMBRUSCKI, S. and BRYAN, G. (1976). South Atlantic diapiric structures. *Bull. Am. Ass. Petrol. Geol.* **60**, 196–212.

LI CHUN-YU, LIU XUEYA, WANG QUAN and ZHANG ZHIMENG (1979). "A tentative contribution to plate tectonics of China." Inst. Geol., Chinese Acad. Geol. Sci. (unpubl.).

LINDGREN, W. (1933). "Mineral Deposits". McGraw-Hill, New York.

LISTER, C. R. B. (1977). Qualitative models of spreading-centre processes, including hydrothermal penetration. *Tectonophysics* **37**, 203-218.

LiVACCARI, R. F. (1979). Reply on Late Cenozoic tectonic evolution of western United States. *Geology* **7**, 371-373.

LIVINGSTONE, D. E. (1973). A plate tectonic hypothesis for the genesis of porphyry copper deposits of the Southern Basin and Range Province. *Earth planet. Sci. Lett.* **20**, 171-179.

LOCARDI, E. and MITTEMPERGHER, M. (1971). Exhalative supergenic uranium, thorium and marcasite occurrences in Quaternary volcanites of central Italy. *Bull. Volcanologique* **35**, 173-184.

LOCK, B. E. (1980). Flat-plate subduction and the Cape Fold Belt of South Africa. *Geology* **8**, 35–39.

LOMBARD, A. F., WARD-ABLE, N. M. and BRUCE, R. W. (1964). The exploration and main geological features of the copper deposit in carbonatite at Loolekop, Palabora Complex. *In* "The Geology of Some Ore Deposits in Southern Africa" (S. H. Haughton, ed.) Vol. 2, 315–337. Geol. Soc. S. Afr., Johannesburg.

LONSDALE, P. (1979). A deep-sea hydrothermal site on a strike-slip fault. *Nature, Lond.* **281**, 531–534.

LORINCZI, G. I. and MIRANDA, J. C. (1978). Geology of the massive sulfide deposits of Campo Morado, Guerrero, Mexico. *Econ. Geol.* **73**, 180–191.

LOWELL, G. R. (1976). Tin mineralisation and mantle hot spot activity in south-eastern Missouri. *Nature, Lond.* **261**, 482–483.

LOWELL, J. D. (1974). Regional characteristics of porphyry copper deposits of the Southwest. *Econ. Geol.* **69**, 601–617.

LOWELL, J. D. and GUILBERT, J. M. (1970). Lateral and vertical alteration-mineralization zoning in porphyry ore deposits. *Econ. Geol.* **65**, 373–408.

MACDONALD, K. C., BECKER, K., SPIESS, F. N. and BALLARD, R. D. (1980). Hydrothermal heat flux of the "black smoke" vents on the East Pacific Rise. *Earth planet. Sci. Lett.* **48**, 1-7.

MACINTYRE, R. M. (1977). Anorogenic magmatism, plate motion and Atlantic evolution. *J. geol. Soc. Lond.* **133**, 175-384.

MACKEVETT, E. A. (1963). Geology and ore deposits of the Bokan Mountain uranium-thorium area, southeastern Alaska. *Bull. U.S. geol. Surv.* **1154**, 1-125.

MACKOWSKY, M. Th. (1975). Comparative petrography of Gondwana and Northern Hemisphere coals related to their origin. *In* "3rd Gondwana Symposium" (K. S. W. Campbell, ed.), 195-220. Aust. National University Press, Canberra.

MARSH, J. S. (1973). Relationships between transform directions and alkaline igneous rock lineaments in Africa and South America. *Earth planets. Sci Lett.* **18**, 317-323.

MARSHAK, R. S. and KARIG, D. E. (1977). Triple junctions as a cause for anomalously near-trench igneous activity between the trench and volcanic arc. *Geology* **5**, 233-236.

MARTIN, H. (1978). The mineralization of the ensialic Damara orogenic belt. *In* "Mineralization in Metamorphic Terranes" (W. J. Verwoerd, ed.), 405-415. Van Schaik, Pretoria.

MASON, D. R. and FEISS, P. G. (1979). On the relationship between whole rock chemistry and porphyry copper mineralization. *Econ. Geol.* **74**, 1506-1510.

MATSUDA, T. and UYEDA, S. (1971). On the Pacific-type orogeny and its model-extension of the paired belts concept and possible origin of marginal seas. *Tectonophysics* **11**, 5-27.

MATTAUER, M. and ETCHECOPAR, A. (1977). Argument en faveur de chevauchements de type Himalayan dans la Chaine Hercynienne du Massif Central Francais. *In* "Colloque sur l'ecologie et geologie de l'Himalaya", 261-267, No. 268. Centre National de la Recherche Scientifique, Paris.

MAUCHER, A. (1965). Die Antimon-Wolfram-Quesckailber Formation und ihre Besiehungern zu magmatismus and Geotektonic. *Freiberger Forschungsh.* **C186**, 173-188.

MAUCHER, A. (1976). The strata-bound cinnabar-stibnite-scheelite deposits. *In* "Handbook of Strata-Bound and Stratiform Ore Deposits" (K. H. Wolf, ed.), Vol. 7, 477-503. Elsevier, Amsterdam.

McCALL, G. J. H. (1963). A reconsideration of certain aspects of the Rangwa and Ruri carbonatite complexes in western Kenya. *Geol. Mag.* **100**, 181-185.

McCARTNEY, W. D. and POTTER, R. F. (1962). Mineralization as related to structural deformation, igneous activity and sedimentation in folded geosynclines. *Can. Min. J.* **83**, 83-87.

McCONNELL, R. B. (1972). Geological development of the rift system of eastern Africa. *Bull. geol. Soc. Am.* **83**, 2549-2572.

McCONNELL, R. B. (1980). A resurgent taphrogenic lineament of Precambrian origin in eastern Africa. *J. geol. Soc. Lond.* **137**, 483-489.

McKELVEY, V. E. (1967). Phosphate deposits. *Bull. U.S. Geol. Surv.* **1252-D**, 1-21.

McMILLAN, R. H. (1978). Genetic aspects and classification of important Canadian uranium deposits. *In* "Uranium Deposits, their Mineralogy and Origin" (M. M. Kimberley, ed.), 187-204. Min. Ass. Can. 3, Univ. Toronto Press, Toronto.

MEGARD, F. and PHILIP, H. (1976). Plio-Quaternary tectono-magmatic zonation and plate tectonics in the Central Andes. *Earth planet. Sci. Lett.* **33**, 231-238.

MEI-ZHONG YAN, YONG-LEE WU and CHANG-YOU LI. (1980). Metallogenic systems of tungsten in Southeast China and their mineralization characteristics. *In* "Granite Magmatism and Related Mineralization" (S. Ishihara and T. Takusumi, eds), 215-221. Min. Geol. Spec. Issue 8, Tokyo, Japan.

MENARD, H. W. (1967). Transitional types of crust under small ocean basins. *J. geophys. Res.* **72**, 140-150.

MENZIES, M. and ALLEN, C. (1974). Plagioclase lherzolite-residual mantle relationships within two eastern Mediterranean ophiolites. *Contr. Miner. Petrol.* **45**, 197-213.

MILLER, C. F. and BRADFISH, L. J. (1980). An inner Cordilleran belt of muscovite-bearing plutons. *Geology* **8**, 412-416.

MILSOM, J. (1978). Discussion on origin of porphyry coppers. *J. geol. Soc.*

Lond. **135,** 457.

MINTER, W. E. L. (1976). Detrital gold, uranium and pyrite concentrations related to sedimentology in the Precambrian Vaal Reef Placer, Witwatersrand goldfields, South Africa. A chronological review of speculations and observations. *Econ. Geol.* **71,** 157–176.

MITCHELL, A. H. G. (1973). Metallogenic belts and angle of dip of Benioff zones. *Nature Phys. Sci.* **245,** 49–52.

MITCHELL, A. H. G. (1974a). Flysch-ophiolite successions: polarity indicators in arc and collision-type orogens. *Nature, Lond.* **248,** 747–749.

MITCHELL, A. H. G. (1974b). Southwest England granites: magmatism and tin mineralization in a post-collision tectonic setting. *Trans. Instn Min. Metall.* **83,** B95–97.

MITCHELL, A. H. G. (1976a). Tectonic settings for emplacement of subduction-related magmas and associated mineral deposits. *Spec. Paps. geol. Ass. Can.* **14,** 3-21.

MITCHELL, A. H. G. (1976b). Southeast Asian tin granites: magmatism and mineralization in subduction and collision-related settings. CCOP Newsletter **3,** 10–14. United Nations, ESCAP, Bangkok.

MITCHELL, A. H. G. (1977). Tectonic settings for emplacement of Southeast Asian tin granites. *Bull. geol. Soc. Malaysia* **9,** 123–140.

MITCHELL, A. H. G. (1978a). Geosynclinal and plate tectonic hypotheses: significance of late orogenic Himalayan tin granites and continental collision. Proc. 11th Comm. Min. Metall. Cong., Hong Kong, Paper 37, 1–13.

MITCHELL, A. H. G. (1978b). The Grampian orogeny in Scotland: arc-continent collision and polarity reversal. *J. Geol.* **86,** 643–646.

MITCHELL, A. H. G. (1979a). Rift, subduction and collision-related tin belts. *Bull. geol. Soc. Malaysia* **11,** 81–102.

MITCHELL, A. H. G. (1979b). Guides to metal provinces in the central Himalayan collision belt; the value of regional stratigraphic correlations and tectonic analogies. *Mem. geol. Soc. China* **3,** 167–194.

MITCHELL, A. H. G. (1981a). Phanerozoic plate boundaries in mainland Southeast Asia, the Himalayas and Tibet. *J. geol. Soc. Lond.* **138,** 109-122.

MITCHELL, A. H. G. (1981b). The Grampian orogeny: almost a fossil Taiwan. *Geol. Surv. Taiwan,* in press.

MITCHELL, A. H. G. and BECKINSALE, R. D. (1981). Mineralization in magmatic rocks at convergent plate margins. *In* "Orogenic Andesites" (R. S. Thorpe, ed.). Wiley, London. In press.

MITCHELL, A. H. G. and BELL, J. D. (1973). Island-arc evolution and related mineral deposits. *J. Geol.* **81,** 381–405.

MITCHELL, A. H. G. and GARSON, M. S. (1972). Relationship of porphyry copper and circum-Pacific tin deposits to palaeo-Benioff zones. *Trans. Instn Min. Metall.* **81,** B10–25.

MITCHELL, A. H. G. and GARSON, M. S. (1976). Mineralization at plate boundaries. *Minerals Sci. Engng* **8,** 129–169.

MITCHELL, A. H. G. and McKERROW, W. S. (1975). Analogous evolution of the Burma orogen and the Scottish Caledonides. *Bull. geol. Soc. Am.* **86,** 305–315.

MITCHELL, A. H. G. and READING, H. G. (1969). Continental margins, geosynclines, and ocean floor spreading. *J. Geol.* **77,** 629–646.

MITCHELL, A. H. G. and READING, H. G. (1978). Sedimentation and tectonics. *In* "Sedimentary Facies and Environments" (H. G. Reading, ed.), 439–476. Blackwell, Oxford.

MITCHELL, A. H. G. and WARDEN, A. J. (1971). Geological evolution of the New Hebrides island arc. *J. geol. Soc. Lond.* **127,** 501–529.

MITCHELL, A. H. G., TIN HLAING and ZAW PE (1978). Structural units and post-Devonian geological history of Burma. Burma Sci. Res. Cong. (unpubl.).

MITCHELL, R. H. (1979). The alleged kimberlite-carbonatite relationship: additional contrary mineralogical evidence. *Am. J. Sci.* **279**, 570–589.

MIYASHIRO, A. (1961). Evolution of metamorphic belts. *J. Petrol.* **2**, 277–311.

MOGHAL, M. Y. (1974). Uranium in Siwalik sandstones, Sulaiman Range, Pakistan. *In* "Formation of uranium ore deposits", 383–403. IAEA, Vienna.

MOLNAR, P. and ATWATER, T. (1978). Interarc spreading and cordilleran tectonics as alternates related to the age of subducted oceanic lithosphere. *Earth planet. Sci. Lett.* **41**, 330-340.

MOLNAR, P. and TAPPONIER, P. (1977). Relation of the tectonics of eastern China to the India-Eurasia collision: application of slip-line field theory to large-scale continental tectonics. *Geology* **5**, 212–216.

MOORE, G. F., CURRAY, J. R., MOORE, D. G. and KARIG, D. E. (1980). Variation in fore-arc structures along the Sunda Arc, eastern Indian Ocean. *In* "Trench and Fore-Arc Sedimentation and Tectonics in Modern and Ancient Subduction Zones (Summary)", 25–36. Geol. Soc. Lond.

MOORE, G. W. (1973). Westward tidal lag as the driving force of plate tectonics. *Geology* **1**, 99–100.

MOORE, J. G., FLEMING, H. S. and PHILLIPS, J. D. (1974). Preliminary model for extrusion and rifting at the axis of the Mid-Atlantic Ridge, 36° 48′ north. *Geology* **2**, 437–440.

MOORE, J. S. and CONNELLY, W. (1979). Tectonic history of the continental margin of Southwestern Alaska: late Triassic to earliest Tertiary. *Prepr. Geol. Soc. Alaska,* H1–H29.

MOORES, E. M. (1969). Petrology and structure of the Vourinos ophiolite complex, northern Greece. *Spec. Paps geol. Soc. Am.* **118**.

MORGAN, W. J. (1972). Plate motions and deep mantle convection. *In* "Studies in Earth and Space Sciences (Hess Volume)" (R. Shagam *et al.,* eds.), *Mem. geol, Soc. Am.* **132**, 7–22.

MORTON, R. D. (1974). Sandstone-type uranium deposits in the Proterozoic strata of northwestern Canada. *In* "Formation of Uranium Ore Deposits", 255–273. IAEA, Vienna.

MORTON, R. D. and BECK, L. S. (1978). The origins of the uranium deposits of the Athabasca region, Saskatchewan, Canada. (Abstr.). *Econ. Geol.* **73**, 1408.

MUNHA, J. (1979). Blue amphiboles, metamorphic regime and plate tectonic modelling in the Iberian pyrite belt. *Contr. Mineral. Petrol.* **69**, 279–289.

MURPHY, R. W. (1973). The Manila Trench – West Taiwan foldbelt: a flipped subduction zone. *Bull. geol. Soc. Malaysia* **6**, 27–42.

NEARY, C. R. and BROWN, M. A. (1980). Chromites from Al'Ays Complex, Int. Saudi Arabia and the Semail Complex, Oman. *In* "Ophiolites", Proc. Ophiolite Symposium, Cyprus, 1979. Geol. Surv. Dept. Nicosia.

NECHAYEVA, I. A. (1965). Apatite-bearing rocks of the Synnyr nepheline-syenite pluton (north Baikal region). *Dokl. Adad. Nauk SSSR* **161**, 165–167.

NEILL, W. M. (1973). Possible continental rifting in Brazil and Angola related to the opening of the South Atlantic. *Nature Phys. Sci.* **245**, 104–107.

NICOLAS, A. (1972). Was the Hercynian orogenic belt of Europe of the Andean type? *Nature, Lond.* **236**, 221–223.

NOBLE, J. A. (1970). Metal provinces of the western United States. *Bull. geol. Soc. Am.* **81**, 1607-1624.

NOBLE, J. A. (1980). Two metallogenic maps for North America. *Geol. Rdsch.* **69**, 594-608.

NORMAN, D. I. (1978). Ore deposits related to the Keweenawan rift. *In*

"Petrology and Geochemistry of Continental Rifts" (E. R. Neumann and I. B. Ramberg, eds), Vol. 1, 245–253. Proc. NATO Advanced Study Inst. Oslo, 1977. Riedel, Holland.

NORMAN, J. W. (1980). Causes of some old crustal failure zones interpreted from Landsat images and their significance in regional exploration. *Trans. Instn Min. Metall.* **89**, B63–72.

NORMARK, R. (1976). Delineation of the main extrusion zone of the East Pacific Rise at 21° N. *Geology* **4**, 681–685.

NOTHOLT, A. J. G. (1979). The economic geology and development of igneous phosphate deposits in Europe and the USSR. *Econ. Geol.* **74**, 339–350.

OBA, N. (1977). Emplacement of granitic rocks in the Outer Zone of Southwest Japan, and geological significance. *Geology* **5**, 383–393.

OBA, N. and MIYAHISA, M. (1977). Relations between chemical composition of granitic rocks and metallization in the Outer Zone of southwest Japan. *Bull. geol. Soc. Malaysia* **9**, 67–74.

OHLE,, E. L. (1980). Some considerations in determining the origin of ore deposits of the Mississippi Valley type–Part II. *Econ. Geol.* **75**, 161–172.

OKRUSCH, M., BUNCH, T. E. and BANK, H. (1976). Paragenesis and petrogenesis of a corundum-bearing marble at Hunza (Kashmir). *Mineral. Deposita* **11**, 278–297.

OLABE, M. A. (1976). Global tectonics and metallogeny of intracontinental rifts and aulacogens. (Abs). 25th Internat. Geol. Cong. Australia 3, 744.

O'NEIL, J. R. and SILBERMAN, M. L. (1974). Stable isotope relations in epithermal Au–Ag deposits. *Econ. Geol.* **69**, 902–909.

O'NIONS, R. K., HAMILTON, P. J. and EVENSEN, N. M. (1980). The chemical evolution of the earth's mantle. *Sci. Am.* 242, No. 5, 91-101.

OXBURGH, E. R. (1972). Flake tectonics and continental collision. *Nature, Lond.* **239**, 202–204.

PAARMA, H. (1970). A new find of carbonatite in north Finland, the Sokli plug in Savukoski. *Lithos* **3**, 129–133.

PAARMA, H. and TALVITIE, J. (1976). Deep fractures—Sokli carbonatite. Univ. Oslo, Dept. Geophysics, Contrib. No. 65.

PACKHAM, G. H. and FALVEY, D. A. (1971). An hypothesis for the formation of marginal seas in the western Pacific. *Tectonophysics* **11**, 79–109.

PALABORA MINING COMPANY LTD. MINE GEOLOGICAL AND MINERALOGICAL STAFF (1976). The geology and the economic deposit of copper, iron and vermiculite in the Palabora Igneous Complex: a brief review. *Econ. Geol.* **71**, 177–192.

PAMIC, J. (1970). Osnovne petroloske karakteristike kromitskog podrucja Dubostic u Bosni. *Geol. Glasnik.* **14**, 135–148.

PARAK, T. (1975). Kiruna iron ores are not "intrusive-magmatic ores of the Kiruna type". *Econ. Geol.* **70**, 1242–1258.

PARK, C. F. (1961). A magnetite "flow" in northern Chile. *Econ. Geol.* **56**, 431–436.

PARK, C. F. and MacDIARMID, R. A. (1964). "Ore deposits". Freeman, San Francisco.

PAVILLON, M. J. (1973). *In* "Some major concepts of metallogeny" (Laboratoire de Geologie Appliquee, Univ. de Paris, France), p. 250. *Mineral. Deposita* **8**, 237–258.

PEARCE, J. A. and GALE, G. H. (1977). Identification of ore-deposition environment from trace-element geochemistry of associated igneous host rocks. *In* "Volcanic Processes in Ore Genesis", 14–24. Instn Min. Metall. and Geol. Soc. Lond. Spec. Publs No. 7.

PENROSE CONFERENCE (1972). Ophiolites. *Geotimes* **17**, 24–25.

PERCHUK, L. L. and VAGANOV, V. I. (1980). Petrochemical and thermodynamic evidence on the origin of kimberlites. *Contr. Mineral. Petrol.* **72**, 219–228.

PEREIRA, J. and DIXON, C. J. (1971). Mineralization and plate tectonics. *Mineral. Deposita* **6**, 404–405.

PETERS, Tj. and KRAMERS, J. D. (1974). Chromite deposits in the ophiolite complex of northern Oman. *Mineral. Deposita* **9**, 253–259.

PETERSEN, U. (1970). Metallogenic provinces in South America. *Geol. Rdsch.* **59**, 834–897.

PETROV , A. J. (1970). Old faults in the eastern part of the Baltic shield and movements along them. *Dokl. Akad. Nauk SSSR* **191**, 56–59.

PEYVE, A. V. and SINITZYN, V. M. (1950). Certain fundamental problems of the doctrine of geosynclines. *Izvest. Akad. Nauk SSR, Ser. Geol* **4**, 28–52.

PHILLIPS, W. J. (1973). Mechanical effects of retrograde boiling and its probable importance in the formation of some porphyry ore deposits. *Trans. Instn Min. Metall.* **82**, B90–98.

PIPER, D. Z. (1980). Japan Geological Survey's Investigation of manganese nodules in the Central Pacific Ocean. *Bull. Office Nat. Res. Sci.* **4**.

PIPER, J. A. D. (1974). Proterozoic crustal distribution, mobile belts and apparent polar movement. *Nature, Lond.* **251**, 381–384.

PITCHER, W. S. (1978). The anatomy of a batholith. *J. geol. Soc. Lond.* **135**, 157–182.

PLAFKER, G. and BRUNE, T. R. (1980). Late Cenozoic subduction—rather than accretion—at the eastern end of the Aleutian Arc. *In* "Trench and Fore-Arc Sedimentation and Tectonics in Modern and Ancient Subduction Zones (Summary)", 29–30. Geol. Soc. Lond.

PLATT, J. W. (1977). Volcanogenic mineralization at Avoca, Co. Wicklow, Ireland, and its regional implications. *In* "Volcanic Processes in Ore Genesis", 163–174. Instn Min. Metall. and Geol. Soc. Lond. Spec. Publs No. 7.

PLIMER, I. R. (1978). Proximal and distal stratabound ore deposits. *Mineral. Deposita* **13**, 345–353.

POULOSE, K. V. (1975). Tourmaline-granites and pegmatites associated with the Thimpu and the Chekha of Bhutan Himalayas. *In* "Recent Geological Studies in the Himalayas—1971" Geol. Surv. India, Misc. Publs No. 24, 256–262.

POWELL, C. Mc. A. (1979). A speculative tectonic history of Pakistan and surroundings: some constraints from the Indian Ocean. *In* "Geodynamics of Pakistan" (A. Farah and K. A. De Jong, eds), 5–24. Geol. Surv. Pakistan, Quetta.

POWELL, C. Mc. A., JOHNSON, B. E. and VEEVERS, J. J. (1980). A revised fit of east and west Gondwanaland. *Tectonophysics* **68**, 13–29.

PRETORIUS, D. A. (1975). The depositional environment of the Witwatersrand goldfields: a chronological review of speculations and observations. *Minerals Sci. Engng* **7**, 18–47.

PRIEM, H. N. A., BOELRIJK, N. A. I. M., HEBEDA, E. H., VERDURMEN, E. A. Th., VERSCHURE, R. H. and BON, E. H. (1971). Granitic complexes and associated tin mineralizations of 'Grenville' age in Rondônia, Western Brazil. *Bull. geol. Soc. Am.* **82**, 1095–1102.

PRYOR, R. N., RHODEN, H. N. and VILLALON, M. (1972). Sampling of Cerro Colorado, Rio Tinto, Spain. *Trans. Instn Min. Metall.* **81**, A143–159.

RAGLAND, P. C. and ROGERS, J. J. W. (1980). Favourable tectonic belts for granitic uranium deposits: Pan-Africa and the southern Appalachians. *J.*

geophys. Explor. **13**, 181–199.

RAIMES, G. L. OFFIELD, T. W. and SANTOS, E. S. (1978). Remote-sensing and subsurface definition of facies and structure related to uranium deposits, Powder River Basin, Wyoming. *Econ. Geol.* **73**, 1706–1723.

RAMBERG, I. B., GRAY, D. F. and RAYNOLDS, R. G. H. (1977). Tectonic evolution of the FAMOUS area of the Mid-Atlantic Ridge, lat. 35° 50' to 37° 20'N. *Bull. geol. Soc. Am.* **88**, 609–622.

RAYBOULD, J. G. (1978). Tectonic controls on Proterozoic stratiform . mineralization. *Trans. Instn Min. Metall.* **87**, B79–86.

RAYNER, R. A. and ROWLANDS, N. J. (1980). Stratiform copper in the late Proterozoic Boorloo Delta, South Australia. *Mineral Deposita,* **15**, 139-150.

REA, D. K. (1975). Model for the formation of topographic features of the East Pacific Rise Crest. *Geology* **3**, 77–80.

REED, B. L. and LANPHERE, M. A. (1973). Alaska-Aleutian range batholith: geochronology, chemistry and relation to circum-Pacific plutonism. *Bull. geol. Soc. Am.* **84**, 2583–2609.

REID, A. R. (1974). Proposed origin for Guianian diamonds. *Geology* **2**, 67–68.

REIS, B. (1970). The use of aeromagnetometry in the determination of deep seated structure and its importance to kimberlite exploration. *Serv. Geol. Min. Luanda* **23**, 11–21.

RENFRO, A. R. (1974). Genesis of evaporite-associated metalliferous deposits: a sabkha process. *Econ. Geol.* **69**, 33–45.

RENTZSCH, J. (1974). The 'Kupferschiefer' in comparison with the deposits of the Zambian Copperbelt. *In* "Gisements Stratiformes et Provinces Cuprifères" (P. Bartholomé, ed.), 235–254. Soc. Geol. Belgique, Liège.

RIGGS, S. R. (1979). Phosphorite sedimentation in Florida—a model phosphogenic system. *Econ. Geol.* **74**, 285–314.

RIMSKAYA-KORSAKOVA, O. M. (1964). Genesis of the Kovdor iron-ore deposits (Kola Peninsula). *Int. Geol. Rev.* **6**, 1735–1746.

RISE PROJECT GROUP (1980). East Pacific Rise: hot springs and geophysical experiments. *Science. N.Y.* **207**, 1421–1433.

ROBERTSON, A. H. F. (1975). Cyprus umbers: basalt-sediment relationships on a Mesozoic ocean ridge. *J. geol. Soc. Lond.* **131**, 511–531.

ROBERTSON, A. H. F. (1977). The Kannaviou Formation, Cyprus: volcanic-lastic sedimentation of a probable late Cretaceous volcanic arc. *J. geol. Soc. Lond.* **134**, 269–292.

ROBERTSON, D. S. (1974). Basal Proterozoic units as fossil time markers and their use in uranium prospection. *In* "Formation of Uranium Ore Deposits", 495–512. IAEA, Vienna.

RODRIGUES, B. (1972). Major tectonic alignments of alkaline complexes in Angola. *In* "African Geology" (T. F. J. Dessauvagie and A. J. Whiteman, eds), 149–153. Unit of Ibaden, Ibaden, Nigeria.

ROGERS, J. J. W., GHUMA, M. A., NAGY, R. M., GREENBERG, J. A. and FULLAGAR, P. D. (1978a). Plutonism in Pan-African belts and the geologic evolution of Northeastern Africa. *Earth planet. Sci. Lett.* **39**, 109–117.

ROGERS, J. J. W., RAGLAND, P. C., NISHIMORI, R. K., GREENBERG, J. K. and HAUCK, S. A. (1978b). Varieties of granitic uranium deposits and favourable exploration areas in the Eastern United States. *Econ. Geol.* **73**, 1539–1555.

RONA, P. A. (1978). Criteria for recognition of hydrothermal mineral deposits in oceanic crust. *Econ. Geol.* **73**, 135–160.

RONA, P. A., HARBISON, R. N., BASSINGER, B. G., SCOTT, R. B. and

NALWALK, A. J. (1976). Tectonic fabric and hydrothermal activity of Mid-Atlantic Ridge crest (lat. 26°N). *Bull. geol. Soc. Am.* **87**, 661–674.

ROSE, P. (1976). Mississippian carbonate shelf margins, western United States. *J. Res. U.S. Geol. Surv.* **4**, 449–466.

ROSSOVSKY, L. N. and NONOVALENKO, S. I. (1976). About South Asian pegmatite belt. *Rep. Acad. Sci. USSR* **229**, No. 3, 695-698.

ROUTHIER, P. (1969). Sur trois principes généraux de la métallogénie et de la recherche minérale. *Mineral. Deposita* **4**, 213–218.

ROUTHIER, P. *et al.*, 1973). Some major concepts of metallogeny (Laboratoire de Geologie Appliquée, Univ. de Paris, France). *Mineral. Deposita* **8**, 237–258.

ROUTHIER, P. (1976). A new approach to metallogenic provinces: the example of Europe. *Econ. Geol.* **71**, 803–811.

ROUTHIER, P. (1980). "Ou sont les metause pour l'arenir? Les provinces metallogenie globale". Mem. Bur. Recherche Geol. Min. no. 105.

ROWLANDS, N. J. (1974). The gitology of some Adelaidean stratiform copper occurrences. *In* "Gisements stratiformes et provinces cupriferes" (P. Bartholomé, (ed.), 419–427. Soc. Geol. Belgique, Liège.

ROWLANDS, N. J. (1980). Discussions and contributions: tectonic controls on Proterozoic stratiform mineralization. *Trans. Instn Min. Metall.* **89**, B167–168.

ROWLANDS, N. J., DRUMMOND, A. J., JARVIS, D. M., WARIN, O. N., KITCH, R. R. and CHUCK, R. G. (1978). Gitological aspects of some Adelaidean stratiform copper deposits. *Min Sci. Engng* **10**, 258–277.

ROWLEY, P. D., LIPMAN, P. W., MEHNERT, H. H., LINDSEY, D. A. and ANDERSON, J. J. (1978). Blue Ribbon Lineament, an east-trending structural zone within the Pioche Mineral Belt of southwestern Utah and eastern Nevada. *J. Res. U. S. Geol. Surv.* **6**, 175–192.

ROZENDAAL, A. (1978). The Gamsberg zinc deposit, Namaqualand. *In* "Mineralization in metamorphic terranes" (W. J. Verwoerd, ed.), 235–265. Van Shaik, Pretoria.

RUIZ, C. AGUILAR, A., EGERT, E., ESPINOSA, W., PEEBLES, F., QUEZADA, R. and SERRANO, M. (1971). Strata-bound copper sulphide deposits of Chile. *In* "International Association of the Genesis of Ore Deposits, papers and Proceedings of the Tokyo–Kyoto Meetings" (Y. Takeuchi, ed), 252–260. Soc. Min. Geol. Japan, Spec. Issue No. 3, Tokyo.

RUSHTON, H. G. (1977). Formation of the Rocky Mountain coking coal deposits. *Can. Min. J.* **98**, 55–58.

RUSSELL, M. J. (1978). Downward-excavating hydrothermal cells and Irish-type ore deposits: importance of an underlying thick Caledonian prism. *Trans. Inst Min. Metall.* **87**, B168–171.

SAMONOV, I. Z. and POZHARISKY, I. F. (1977). Deposits of copper. *In* "Ore Deposits of the USSR", (V. I. Smirnov, ed.), Vol. 2, 106–181. Pitman, London.

SANGSTER, D. F. (1970). Metallogenesis of some Canadian lead-zinc deposits in carbonate rocks. *Geol. Ass. Can. Proc.* **22**, 27-36.

SANGSTER, D. F. (1976a). Carbonate-hosted lead-zinc deposits. *In* "Handbook of Stratabound and Stratiform Ore Deposits" (K. H. Wolf, ed.), Vol. 6, 445–456. Elsevier, Amsterdam.

SANGSTER, D. F. (1976b). Possible origin of lead in volcanogenic marine sulphide deposits of calc-alkaline affiliation. *Spec. Pap. Geol. Ass. Can.* **14**, 103-104.

SATO, T. (1976). Origin of the green tuff metal province of Japan. *Spec. Pap. Geol. Ass. Can.* **14**, 105-120.

SATO, T. (1977). Kuroko deposits: their geology, geochemistry and origin. *In* "Volcanic Processes in Ore Genesis", 153–161. Instn. Min. Metall. and Geol. Soc. Lond. Spec. Publ. No. 7.

SAWKINS, F. J. (1972). Sulfide ore deposits in relation to plate tectonics. *J. Geol.* **80**, 377–397.

SAWKINS, F. J. (1976a). Widespread continental rifting: some considerations of timing and mechanism. *Geology* **4**, 427–430.

SAWKINS, F. J. (1976b). Metal deposits related to intracontinental hotspot and rifting environments. *J. Geol.* **84**, 653–671.

SAWKINS, F. J. (1976c). Massive sulphide deposits in relation to geotectonics. *Spec. Pap. geol. Ass. Can.* **14**, 221–240.

SCHEIBNER, E. and MARKHAM, N. L. (1976). Tectonic setting of some strata-bound massive sulphide deposits in New South Wales, Australia. *In* "Handbook of Strata-bound and Stratiform Ore Deposits" (K. H. Wolf, ed.), 55–77. Elsevier, Amsterdam.

SCHILLING, J. G. (1973). Iceland mantle plume: geochemical evidence along Reykjanes Ridge. *Nature, Lond.* **242**, 565–571.

SCHLANGER, S. O. and JENKYNS, H. C. (1976). Cretaceous oceanic anoxic events: causes and consequences. *Geol. Mijn.* **55**, 179–184.

SCHOLL, D. W. and MARLOW, M. S. (1974). Sedimentary sequence in modern Pacific trenches and the deformed circum-Pacific eugeosyncline. *In* "Modern and Ancient Geosynclinal Sedimentation" (R. H. Doft Jr and R. H. Shaver, eds). *Spec. Publs. econ. Palaeont. Miner.* **19**, 193-211.

SCHOLL, D. W. and VALLIER, T. L. (1980). Subduction erosion at modern and ancient underthrust Pacific margins: an evaluation of the evidence. *In* "Trench and Fore-Arc Sedimentation and Tectonics in Modern and Ancient Subduction Zones (Summary)", 32–33. Geol. Soc. Lond.

SCHOLL, D. W., HUENE, R. VON, VALLIER, T. L. and HOWELL, D. G. (1980). Sedimentary masses and concepts about tectonic processes at under-thrust ocean margins. *Geology* **8**, 564–568.

SCHOLZ, C. H., BARAZANGI, M. and SBAR, M. L. (1971). Late Cenozoic evolution of the Great Basin, western United States, as an ensialic interarc basin. *Bull. geol. Soc. Am.* **82**, 2979–2990.

SCHUILING, R. D. (1967). Tin belts around the Atlantic Ocean: some aspects of the geochemistry of tin. *In* "A Technical Conference on Tin", 531–547. International Tin Council, London.

SCOTT, B. (1976). Zinc and lead mineralization along the margins of the Caledonian orogen. *Trans. Instn Min. Metall.* **85**, B200-204.

SCOTT, M. R. SCOTT, R. B., RONA, P. A., BUTLER, L. W. and NALWARK, A. J. (1974). Rapidly accumulating manganese deposit from the median valley of the Mid-Atlantic Ridge. *Geophys. Res. Lett.* **1**, 355–358.

SEARLE, D. L. (1972). Mode of occurrence of cupriferous pyrite deposits of Cyprus. *Trans. Instn Min. Metall.* **81**, B189-197.

SENGOR, A. M. C. (1976). Collision of irregular continental margins: implications for foreland deformation of Alpine-type orogens. *Geology* **4**, 779–782.

SENGOR, A. M. C., BURKE, K. and DEWEY, J. F. (1978). Rifts at high angles to orogenic belts: tests for their origin and the Upper Rhine Graben as an example. *Am. J. Sci.* **278**, 24-40.

SHARP, J. M. (1978). Energy and momentum transport model of the Ouachita Basin and its possible impact on formation of economic mineral deposits. *Econ. Geol.* **73**, 1057–1068.

SHATSKI, N. S. (1955). On the origin of the Pachelma trough. *Mosk. O-va. Lyubit. Prir. Byull. Geol. Sec.* **5**, 5–26.

SHATSKI, N. S. (1947). Structural correlations of platforms and geosynclinal folded regions. *Akad. Nauk SSSR Izv. Geol. Ser.* **5**, 37–56.

SHAWE, D. R., HITE, R. J. and INTHUPUTI, B. (1975). Potential for sandstone-type uranium deposits in Jurassic rocks, Khorat Plateau, Thailand. *Econ. Geol.* **70**, 538–541.

SHCHEGLOV, A. D. (1976). "Principles of Metallogenic Analyses." Hyedra, Moscow.

SHELDON, R. P. (1964). Palaeolatitudinal and palaeogeographic distribution of phosphorite. *In* Geological Survey Research. *Prof. Pap. U. S. geol. Surv.* 501–C, 106–113.

SHEPPARD, S. M. F. (1977). Identification of the origin of ore-forming solutions by the use of stable isotopes. *In* "Volcanic Processes in Ore Genesis". 25–41. Instn Min. Metall. and Geol. Soc. Lond. Spec. Publs No. 7.

SHERIDAN, R. E. (1974). Atlantic continental margin of North America. *In* "The Geology of Continental Margins" (C. A. Burke and C. L. Drake, eds), 391–407. Springer-Verlag, New York.

SHOEMAKER, E. M. (1978). Induced suction and lithospheric plate pull. *Nature, Lond.* **272**, 241–242.

SIESSER, W. G. (1976). Native copper in DSDP sediment cores from the Angola Basin. *Nature, Lond.* **263**, 308-0309.

SILLITOE, R. H. (1970). South American porphyry copper deposits and the new global tectonics. Resumenes Primer Congr. Latinoamericano Geol., Lima, Peru, 254–256.

SILLITOE, R. H. (1972a). Formation of certain massive sulphide deposits at sites of sea-floor spreading. *Trans. Instn Min. Metall.* **81**, B141–148.

SILLITOE, R. H. (1972b). Relation of metal provinces in western America to subduction of oceanic lithosphere. *Bull. geol. Soc. Am.* **83**, 813–818.

SILLITOE, R. H. (1974). Tin mineralization above mantle hot spots. *Nature, Lond.* **248**, 497–499.

SILLITOE, R. H. (1975). Subduction and porphyry copper deposits in south-western North America–a reply to recent objections. *Econ. Geol.* **70**, 1474–1477.

SILLITOE, R. H. (1976a). A reconnaissance of the Mexican porphyry copper belt. *Trans. Instn Min. Metall.* **85**, B170–189.

SILLITOE, R. H. (1976b). Andean mineralization: a model for the metallogeny of convergent plate margins. *Spec. Pap. geol. Ass. Can.* **14**, 59-100.

SILLITOE, R. H. (1977). Metallic mineralization affiliated to subaerial volcanism: a review. *In* "Volcanic Processes in Ore Genesis", 99–116. Instn Min. Metall. and Geol. Soc. Lond. Spec. Publs. No. 7.

SILLITOE, R. H. (1978). Metallogenic evolution of a collisional mountain belt in'Pakistan: a preliminary analysis. *J. geol. Soc. Lond.* **135**, 377–387.

SILLITOE, R. H. (1979). Some thoughts on gold-rich porphyry copper deposits. *Mineral. Deposita* **14**, 161–174.

SILLITOE, R. H. (1980a). Types of porphyry deposits. *Min. Mag.* June, 550–551.

SILLITOE, R. H. (1980b). Are porphyry copper and Kuroko-type massive sulphide deposits incompatible? *Geology* **8**, 11-14.

SILLITOE, R. H. and SAWKINS, F. J. (1971). Geologic, mineralogic and fluid inclusion studies relating to the origin of copper-bearing tourmaline breccia pipes, Chile. *Econ. Geol.* **66**, 1028-1041.

SILVER, E. A. (1971). Transitional tectonics and late Cenozoic structure of the continental margin off northernmost California. *Bull. geol. Soc. Am.* **82**, 1–22.

SILVER. E. A. and MOORE, J. C. (1978). The Molluca Sea collision zone, Indonesia. *J. geophys. Res.* **83**, 1681–1691.

SIMPSON, P. R., BROWN, G. C., PLANT, J. and OSTLE, D. (1979). Uranium mineralization and granite magmatism in the British Isles. *Phil. Trans. R. Soc. Lond.* **291A**, 133–160.

SKILBECK, J. and WHITEHEAD, J. A. (1978). Formation of discrete islands in linear island chains. *Nature, Lond,* **272**, 499–501.

SLEEP, N. and TOKSOZ, M. N. (1971). Evolution of marginal basins. *Nature, Lond.* **233**, 548-550.

SMIRNOV, V. I. (1968). The sources of ore-forming fluids. *Econ. Geol.* **63**, 380-389.

SMIRNOV, V. I. (1977). Preface. *In* "Ore deposits of the USSR" (V. I. Smirnov, ed.), Vol. 1. Pitman, London.

SMITH, A. G. and WOODCOCK, N. H. (1976). Emplacement model for some 'Tethyan' ophiolites. *Geology* **4**, 653–656.

SMITH, D. B. (1979). Rapid marine transgressions of the Upper Permian Zechstein Sea. *J. geol. Soc. Lond.* **136**, 155–156.

SMITH, D. B. and CROSBY, A. (1979). The regional and stratigraphic context of Zechstein 3 and 4 potash deposits in the British sector of the southern North Sea and adjoining land areas. *Econ. Geol.* **74**, 397–408.

SMITH, R. B. and CHRISTIANSEN, R. L. (1980). Yellowstone Park as a window on the earth's interior. *Sci. Am.* **242**, 84-95.

SMITHERINGALE, W. G. (1972). Low-potash Lush's Bight tholeiites: ancient oceanic crust in Newfoundland. *Can. J. Earth Sci.* **9**, 574-583.

SNELGROVE, A. K. (1971). Metallogeny and the new global tectonics. *Min. Res. Expl. Inst. Turkey, Bull.* **76**, 130–149.

SNYDER, W. S. (1978). Manganese deposited by submarine hot springs in chert-greenstone complexes, western United States. *Geology* **6**, 741–744.

SOE WIN. (1968). The application of geology to the mining of jade. *Union Burma J. Sci. Tech.* **1**, 445–456.

SOLOLVIEV, S. L., TOUEZOV, I. K. and VASILIEV, B. I. (1977). The structure and origin of the Okhotsk and Japan Sea abyssal depressions according to new geophysical and geological data. *Tectonophysics* **37**, 153-166.

SOLOMON, M. (1976). "Volcanic" massive sulphide deposits and their host rocks – a review and an explanation. *In* "Handbook of Strata-bound and Stratiform Ore Deposits" (K. H. Wolf, ed.), Vol. 6, 21–54. Elsevier, Amsterdam.

SOLOMON, M. and WALSHE, J. L. (1979). The formation of massive sulfide deposits on the sea floor. *Econ. Geol.* **74**, 797-813.

SPOONER, E. T. C. (1977). A hydrodynamic model for the origin of the ophiolitic cupriferous pyrite ore deposits of Cyprus. *In* "Volcanic Processes in Ore Genesis", 58-71. Instn Min. Metall. and Geol. Soc. Lond. Spec. Publs No. 7.

SPOONER, E. T. C. and BRAY, C. J. (1977). Hydrothermal fluids of seawater salinity in ophiolitic sulphide ore deposits in Cyprus. *Nature, Lond.* **266**, 808-812.

SPOONER, E. T. C., BECKINSALE, R. D., ENGLAND, P. C. and SENIOR, A. (1977a). Hydration, [18]O enrichment and oxidation during ocean floor hydrothermal metamorphism of ophiolitic metabasic rocks from E. Liguria, Italy. *Geochim. cosmochim. Acta* **41**, 857–871.

SPOONER, E. T. C., CHAPMAN, H. J. and SMEWING, J. D. (1977b). Strontium isotopic contamination and oxidation during ocean floor hydrothermal metamorphism of the ophiolitic rocks on the Troodos Massif, Cyprus. *Geochim. cosmochim. Acta* **41**, 873–890.

STANTON, R. L. (1955). Lower Palaeozoic mineralization near Balhurst, New South Wales. *Econ. Geol.* **50**, 681–714.

STANTON, R. L. (1960). General features of the conformable pyritic ore bodies. *Trans. Can. Inst. Min. Metall.* **63**, 22–27.

STANTON, R. L. (1972). "Ore Petrology". McGraw-Hill, New York.

STILLE, H. (1936). Wege und Ergebnisse der geologischtektonischen Forschung. *In* "25 Jahre Kaiser Wilhelm Gesellschaft, 2", 77–97.

STILLE, H. (1940). "Einfuhrung in den Bau Nordamerikas." Borntraeger, Berlin.

STILLMAN, C. J., FUSTER, J. M., BENNELL-BAKER, M. S., MUNOZ, M., SMEWING, J. D. and SAGREDO, N. (1975). Basal complex of Fuerteventura (Canary Islands) is an oceanic intrusive complex with rift-system affinities. *Nature, Lond,* **257**, 469.

STOCKLIN, J. (1980). Geology of Nepal and its regional frame. *J. geol. Soc. Lond.* **137**, 1–34.

STRACKE, K. J., FERGUSON, J. and BLACK, L. P. (1979). Structural setting of kimberlites in south-eastern Australia. *In* "Kimberlites, Diatremes, and Diamonds: Their Geology, Petrology, and Geochemistry" (F. R. Boyd and H. O. A. Meyer, eds), 71–91. Proc. Second Internat. Kimberlite Conf. 1. Am. Geophys. Union, Washington.

STRAUSS, G. T. and MADEL, J. (1974). Geology of massive sulphide deposits in the Spanish-Portuguese Pyrite Belt. *Geol. Rundschau* **63**, 191–211.

STRAUSS, G. T., MADEL, J. and ALONSO, F. D. (1977). Exploration practice for strata-bound volcanogenic sulphide deposits in the Spanish-Portuguese Pyrite Belt: geology, geophysics and geochemistry. *In* "Time- and Strata-bound Ore Deposits" (D. D. Klemm and H-J. Schneider, eds), 55–93. Springer-Verlag, Berlin.

STRONG, D. F. (1972). Sheeted diabases of central Newfoundland: new evidence for Ordovician sea-floor spreading. *Nature, Lond.* **235**, 102–104.

STUCKLESS, J. S. and NKOMO, I. T. (1978). Uranium-lead isotope systematics in uraniferous alkali-rich granites from the Granite Mountains, Wyoming: implications for uranium source rocks. *Econ. Geol.* **73**, 427–441.

STUMPFL, E. F. (1966). On the occurrence of native platinum with copper sulphides at Congo Dam, Sierra Leone. *Overseas Geol. Miner. Resour.* **10**, 1–10.

STUMPFL, E. F., CLIFFORD, T. N., BURGER, A. J. and ZYL, F. VAN (1976). The copper deposits of the O'okiep District, South Africa: new data and concepts. *Mineral. Deposita* **11**, 46–70.

SUPPE, J. and LIOU, J. G. (1979). Tectonics of the Lichi melange and east Taiwan ophiolite. *Mem. geol. Soc. China* **3**, 147–154.

SVISERO, D. P., MEYER, H. O. A. and TSAI, H-M. (1976). Kimberlites in Brazil: an initial report. *In* "Kimberlites, Diatremes, and Diamonds: Their Geology, Petrology and Geochemistry" (F. R. Boyd and H. O. A. Meyer, eds), 92–100. Proc. Second Internat. Kimberlite Conf., 1. Am. Geophys. Union, Washington.

SWANSON, E. R., KEIZER, R. P. and CLABAUGH, S. E. (1978). Tertiary volcanism and caldera development near Durango City, Sierre Madre Occidental, Mexico. *Bull. geol. Soc. Am.* **89**, 1000–1012.

TAHIRKHELI, R. A., MATTAUER, M., PROUST, F. and TAPPONIER, P. (1979). The India-Eurasia suture zone in northern Pakistan: synthesis and interpretation of recent data at plate scale. *In* "Geodynamics of Pakistan" (A. Farah and H. De Jong, eds), 125–130. Geol. Surv. Pakistan, Quetta.

TALALOV, V. A. (1977). Main features of magmatism and metallogeny of the Nepalese Himalayas. *In* "Colloque sur l'ecologie et geologie de la Himalaya" No. 268, 409-430. Centre National de la Recherche Scientifique, Paris.

TAPPONIER, P. and MOLNAR, P. (1977). Active faulting and tectonics in China. *J. geophys. Res.* **82**, 2905–2930.

TARNEY, J. and WINDLEY, B. F. (1979). Continental growth, island arc accretion and the nature of the lower crust – a reply to S. R. Taylor and S. M. McLennan. *J. geol. Soc. Lond.* **136**, 501–504.

TATSCH, J. H. (1976). "Copper Deposits: Origin, Evolution and Present Characteristics." Tatsch, Sudbury.

TATSUMI, T., TAKAGI, Y. and OTAGAKI, T. (1972). Geology of the Kuroko deposits. MMIJ-AIME Joint Meeting, Soc. Petroleum Engrs. Preprint TIbl.

TAYLOR, R. P. (1979). Topsails igneous complex–further evidence of middle Palaeozoic epeirogeny and anorogenic magmatism in the northern Appalachians. *Geology* **7**, 488–490.

TEGGIN, D. E. (1975). The granites of northern Thailand. PhD Thesis (unpubl.). Univ. Manchester.

THAYER, T. P. (1969). Gravity differentiation and magmatic re-emplacement of podiform chromite deposits. *In* "Magmatic Ore Deposits" (H. D. B. Wilson, ed.), 132–146. Econ. Geol. Mon. 4.

THAYER, T. P. (1970). Chromite segregations as petrogenetic indicators. *Spec. Publs Geol. Soc. S. Afr.* **1**, 380–390.

THIESSEN, R., BURKE, K. and KIDD, W. S. F. (1979). African hotspots and their relation to the underlying mantle. *Geology* **7**, 263–266.

THOMPSON, G. and MELSON, W. G. (1972). The petrology of oceanic crust across fracture zones in the Atlantic Ocean: evidence of a new kind of sea-floor spreading. *J. Geol.* **80**, 526–538.

THOMPSON, T. L. (1976). Plate tectonics in oil and gas exploration of continental margins. *Bull. Am. Ass. Petrol. Geol.* **60**, 1463–1501.

THURLOW, J. G. (1977). Occurrence, origin and significance of mechanically transported sulphide ores at Buchans, Newfoundland. *In* "Volcanic Processes in Ore Genesis", 127. Instn Min. Metall. and Geol. Soc. Lond. Spec. Publs No. 7.

THURLOW, J. G., SWANSON, E. A. and STRONG, D. F. (1975). Geology and lithogeochemistry of the Buchans polymetallic sulfide deposits, Newfoundland. *Econ. Geol.* **70**, 130–144.

TISCHLER, S. E. and FINLOW-BATES, T. (1980). Plate tectonic processes that governed the mineralization of the Eastern Alps. *Mineral. Deposita* **15**, 19–34.

TOKSOZ, M. N. and BIRD, P. (1977). Tectonophysics of the continuing Himalayan orogeny. *In* "Colloque sur l'ecologie et geologie de la Himalaya", No. 268, 443–448. Centre National de la Recherche Scientifique, Paris.

TURCOTTE, D. L. and OXBURGH, E. R. (1973). Mid-plate tectonics. *Nature, Lond.* **244**, 337–339.

TURCOTTE, D. L. and OXBURGH, E. R. (1978). Intra-plate volcanism. *Phil. Trans. R. Soc. Lond.* **288A**, 561–579.

TURNEAURE, F. S. (1955). Metallogenetic provinces and epochs. *Econ. Geol.* 50th anniv. vol., 38–98.

TURNEAURE, F. S. (1971). The Bolivian tin-silver province. *Econ. Geol.* **66**, 215–225.

TURNER, D. C. and BOWDEN, P. (1979). The Ningi-Burra complex, Nigeria: dissected calderas and migrating magma centres. *J. geol. Soc. Lond.* **136**, 105–119.

TURNER, D. C. and WEBB, P. K. (1974). The Daura igneous complex, N. Nigeria: a link between the Younger Granite district of Nigeria and S. Niger. *J. geol. Soc. Lond.* **130**, 71–77.

TURNER, W. M. (1973). The Cyprian gravity nappe and the autochthonous basement of Cyprus. *In* "Gravity and Tectonics" (K. A. De Jong and R.

Scholten, eds), 287–301. Wiley, New York.

UDAS, G. R. and MAHADEVAN, T. M. (1974). Controls and genesis of uranium mineralization in some geological environments in India. *In* "Formation of Uranium Ore Deposits", 425–436. IAEA, Vienna.

UENO, H. (1975). Duration of the Kuroko mineralization episode. *Nature, Lond.* **253**, 428–429.

UKPONG, E. E. and OLABE, M. A. (1979). Geochemical surveys for lead-zinc mineralization, southern Benue Trough, Nigeria. *Trans. Instn Min. Metall.* **88**, B81–92.

ULMER, G. C. (1969). Experimental investigations of chromite spinels. *In* "Magmatic Ore Deposits" (H. D. B. Wilson, ed.), 114–131. Econ. Geol. Mon. 4.

UMGROVE, J. H. F. (1949). "Structural History of the East Indies". Cambridge University Press, London.

UNITED NATIONS. (1980). "Studies in East Asian Tectonics and Resources (SEATAR)." CCOP-IOC, Bangkok.

UPADHYAY, H. D. and STRONG, D. F. (1973). Geological setting of the Betts Cove copper deposits, Newfoundland: an example of ophiolite sulfide mineralization. *Econ. Geol.* **68**, 161–167.

UPADHYAY, H. D., DEWEY, J. F. and NEALE, E. R. W. (1971). The Betts Cove ophiolite complex, Newfoundland: Appalachian oceanic crust and mantle. *Proc. geol. Ass. Can.* **24**, 27–34.

URABE, T. and SATO, T. (1978). Kuroko deposits of the Kosaka Mine, Northeast Honshu, Japan—products of submarine hot springs on Miocene sea floor. *Econ. Geol.* **73**, 161–179.

UYEDA, S. and MIYASHIRO, A. (1974). Plate tectonics and the Japanese Islands. *Bull. geol. Soc. Am.* **85**, 1159–1170.

UYEDA, S. and NISHIWAKI, C. (1980). Stress field, metallogenesis and mode of subduction. *Spec. Pap. geol. Ass. Can.* **20**.

VAIL, J. R. (1978). Further data on the alignment of basic igneous intrusive complexes in southern and eastern Africa. *Trans. geol. Soc. S. Afr.* **62**, 87–92.

VALLÉE, M. and DUBUC, F (1970). The St. Honoré carbonatite complex, Quebec. *Trans. Can. Inst Min. Metall.* **73**, 346–356.

VARTIAINEN, H. and PAARMA, H. (1979). Geological characteristics of the Sokli carbonatite complex, Findland. *Econ. Geol.* **74**, 1296–1306.

VEEVERS, J. J. and COTTERILL, D. (1976). Western margin of Australia: a Mesozoic analog of the East African rift system. *Geology* **4**, 713–717.

VERWOERD, W. J. (1966). South African carbonatites and their probable mode of origin. *Ann. Univ. Stellenbosch* **66**, Ser. A, No. 2, 121–233.

VINE, F. J. and MATTHEWS, D. H. (1963). Magnetic anomalies over oceanic ridges. *Nature, Lond.* **199**, 941–949.

VOKES, F. M. (1968). Regional metamorphism of the Palaeozoic geosynclinal sulphide ore deposits of Norway. *Trans. Inst. Min. Metall.* **77**, B53–59.

VOKES, F. M. and GALE, G. H. (1976). Metallogeny relatable to global tectonics in southern Scandinavia. *Spec. Pap. geol. Ass. Can.* **14**, 413–441.

WALKER, W. (1976). Eras, mobile belts and metallogeny. *Spec. Pap. geol. Ass. Can.* **14**, 517–557.

WALTHAM, A. C. (1968). Classification and genesis of some massive sulphide deposits in Norway. *Trans. Inst Min. Metall.* **77**, B153–161.

WATTERS, B. R. (1976). Possible late Precambrian subduction zone in South West Africa. *Nature, Lond* **259**, 471–473.

WEDEPOHL, H. H. (1971). "Kupferschiefer" as a prototype of syngenetic sedimentary ore deposits. *In* "Proceedings IMA-IAGOD Meeting" (Y.

Takewchi, ed.). Vol. 1970. IAGOD.

WEIBLEN, P. W., COOPER, R. J. and CHURCHILL, R. K. (1978). Relationship between mineralization and structure in the Duluth Complex. *Econ. Geol.* 73, 316.

WELLMAN, P. and McDOUGALL, I. (1974). Cenozoic igneous activity in eastern Australia. *Tectonophysics* 23, 49–65.

WELLS, M. K. (1962). Structure and petrology of the Freetown layered basic complex of Sierra Leone. *Bull. Overseas Geol. Miner. Resour. Suppl.* 4.

WHITE, A. J. R. and CHAPPELL, B. W. (1977). Ultrametamorphism and granite genesis. *Tectonophysics* 43, 7–22.

WHITE, D. E. (1968). Environments of generation of some base-metal ore deposits. *Econ. Geol.* 63, 301–335.

WHITE, D. E., MUFFLER, L. J. P. and TRUESDELL, A. H. (1971). Vapor-dominated hydrothermal systems compared with hot water systems. *Econ. Geol.* 66, 75–97.

WHITE, D. E., BARNES, I. and O'NEIL, J. R. (1973). Thermal and mineral water of nonmeteoric origin, California coast ranges. *Bull. geol. Soc. Am.* 84, 547–560.

WHITE, W. (1968). Ore deposits of the Western United States, 1933–1967. *Am. Inst. Min. Metall. Engng* 303.

WHITEHEAD, A., NAYLOR, D., PEGRUM, R. and REES, G. (1975). North Sea troughs and plate tectonics. *Tectonophysics* 26, 39–54.

WHITEMAN, A. J. (1971). "The Geology of the Sudan Republic." Clarendon Press, Oxford.

WHITNEY, J. A., JONES, L. M. and WALKER, R. L. (1976). Age and origin of the Stone Mountain Granite, Lithonia District, Georgia. *Bull. geol. Soc. Am.* 87, 1067–1077.

WILLIAMS, D. (1962). Further reflections on the origin of the porphyries and ores of Rio Tinto, Spain. *Trans. Instn Min. Metall.* 71, B265–B266.

WILLIAMS, H. R. and WILLIAMS, R. A. (1977). Kimberlites and plate-tectonics in West Africa. *Nature, Lond.* 270, 507–508.

WILSON, J. T. (1963). A possible origin of the Hawaiian Islands. *Can. J. Phys.* 41, 863–870.

WILSON, J. T. (1965a). A new class of faults and their bearing on continental drift. *Nature, Lond.* 207, 343–347.

WILSON, J. T. (1965b). Evidence from ocean islands suggesting movement in the earth. *In* "Symposium on continental drift". *Phil. Trans. R. Soc. Lond.* 258A, 145–167.

WILSON, J. T. (1968). Static or mobile earth: The current scientific revolution. *Proc. Am. Phil. Soc.* 112, 309–320.

WILSON, R. C. L. and WILLIAMS, C. A. (1979). Oceanic transform structures and the development of Atlantic continental margin sedimentary basins—a review. *J. geol. Soc. Lond.* 136, 311–320.

WINKLER, H. G. F. (1967). "Petrogenesis of Metamorphic Rocks." Springer-Verlag, New York.

WOLF, K. H. (1976). Conceptual models in geology. *In* "Handbook of Strata-Bound and Stratiform Ore Deposits" (K. H. Wolf, ed.), Vol. 1. 11–78. Elsevier, Amsterdam.

WOODCOCK, J. R. and HOLLISTER, J. F. (1978). Porphyry molybdenite deposits of the North American Cordillera. *Minerals Sci. Engng* 10, 3–18.

WORST, B. G. (1960). The Great Dyke of Southern Rhodesia. *Bull. geol. Surv. S. Rhod.* 47.

WORTEL, M. J. R. and VLAAR, N. J. (1978). Age-dependent subduction of the ocean lithosphere beneath western North America. *Phys. Earth planet. Int.* **17**, 201–208.

WRIGHT, J. B. (1976). Fracture systems in Nigeria and initiation of fracture zones in the South Atlantic. *Tectonophysics* **34**, T43–47.

WYLLIE, P. J. (1967). Kimberlites. *In* "Ultramafic and Related Rocks" (P. J. Wyllie, ed.). Wiley, New York.

WYLLIE, P. J. and HUANG, W. L. (1975). Influence of mantle CO_2 in the generation of carbonatites and kimberlites. *Nature, London.* **257**, 297–299.

WYLLIE, P. J. and HUANG, W.L. (1976). Carbonation and melting reactions in the system $CaO\text{-}MgO\text{-}SiO_2\text{-}CO_2$ at mantle pressure with geophysical and petrological applications. *Contr. Mineral. Petrol.* **54**, 79–107.

YOUNG, R. B. (1917). "The Banket of the South African Goldfields." Gurney and Jackson, London.

ZIEGLER, P. A. (1975). North Sea Basin history in the tectonic framework of North-Western Europe. *In* "Petroleum and the Continental Shelf of North West Europe" (A. W. Woodland, ed.), Vol. 1, 131–148. Applied Science, Essex.

ZIEGLER, P. A. (1978). Northwestern Europe: tectonics and basin development. *Geol. Mijn.* **57**, 589–626.

ZIEGLER, W. H. (1975). Outline of the geological history of the North Sea. *In* "Petroleum and the Continental Shelf of North West Europe" (A. W. Woodland, ed.), Vol. 1. 165–187. Applied Science, Essex.

ZOBACK, M. L. and THOMPSON, G. A. (1978). Basin and Range rifting in northern Nevada: clues from a mid-Miocene rift and its subsequent offsets. *Geology* **6**, 111–116.

ZUFFARDI, P. (1977). Ore/mineral deposits related to the Mesozoic ophiolites in Italy. *In* "Time- and Strata-bound Ore Deposits" (D. D. Klemm and H-J. Schneider, eds), 314–323. Springer-Verlag, Berlin.

Index of Localities

A

Abakiliki, Nigeria, 82, *322*
Abu Ghalaga, Egypt, 317
Acupan, Philippines, 208
Adelaide Geosyncline, 67, 70, 77
Adriatic Sea, *269*, 290
Afar, Eritrea, 125
Africa, *16*, 27
Afrikanda, USSR, 36, *37*
African Platform, 298
Ahaggar, Africa, 109
Air, Niger, *31, 32*, 109
Ajo, USA, 319
Aleutian Arc, 160, 177, 178, 181, 206
Aleutian Trench, 160, 163
Aljustel, Portugal, 195
Alligator River, Australia, 109, 110, *111*
Alno, Sweden, 60
Alpine Chains 8, *9*, 252
Alpine Fault, New Zealand, 306
Alps, 48, 104, 269, 290
Altiplano, Bolivia, 219, 222, *223*
Alto-San Franciscana Basin, Brazil, 313
Altyn Tagh, Tibet, 299, 306
Amadeus Basin, Australia, 109
Amazon Basin, 83
Amazon Fracture Zone, 51, *82*
Andaman Sea, *42, 161, 204*, 221, 225, *244, 246*, 247, 249, 305, 336
Andean Arc, 202, 203, 246, 338
Andean Cordillera, *145*, 229
Andes, 20, 181, 186, *223*, 339
Angola, 70, 310, *312*
Angola Basin, 71
Antamok, Philippines, 208
Antarctic, 233
Antarctic Ridge, 314
Appalachian Basin, *291*
Appalachians, 187, *213*, 270

Appenines, Italy, *123*, 142, 217, 251, 332
Apusini Mts., 243
Arabian Platform, 175
Arabian Shield, 181
Aravalli, India, 95, *255*
Araxa, Brazil, 311, 313
Ardennes, 5
Arika Basin, 214
Asmara, Ethiopia, 199
Athabasca Basin, 109, *110*
Athapuscow, Canada, 24, 46, *47*, 48, 50, *79*, 80
Atlantic Coastal Plain, 97
Atlantis II Deep, *126, 316*
Austral Chain, 152, *153*
Avoca, Eire, 192, 199
Aznacallar, Spain, 195

B

Baguio, Philippines, 209, *210*
Baja California, 93, 97
Baikal Rift, 50, 60, 298
Banda Arc, *271*
Bangka Island, Indonesia, 277, 279
Basin and Range Province, North America, 41, 221, 236, 237, 238, *240*, 241
Bas Ranas, Egypt, 317
Batchawan, Ontario, 67
Bathurst, New Brunswick, 192, *201*
Bathurst, New South Wales, 192
Batton Trough, Australia, *78*
Bauer Deep, Pacific, 125
Bawdwin, Burma, 193, 321
Bay of Bengal, *10*
Bay of Islands, Newfoundland, *139*, 141, 262
Benfontein, S. Africa, 64

Index of Mineralization in Relation to Tectonic Setting